Undergraduate Texts in Mathematics

Editors

F. W. Gehring
P. R. Halmos

Advisory Board

C. DePrima
I. Herstein

Undergraduate Texts in Mathematics

continued after Index

Kennan T. Smith

Primer of
Modern Analysis

(Directions for Knowing All Dark Things,
Rhind Papyrus, 1800 B.C.)

Springer-Verlag
New York Berlin Heidelberg Tokyo

Kennan T. Smith
Mathematics Department
Oregon State University
Corvallis, Oregon 97331
U.S.A.

AMS Subject Classification: 26-01, 28-01

Library of Congress Cataloging in Publication Data
Smith, Kennan T., 1926–
    Primer of modern analysis.
    (Undergraduate texts in mathematics)
    Includes index.
    1. Mathematical analysis.   I. Title. II. Series.
QA300.S77   1983   515   83–538

The original version of this book was published by Bogden & Quigley, Inc.,
Publishers, in 1971.

Printed and bound by Halliday Lithograph, West Hanover, MA.
Printed in the United States of America.

9 8 7 6 5 4 3 2 1

ISBN 0-387-90797-1   Springer-Verlag New York Berlin Heidelberg Tokyo
ISBN 3-540-90797-1   Springer-Verlag Berlin Heidelberg New York Tokyo

To J.

Preface

This book discusses some of the first principles of modern analysis. It can be used for courses at several levels, depending upon the background and ability of the students.

It was written on the premise that today's good students have unexpected enthusiasm and nerve. When hard work is put to them, they work harder and ask for more. The honors course (at the University of Wisconsin) which inspired this book was, I think, more fun than the book itself. And better. But then there is acting in teaching, and a typewriter is a poor substitute for an audience. The spontaneous, creative disorder that characterizes an exciting course becomes silly in a book. To write, one must cut and dry. Yet, I hope enough of the spontaneity, enough of the spirit of that course, is left to enable those using the book to create exciting courses of their own.

Exercises in this book are not designed for drill. They are designed to clarify the meanings of the theorems, to force an understanding of the proofs, and to call attention to points in a proof that might otherwise be overlooked. The exercises, therefore, are a real part of the theory, not a collection of side issues, and as such nearly all of them are to be done. Some drill is, of course, necessary, particularly in the calculation of integrals.

Those using the book should not feel obliged to do every proof. It is more important for teachers to explain the theorems well and to show how they are used, and why they are interesting, than to spend all the time on proofs. This is one place where the teacher has an advantage over the author. He can choose proofs that seem to him exciting or illuminating, and skip some of the others. The author, however, must do nearly all. In this book I have omitted only the proof of Fubini's theorem—in favor of a long list of applications.

Many topics in the mathematics curriculum find their best use in the calculus of several variables: for example, much linear algebra, much topology, much measure theory, and so forth. Usually students learn them as separate topics. As a result, they understand these subjects narrowly and apply them poorly. I have therefore done quite a bit of linear algebra, topology, and measure theory—but always with the applications in mind and following close behind. The result should be that students will understand *both* sides much better.

Part I begins with a half intuitive–half rigorous discussion of applications, chosen to arouse interest and to show the need for a precise and general theory, and then develops this theory for functions of one variable. Unusual features include the solid treatment of Taylor's formula, the discussion of real analytic functions, and the Weierstrass approximation theorem.

In Part II the differential properties of functions of several variables are studied. There is some background on metric and vector spaces, but the bulk of this part deals with applications of the implicit-function theorem to the study of surfaces and manifolds, tangent and normal planes, maximum and minimum problems in several variables and on manifolds, and so forth. Various interesting sidelights, such as the derivation of Kepler's laws of planetary motion and mini–max descriptions of eigenvalues, are included.

In Part III the integration and differentiation of measures are studied. The Lebesgue theory of integration is developed in the simple, yet perfectly general, abstract setting of outer measures, and applied in many and diverse situations, such as integration in $\mathbf{R}^n$, summation of multiple power series, and Sard's theorem on regular values of differentiable functions. The Lebesgue theory of differentiation is presented for regular Borel measures on $\mathbf{R}^n$ and used, for example, in establishing the formulas for change of variable in multiple integrals. The theory of differentiation leads naturally to the study of surface area via the area measures of Hausdorff. In the final chapter I discuss the Brouwer degree of maps of spheres and its applications, developing the degree from the analytic point of view suggested by John Milnor.

Theorems, Definitions, etc., are numbered within each chapter and section. Thus, Theorem 6.3 of Chapter 8 is found in Section 6 of Chapter 8. Theorem 6.3 without any chapter reference is found in Section 6 of the chapter in which the reference is made. The chapter number and title are printed in the upper left-hand corner of each double-page spread.

The index lists most of the terms and symbols that are used and the page or pages on which they are defined. The symbols occur ahead of the terms beginning with the same letter. Thus, $|A|$ and α_m occur at the head of the a's.

I wish to thank my colleagues at Oregon State University and at the University of Oregon who read and commented upon earlier versions of the manuscript. These include Professors P. M. Anselone, D. S. Carter, R. B. Guenther, B. Petersen, and, particularly, R. M. Koch. Professor Norton Starr of Amherst College also read an earlier version of the manuscript and made suggestions. In addition, I wish to thank Professor D. C. Rung of The Pennsylvania State University for suggesting the title. Finally, I wish to praise Mr. Edward J. Quigley, who is a new publisher, but a good one.

It is fitting to end this preface with advice to the reader from the creator and patron saint of calculus. The following statement came in answer to the question of how he had made his famous discoveries:

Isaac Newton

"By always thinking about them, I keep the subject constantly before me and wait till the first dawnings open little by little into the full light."

K. T. S.

PREFACE TO THE
SPRINGER EDITION

Rademacher's theorem on the differentiability of Lipschitz functions has been added. Applications of Rademacher's theorem and the Brouwer degree to changes of variable in multiple integrals have been added. The main addition, however, is a chapter on the results of Hestenes, Seeley, and Adams–Aronszajn–Smith on extension of differentiable functions of various kinds across Lipschitz graphs. A construction is given for a single extension operator which applies to functions of class C^m, functions of class C^m with bounded derivatives, functions of class C^m with Hölder continuous derivatives, and to Sobolev functions. It applies to many other function classes as well, but these are the ones discussed explicitly. The discussion of the Sobolev spaces requires a minimal knowledge of L^p spaces (mainly the Hölder and Minkowski inequalities). The theorems cover polyhedral domains, so they are of use in the numerical study of partial differential equations, as well as of theoretical interest.

K. T. S.

Contents

PART I

1 } Applications

1   TANGENT LINES

The origin of calculus was the problem of finding the tangent to a curve. Like most geometric problems, this has an immediate appeal and is very tricky. What is a curve? What is the tangent line? From a straight geometrical point of view both questions are almost impossible.

The thing to do with impossible questions is to avoid them. In the first place, we shall not consider an arbitrary curve but rather the graph of a function. In the second place, we shall not attempt a geometric definition of the tangent line but shall use geometric intuition to come to an analytic definition. This has several advantages. The analytic definition is fairly easy to give. The notion that emerges is relevant not only to the tangent line, but also to other problems where the tangent line has no role. Finally, in an analytic setting the power of arithmetic and algebra can be brought to bear.

Let f be a real-valued function defined on an interval I, and let a be an interior point of I (i.e., not an end point). We ask for the tangent line to the graph of f at the point (a, b), $b = f(a)$ (Figure 1).

A straightforward preliminary notion is that of a chord through the point (a, b). It is simply a line through this point and some other point (x, y) of the graph. Geometric intuition says that the tangent line should be the limit of the chord as the point x approaches the point a.

The idea of the limit of a family of lines may seem as nebulous as that of the tangent line itself. The trick is to replace each line by a number and to deal with a limit of numbers instead. The number to use gives the direction of the line. It is called the *slope*.

DEFINITION
1.1

Let L be the line passing through the two points (a, b) and (x, y), $a \neq x$.
The slope of L is the number $m = (y - b)/(x - a)$.

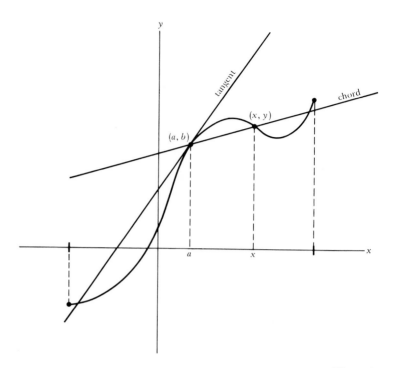

Figure 1

Exercise 1   For the definition to make sense x must be different from a.   What condition does this impose on the line L?

Elementary trigonometry shows that the slope is the tangent of the counter-clockwise angle from the positive x axis to L.   It is independent of the particular points (a, b) and (x, y) chosen on L.

Exercise 2   Show that the slope is independent of the points (a, b) and (x, y) chosen on L by using similar triangles.

Exercise 3   Find the equation of the line passing through the points $(-1, 2)$ and $(3, 6)$. [*Hint:* Calculate the slope in two ways—first by using the two points $(-1, 2)$ and $(3, 6)$, and then by using the two points $(-1, 2)$ and (x, y).]

If (a, b) and (x, y) are points on the graph of f, then $b = f(a)$ and $y = f(x)$, so the slope of the chord joining them is

$$\frac{y - b}{x - a} = \frac{f(x) - f(a)}{x - a}.$$

According to our intuitive geometric reasoning, the slope of the tangent line to the graph of f at the point (a, b) should be the limit of these numbers as x approaches a.

2   DERIVATIVES

The limit of the numbers

$$\frac{f(x) - f(a)}{x - a}$$

as x approaches a is called the *derivative* of f at the point a. It is written $f'(a)$.

The result of Section 1 is that the tangent line to the graph of f at the point (a, b) is the line passing through this point with slope $f'(a)$. According to the definition of slope, the equation of the tangent line is therefore

$$\frac{y - b}{x - a} = f'(a).$$

Of course, the definition of a limit of numbers is lacking. Intuitively, the limit of $g(x)$ as x approaches a is the number l, if $g(x)$ is as close as we please to l for every x that is close enough to a. To be useful in real proofs the definition must be given a precise quantitative form. Note that the distance between two numbers z and w is $|z - w|$.

DEFINITION 2.1   *The limit of $g(x)$ as x approaches a is the number l, written $\lim_{x \to a} g(x) = l$, if for each positive number ϵ there is a positive number δ such that $|g(x) - l| < \epsilon$ whenever $|x - a| < \delta$ and $x \neq a$.*

The definition would seem to fit the intuitive idea of limit, but its real significance must come out of the results that can be obtained from it. Before taking these up (in most of the rest of the book), let us look at some examples in which the value of the limit is pretty clear.

First, let $f(x) = x^2$. Then

$$\frac{f(x) - f(a)}{x - a} = \frac{x^2 - a^2}{x - a} = x + a.$$

When x is close to a, $x + a$ is close to $a + a$, so the limit is $2a$; that is, $f'(a) = 2a$.

Example   Find the tangent line to the curve $y = x^2$ at the point $(2, 4)$.

The slope is $f'(2) = 4$, and the equation is

$$\frac{y - 4}{x - 2} = 4 \quad \text{or} \quad y = 4x - 4.$$

Next, let $f(x) = x^3$. Then

$$\frac{f(x) - f(a)}{x - a} = \frac{x^3 - a^3}{x - a} = x^2 + xa + a^2.$$

When x is close to a, x^2 is close to a^2 and xa is close to a^2. Thus, the sum is close to $3a^2$ and $f'(a) = 3a^2$.

Let $f(x) = x^n$, where n is any positive integer. Then

$$\frac{f(x) - f(a)}{x - a} = \frac{x^n - a^n}{x - a} = x^{n-1} + x^{n-2}a + x^{n-3}a^2 + \cdots + xa^{n-2} + a^{n-1}.$$

To see this call the right side R and consider $(x - a)R = xR - aR$. Each term in xR cancels with the previous one in aR, so all terms cancel except the first one in xR, which is x^n, and the last one in aR, which is a^n. It looks like this:

$$xR = x^n + x^{n-1}a + x^{n-2}a^2 + \cdots + x^2a^{n-2} + xa^{n-1},$$
$$aR = \qquad x^{n-1}a + x^{n-2}a^2 + \cdots \cdots \cdots + xa^{n-1} + a^n.$$

In the difference each term cancels with the one above or below it, leaving only $x^n - a^n$.

The limit is a sum of n terms each of which is a^{n-1}. Therefore, $f'(a) = na^{n-1}$.

THEOREM 2.2   *If $f(x) = x^n$, where n is a positive integer, then $f'(x) = nx^{n-1}$.*

Exercise 1   When n is a negative integer the same formula holds at any point $x \neq 0$. Try to fashion an intuitive proof based on the one above. Discuss also the case $n = 0$.

The common functions occur as combinations, such as sums, products, and quotients, of a certain few functions, such as x^n, $\sin x$, $\cos x$, a^x, and $\log x$. To calculate the derivative of any common function, what is needed is the calculation for each one of the few functions, and then some rules to deal with combinations. The special calculations, even more than the general rules, involve points of considerable interest and difficulty. They are carried out in Chapter 2, as are the proofs of the general rules. Here we shall state without proof one simple general rule that can be used in conjunction with Theorem 2.2 and Exercise 1 to illustrate the developing theory.

THEOREM 2.3   *If $f(x) = \alpha g(x) + \beta h(x)$, where α and β are real numbers, then $f'(x) = \alpha g'(x) + \beta h'(x)$.*

Example   Let $f(x) = 3x^2 - (8/x)$. If $g(x) = x^2$, then by Theorem 2.2, $g'(x) = 2x$; if $h(x) = x^{-1}$, then by Exercise 1, $h'(x) = -x^{-2}$. Therefore,

$$f'(x) = 3 \cdot 2x + (-8)(-1)x^{-2} = 6x + \frac{8}{x^2}.$$

3   MAXIMUM AND MINIMUM PROBLEMS

One of the intriguing applications of the derivative comes in finding the maximum and minimum values of a function and the points where they occur.   The geometric idea is that if f has a maximum or minimum at the point a, then the tangent line to the graph at the point $(a, f(a))$ should be horizontal.   In other words, its slope is 0, or, in still other words, $f'(a) = 0$.

   This is apparent geometrically, but it can be looked at analytically, too. Suppose that f has a minimum at a.   Then $f(x) \geq f(a)$ for every point x, which means that the quotient

$$g(x) = \frac{f(x) - f(a)}{x - a}$$

is ≥ 0 if $x > a$ and is ≤ 0 if $x < a$.   Let x approach a but be always $>a$.   The limit $f'(a)$ must be ≥ 0, since it is the limit of numbers that are all ≥ 0.   Now let x approach a but be always $<a$.   This time $f'(a)$ must be ≤ 0, since it is the limit of numbers that are all ≤ 0.   Thus, $f'(a) \geq 0$ and $f'(a) \leq 0$, which leaves only $f'(a) = 0$.

THEOREM 3.1   *If f has a maximum or minimum at a and if $f'(a)$ exists, then $f'(a) = 0$.*

Now let us give a real proof using the formal Definition 2.1 of limit.

Proof   Suppose that f has a minimum at a.   We assume that $f'(a) > 0$ and derive a contradiction.   [The contradiction is similar if we assume that $f'(a) < 0$.]

   In Definition 2.1 take $\epsilon = \frac{1}{2}f'(a)$ and find the corresponding positive number δ such that

$$|g(x) - f'(a)| < \epsilon = \tfrac{1}{2}f'(a) \qquad \text{if} \quad |x - a| < \delta \quad \text{and} \quad x \neq a.$$

Then

$$g(x) \geq f'(a) - \tfrac{1}{2}f'(a) > 0 \qquad \text{if} \quad |x - a| < \delta \quad \text{and} \quad x \neq a,$$

whereas we saw above that $g(x) \leq 0$ for every $x < a$ when f has a minimum at a.

Example   A cylindrical barrel is to contain 1 ft^3 of whiskey.   What should be the dimensions so that the barrel is built with the least amount of wood?

   If x is the radius of the barrel and h is the height, then the volume is $\pi x^2 h$, so

$$1 = \pi x^2 h \qquad \text{and} \qquad h = \frac{1}{\pi x^2}.$$

The amount of wood used is essentially the surface area of the barrel, which is the area of the top plus the area of the bottom plus the area of the cylindrical side. Thus,

$$\text{area} = \pi x^2 + \pi x^2 + 2\pi x h = 2\pi x^2 + \frac{2}{x}.$$

Therefore, the problem is to find the value of x at which the function

$$f(x) = 2\pi x^2 + \frac{2}{x}$$

is minimum.

By Theorems 2.2 and 2.3 (the same calculation as in the last section) we have

$$f'(x) = 4\pi x - \frac{2}{x^2}.$$

If f has a minimum at a, then $f'(a) = 0$. Hence, $2\pi a^3 = 1$, or

$$a = (2\pi)^{-1/3} \qquad \text{and} \qquad h = \frac{1}{\pi a^2} = 2^{2/3}\pi^{-1/3} = 2a.$$

The legitimate conclusion of this is that if the problem does have a solution, then the best barrel is the one whose height is twice its radius. But it is not at all clear that the problem does have a solution. Perhaps the function f does not have a minimum. (Note that it certainly does not have a maximum.) It could well be that there is no best barrel for a cubic foot of whiskey!

To settle this kind of question (from an amoral point of view, of course) we shall have to prove a theorem to the effect that under the right conditions a function must have a minimum or a maximum.

Exercise 1   What are the right dimensions to make a rectangular field that contains 100 yd² of grass using the least amount of fencing?

Exercise 2   What is the shortest distance from the point $(18, 0)$ to the curve $y = x^2$? Where is the closest point on the curve?

4   VELOCITY AND ACCELERATION

A physical problem, apparently unrelated to the geometrical problem of tangent lines, is the motion of an object along a straight line. An example is a falling body.

Let coordinates be chosen on the line, and let $s(t)$ be the coordinate of the object at the time t. In elementary physics the average velocity over the time interval from $t = a$ to $t = x$ is (by definition) the difference between the final and initial positions divided by the length of the time interval. Thus,

$$\text{average velocity} = \frac{s(x) - s(a)}{x - a}. \tag{1}$$

It is plain then how to define the velocity *at the time* $t = a$. It is the limit of the average velocity as the time interval goes to 0. In other words, the velocity at the time $t = a$ is the derivative $s'(a)$.

In this context it is natural to consider the velocity function, the function v whose value at any time t is the velocity at that time. We have

$$v(t) = s'(t) \qquad \text{for each } t. \tag{2}$$

In elementary physics the average acceleration of the object over the time interval from $t = a$ to $t = x$ is (by definition) the difference between the final and initial velocities divided by the length of the time interval:

$$\text{average acceleration} = \frac{v(x) - v(a)}{x - a}. \tag{3}$$

The acceleration at the time $t = a$ is the limit of the average acceleration as the time interval goes to 0. Thus, the acceleration at the time $t = a$ is the derivative $v'(a)$.

The acceleration is the derivative of the derivative of s, which is called the second derivative of s and is written s''.

Again, it is natural to consider the acceleration function a whose value at any time t is the acceleration at that time:

$$a(t) = v'(t) = s''(t) \qquad \text{for each } t. \tag{4}$$

Example   A stone is dropped from the top of a 100-ft tree. When does it hit the ground?

What is known is the total force that acts on the stone. There is the force of gravity pulling the stone down and the air resistance pushing the stone up. Knowing these two forces, we must solve the problem.

The basis for the solution is the law of physics, the famous second law of Newton, which states that the acceleration of an object is proportional to the force acting on it. In other words, there is a constant c such that if $a(t)$ is the acceleration at time t and $f(t)$ is the force acting on the object at time t, then

$$a(t) = cf(t) \qquad \text{for each } t. \tag{5}$$

(The constant c is determined by the units in which the acceleration and force are measured.)

In our present case the air resistance is nearly negligible, and we shall neglect it.   The force of gravity is nearly constant.   (It depends on the distance between the stone and the center of the earth which varies only 100 ft during the fall.)   We shall assume that it is constant.   Therefore, according to formula (5), the acceleration is constant.   This constant, usually called g, has the value of about 32 ft/sec/sec.

To proceed we have to choose the coordinates.   The line of motion is the line from the top of the tree to the center of the earth.   Let the coordinates on this line be such that the origin is at the surface of the earth and the positive direction on the line is upward.   Let the time be measured from the moment the stone is dropped.   Then the information we have is that

$$v'(t) = a(t) = -32 \qquad \text{and} \qquad v(0) = 0.$$

The condition $v(0) = 0$ says that the stone has velocity 0 at the moment it is dropped.   If it were thrown down with a speed v_0, then the condition would be $v(0) = -v_0$.   The minus sign here and in the acceleration come from the fact that they are directed downward, while the positive direction on the line is upward.

Now we see what the problem is: to determine the function v from its derivative and its value at one point.   Once this is done the function s is to be determined from similar information.

An obvious question occurs.   To what extent is a function determined by its derivative? The answer is as follows.

THEOREM 4.1   *Two functions have the same derivative at each point of an interval if and only if they differ by a constant.*

Part of the theorem is easy.   If $f = g - h$, then by Theorem 2.3, $f' = g' - h'$.   If f is constant, then $f' = 0$, so $g' = h'$.   The other part is not so easy.   It is proved in Section 10 of Chapter 2.

Now let us return to the problem of the stone.   We have $v'(t) = -32$, and we know from Theorem 2.2 that the derivative of $-32t$ is also -32.   According to the present theorem, we must have $v(t) = -32t + c$ for some constant c. The value of c is determined by the condition $v(0) = 0$.   Indeed, $0 = v(0) = -32 \cdot 0 + c$.   Therefore, $c = 0$, and

$$v(t) = -32t. \tag{6}$$

Now for s.   We have $s'(t) = -32t$, and we know from Theorem 2.2 that the derivative of $-16t^2$ is also $-32t$.   According to the present theorem we must have $s(t) = -16t^2 + d$ for some constant d.   This time $s(0) = 100$, so $100 = s(0) = -16 \cdot 0 + d$.   Therefore, $d = 100$, and

$$s(t) = -16t^2 + 100. \tag{7}$$

When does the stone hit the ground? At the time t, when $s(t) = 0$ we have

$$-16t^2 + 100 = 0 \qquad \text{or} \qquad t = \tfrac{5}{2}.$$

With what velocity does it hit the ground? With the velocity

$$v(\tfrac{5}{2}) = -32\tfrac{5}{2} = -80.$$

5 AREA

The problem is to find the area under the graph of a function.

To be more precise, let f be a nonnegative function on an interval I, and let a and b be two points of I with $a \leq b$. We want to find the area of the set that is under the graph of f, above the x axis, and between the lines $x = a$ and $x = b$—that is, of the set

$$\{(x, y) : a \leq x \leq b \text{ and } 0 \leq y \leq f(x)\}.$$

Let $\int_a^b f$ denote the area of this set.

The number $\int_a^b f$ is called the integral of f from a to b. The symbol $\int$ is designed to be a peculiar letter S, standing for sum. When the integral is defined properly (Chapter 3), it will appear as a limit of sums associated with the function f, and area will be only one of many interpretations that can be given to it.

It is not clear that the problem of area makes sense. Rectangles, triangles, etc., have areas, but there is little reason to believe that such general sets do. One thing is clear, however: If the area does make sense, then it ought to satisfy the following two conditions:

A. *If $m \leq f(x) \leq M$ on $a \leq x \leq b$, then*

$$m(b - a) \leq \int_a^b f \leq M(b - a).$$

B. *If $a \leq b \leq c$, then*

$$\int_a^c f = \int_a^b f + \int_b^c f.$$

The first condition says that if a set contains a rectangle, its area is larger than or equal to the area of the rectangle, whereas if it is contained in a rectangle, its area is smaller than or equal to the area of the rectangle. The second condition says that if a set is cut into two parts by a vertical line, the area is the sum of the areas of the parts.

In fact, it is not possible to define the area so that these two simple conditions hold unless some restriction is put on the function f.   One natural restriction is that f be continuous at each point in the following sense:

DEFINITION 5.1   *The function f on the interval I is continuous at the point $a \in I$ if $\lim_{x \to a} f(x) = f(a)$.*

There is a detailed discussion of continuity in Chapters 2 and 3.   For the present it suffices to say that all the common functions are continuous except at certain quite obvious points.   For instance, the function $f(x) = 1/x$ is continuous at every point except 0—and there is no way to define it at 0 so that it becomes continuous there.   The same is true of the function $f(x) = \sin(1/x)$.

Exercise 1   Draw graphs of the functions $f(x) = 1/x$ and $f(x) = \sin(1/x)$.

Exercise 2   If f has a derivative at the point a, then f is continuous at a.

(Don't worry if there is difficulty with this one.   The proof appears in Section 3 of the next chapter.   The statement is given mainly to bear out the contention that the common functions really are continuous.)

The theory of area works very well for continuous functions.

THEOREM 5.2   *Let f be continuous at each point of the interval I.*
    *(a) For any two points a and b of I it is possible to define $\int_a^b f$ so that conditions **A** and **B** above hold.*
    (b) Let a be a fixed point of I, and for each $x \in I$ define $F(x) = \int_a^x f$. Then $F'(x) = f(x)$ for each $x \in I$.
    (c) Let G satisfy $G'(x) = f(x)$ for each $x \in I$.   Then for any two points a and b of I, $\int_a^b f = G(b) - G(a)$.

Part (c) of this astonishing theorem is what permits the calculation of integrals.

Example 1   Find the area under the curve $y = x^3$ between $x = 0$ and $x = 2$.

According to the theorem, we should look for a function whose derivative is x^3.   One such is $G(x) = x^4/4$, so

$$\text{area} = G(2) - G(0) = 4.$$

Part (a) of the theorem is not easy to prove.   Parts (b) and (c) can be proved now, but part (a) is postponed to Chapter 4.   [Logically, however, parts (b) and (c) do not make sense without part (a) to show that they do.]

It is technically convenient to use the symbol $\int_a^b f$ also when $a > b$. In this case it is defined to be $-\int_b^a f$.

Exercise 3 Show that condition B holds, that is, that

$$\int_a^c f = \int_a^b f + \int_b^c f$$

no matter what the relative positions of the three points.

Proof of Part (b)   This is a situation that requires the quantitative definition of limit. It must be shown that for every positive number ϵ there is a positive number δ such that

$$\left| \frac{F(x) - F(b)}{x - b} - f(b) \right| \le \epsilon \qquad \text{if } |x - b| < \delta \quad \text{and} \quad x \ne b,$$

which is the same as

$$f(b) - \epsilon \le \frac{F(x) - F(b)}{x - b} \le f(b) + \epsilon \qquad \text{if } |x - b| < \delta \quad \text{and} \quad x \ne b. \quad (1)$$

According to the definition of F and condition B,

$$F(x) - F(b) = \int_a^x f - \int_a^b f = \int_a^x f + \int_b^a f = \int_b^x f.$$

Therefore, formula (1) is the same as

$$f(b) - \epsilon \le \frac{1}{x - b} \int_b^x f \le f(b) + \epsilon \qquad \text{if } |x - b| < \delta \quad \text{and} \quad x \ne b. \quad (2)$$

Let $\epsilon > 0$ be given. Use the fact that f is continuous at b to find $\delta > 0$ such that $|f(y) - f(b)| \le \epsilon$ if $|y - b| < \delta$, hence such that

$$f(b) - \epsilon \le f(y) \le f(b) + \epsilon \qquad \text{if } |y - b| < \delta. \quad (3)$$

This is the δ required in formula (2). Indeed, let $|x - b| < \delta$ and $x > b$. If y is in the interval between b and x, then $|y - b| < \delta$; so inequality (3) holds. Therefore, by condition A,

$$(f(b) - \epsilon)(x - b) \le \int_b^x f \le (f(b) + \epsilon)(x - b).$$

Division by $x - b$ gives inequality (2).

Exercise 4 In the final paragraph it is assumed that $x > b$. What happens when $x < b$?

Proof of Part (c)   This one is easy now that part (b) is established. Indeed, G and F have the same derivative. According to Theorem 4.1, they must differ by a constant; that is, $G(x) = F(x) + \alpha$. In the difference $G(b) - G(a)$ the constant α cancels out, so

$$G(b) - G(a) = F(b) - F(a) = \int_a^b f - \int_a^a f = \int_a^b f.$$

The integral shows up in a great variety of mathematical and physical problems.

Example 2   A spring has a length of 1 ft when it is unstretched. Find the work done in stretching it to a length of 2 ft.

First consider the physical background. Work is done when a force acts through a distance. When the force is constant, the work is by definition the product of the force and the distance.

In the present case the force is not constant. A characteristic of springs is that the force is proportional to the amount of stretching. Let the spring be anchored at the origin and stretched along the x axis, and let $f(x)$ be the force when the unanchored end is at the point x. The fact that the force is proportional to the amount of stretching means that there is a constant c such that $f(x) = c(x - 1)$. The constant c is a quantity associated with the particular spring, which is determined by experiment. Let us suppose that $c = -1$, so $f(x) = -(x - 1)$.

Exercise 5   Why is c negative?

In general, let $W_a^b(f)$ be the work done by the force f as it acts through the interval from a to b. Conditions A and B are clearly satisfied by W. The first says that if the force is everywhere $\geq m$, then the work done is $\geq$ that done by the constant force m, whereas if the force is everywhere $\leq M$, then the work done is $\leq$ that done by the constant force M. The second says that if b is between a and c, then the work done over the interval from a to c is the sum of the work done from a to b and the work done from b to c.

Theorem 5.2 was proved solely on the basis of conditions A and B. Therefore, if $G'(x) = f(x)$, then

$$W_a^b(f) = G(b) - G(a) = \int_a^b f.$$

In our particular case, $G(x) = -\tfrac{1}{2}x^2 + x$ satisfies $G'(x) = f(x)$, so the solution of the problem is

$$\text{work} = G(2) - G(1) = -\tfrac{1}{2}.$$

Exercise 6   Is this really the solution, or should the solution be $+\tfrac{1}{2}$?

2 Calculation of Derivatives

1 LIMITS

The statement $\lim_{x \to a} g(x) = l$ has been discussed in two cases. In the first, the definition of the derivative, g is the quotient

$$g(x) = \frac{f(x) - f(a)}{x - a},$$

where f is a function defined on some interval and a is an interior point of the interval (i.e., not an end point). In this case g is defined at every point sufficiently close to a, except for a itself.

In the second case, the definition of continuity, g is a function defined on some interval I and a is a point of I, quite possibly an end point.

In general g is a function defined on some set S, a is a point that may or may not belong to S, and l is a number. There are two ideas to be expressed. The first is that there are points of S as close as we please to a. The second is that $g(x)$ is as close as we please to l if x is in S and is close enough to a.

These ideas are relevant in a wide range of situations. There is no reason that the set S on which g is defined must be a set of real numbers, or that the values of g must be real. What is necessary is that there be a distance, so that it makes sense to say that x is close to a and that $g(x)$ is close to l. For instance, either set could be the plane, or the three-dimensional space. This general point of view will be necessary in the end, but for the time being it will be simpler to stick to real-valued functions defined on sets of real numbers.

DEFINITION 1.1

Let g be a real-valued function on a set S of real numbers.   Let a and l be real numbers.   The statement

$$\lim_{\substack{x \to a \\ x \in S}} g(x) = l$$

means that

(a) *For each positive number δ there is at least one point $x \in S$ with $|x - a| < \delta$.*

(b) *For each positive number ϵ there is a positive number δ such that if $|x - a| < \delta$ and $x \in S$, then $|g(x) - l| < \epsilon$.*

THEOREM 1.2

If a limit exists, it is unique.

Proof

Suppose that

$$\lim_{\substack{x \to a \\ x \in S}} g(x) = l \qquad \text{and} \qquad \lim_{\substack{x \to a \\ x \in S}} g(x) = m.$$

Let ϵ be a positive number.   (We shall see how small to take it at the end.)   Find $\delta_1 > 0$ so that

$$|g(x) - l| < \epsilon \qquad \text{if } |x - a| < \delta_1 \quad \text{and} \quad x \in S,$$

and find $\delta_2 > 0$ so that

$$|g(x) - m| < \epsilon \qquad \text{if } |x - a| < \delta_2 \quad \text{and} \quad x \in S.$$

If δ is the smaller of the two numbers δ_1 and δ_2, then by condition (a) in the definition there is at least one point $x \in S$ with $|x - a| < \delta$.   For this point x we have

$$|l - m| \leq |l - g(x)| + |g(x) - m| < \epsilon + \epsilon = 2\epsilon.$$

If $l \neq m$, then we can take $\epsilon \leq \frac{1}{2}|l - m|$ and obtain the contradiction

$$|l - m| < 2\epsilon \leq |l - m|.$$

It is not true, of course, that a limit always exists.   Consider the function $1/x$ on the set $S = \{x : x \neq 0\}$.   It is clear that for every interval I containing 0 this function is unbounded on $I \cap S$.   On the other hand, we have the following theorem.

THEOREM 1.3

If the limit

$$\lim_{\substack{x \to a \\ x \in S}} g(x)$$

exists, then there is an interval I with center a such that g is bounded on $I \cap S$.

Proof   Let l be the limit and find $\delta > 0$ corresponding to $\epsilon = 1$. If I is the interval $\{x : |x - a| < \delta\}$, then

$$|g(x)| \leq |g(x) - l| + |l| < 1 + |l| \qquad \text{for } x \in I \cap S.$$

It is not true either that a limit always exists when the function g is bounded. Give a couple of examples.

THEOREM   *If $g(x) \geq 0$ everywhere on S and if*
1.4

$$\lim_{\substack{x \to a \\ x \in S}} g(x)$$

exists, then

$$\lim_{\substack{x \to a \\ x \in S}} g(x) \geq 0.$$

Proof   Suppose that the limit l is negative, let $\epsilon = -l$, and find the corresponding δ. By condition (a) there is at least one point $x \in S$ with $|x - a| < \delta$. For this point we have

$$g(x) \leq l + |g(x) - l| < l - l = 0,$$

while by hypothesis $g(x) \geq 0$.

Exercise 1   Suppose that

$$\lim_{\substack{x \to a \\ x \in S}} g(x)$$

exists and that there is an interval I with center a such that $\alpha \leq g(x) \leq \beta$ for every $x \in I \cap S$. Then

$$\alpha \leq \lim_{\substack{x \to a \\ x \in S}} g(x) \leq \beta.$$

There are some special cases that are particularly common and useful and have their own particular names.

DEFINITION   *When $S = I - \{a\}$, where I is an interval and a is an interior point (i.e.,*
1.5   *not an end point), we write*

$$\lim_{\substack{x \to a \\ x \neq a}} g(x) \qquad \text{for} \qquad \lim_{\substack{x \to a \\ x \in S}} g(x).$$

When $S = I - \{a\}$, and a is the right end point of I, we write

$$\lim_{\substack{x \to a \\ x < a}} g(x),$$

and call the limit the left-hand limit.

When $S = I - \{a\}$ and a is the left end point of I, we write

$$\lim_{\substack{x \to a \\ x > a}} g(x),$$

and call the limit the right-hand limit.

Exercise 2   The limit

$$\lim_{\substack{x \to a \\ x \neq a}} g(x)$$

exists if and only if both the left- and right-hand limits exist and are equal. (Then, of course, they are equal to the limit.)

2   LIMITS AND DERIVATIVES

Let f be a real-valued function defined on an interval I and let a be an interior point of I.

DEFINITION 2.1   *If the limit*

$$\lim_{\substack{x \to a \\ x \neq a}} \frac{f(x) - f(a)}{x - a}$$

exists, then f is differentiable at the point a. The value of the limit is called the derivative of f at a, or $f'(a)$.

The left- and right-hand derivatives are defined in the same way with limit replaced by left- or right-hand limit.

Of course, the derivative does not always exist.

Exercise 1   The derivative exists if and only if the left- and right-hand derivatives both exist and are equal.

Exercise 2   Let $f(x) = |x|$. The left- and right-hand derivatives exist at every point. They are equal at every point except 0 and are different at 0.

Let us calculate the derivative of the function $f(x) = x^n$, where n is a positive integer. We have seen in Section 2 of Chapter 1 that

$$\frac{f(x) - f(a)}{x - a} = x^{n-1} + x^{n-2}a + x^{n-3}a^2 + \cdots + xa^{n-2} + a^{n-1}, \quad (1)$$

so we have to calculate the limit of this sum. In Section 2 of Chapter 1 we reasoned that each term in the sum has the limit a^{n-1} and that there are n such terms, so the limit should be na^{n-1}. The justification of this reasoning calls for a theorem on limits of sums and products.

THEOREM 2.2

Let

$$\lim_{\substack{x \to a \\ x \in S}} g(x) = l \quad \text{and} \quad \lim_{\substack{x \to a \\ x \in S}} h(x) = m.$$

(a) *If $f = g + h$, then*

$$\lim_{\substack{x \to a \\ x \in S}} f(x) = l + m.$$

(b) *If $f = gh$, then*

$$\lim_{\substack{x \to a \\ x \in S}} f(x) = lm.$$

(c) *If $f = g/h$, then*

$$\lim_{\substack{x \to a \\ x \in S}} f(x) = l/m \quad \text{provided } m \neq 0.$$

[In these statements it is assumed tacitly that f, g, and h are all defined on the same set S. In part (c) this requires that $h(x) \neq 0$ for all $x \in S$. However, see the exercises.]

The theorem is applied in the following way. Part (b) shows that the limit of each term in (1) is a^{n-1}, and part (a) shows that the limit of the sum is na^{n-1}. This is not quite fair, since the theorem deals with the sum and product of two functions, while here there are sums and products of several. (The typical term in the sum is $x^{n-k-1}a^k$, which should be thought of as a product of $n - 1$ factors, k of them equal to a and $n - k - 1$ of them equal to x. Each factor obviously has the limit a, so there are $n - 1$ factors each with the limit a.) The case of several functions follows easily from the case of two, with the result that *the limit of a sum is the sum of the limits, and the limit of a product is the product of the limits*, no matter how many functions are involved.

Proof of the Theorem

The idea is always to estimate the quantity that must be *proved* to be small by means of those that are *known* to be small.

First take part (a). The quantity that must be proved to be small is $|f(x) - (l + m)|$, and those that are known to be small are $|g(x) - l|$ and $|h(x) - m|$. In this case we have the estimate

$$|f(x) - (l + m)| = |g(x) - l + h(x) - m| \leq |g(x) - l| + |h(x) - m|.$$

If ϵ is a given positive number, then $\epsilon/2$ is also a positive number; so we can find a positive number δ such that

$$|g(x) - l| < \epsilon/2 \quad \text{and} \quad |h(x) - m| < \epsilon/2$$

$$\text{whenever } |x - a| < \delta \quad \text{and} \quad x \in S.$$

Then

$$|f(x) - (l + m)| < \epsilon/2 + \epsilon/2 = \epsilon \qquad \text{whenever } |x - a| < \delta \quad \text{and} \quad x \in S.$$

Strictly speaking, we should find first a δ_1 for g and a δ_2 for h, and then take δ to be the minimum of δ_1 and δ_2. Usually some of these intermediate steps are skipped, and δ is chosen so as to satisfy several conditions simultaneously.

Part (b) is more complicated. In this case the quantity that must be proved to be small is $|f(x) - lm|$, and those that are known to be small are again $|g(x) - l|$ and $|h(x) - m|$. There is a trick that is almost always used with products, which is to add and subtract the same number, in this case the number $lh(x)$. We have

$$\begin{aligned} |f(x) - lm| &= |g(x)h(x) - lh(x) + lh(x) - lm| \\ &\leq |g(x) - l|\,|h(x)| + |l|\,|h(x) - m|. \end{aligned}$$

The term $|l|\,|h(x) - m|$ is not at all troublesome. [If $|h(x) - m|$ is small, then so is $|l|\,|h(x) - m|$.] The term $|g(x) - l|\,|h(x)|$ could be. It is conceivable that although $|g(x) - l|$ is small, $|h(x)|$ is big, so that the product is not small. This is covered by Theorem 1.3.

We proceed as follows. Let ϵ be a given positive number. First choose a positive number δ_0 and a positive number M so that $|h(x)| \leq M$, whenever $|x - a| < \delta_0$ and $x \in S$. Then $\epsilon/(|l| + M)$ is also a positive number, so we can find a positive number δ_1 such that

$$|g(x) - l| < \frac{\epsilon}{|l| + M} \qquad \text{and} \qquad |h(x) - m| < \frac{\epsilon}{|l| + M}$$

$$\text{whenever } |x - a| < \delta_1 \quad \text{and} \quad x \in S.$$

If δ is the minimum of δ_0 and δ_1 and if $|x - a| < \delta$ and $x \in S$, then

$$|f(x) - lm| < \frac{\epsilon}{|l| + M} M + |l| \frac{\epsilon}{|l| + M} = \epsilon.$$

In doing part (c) we can take account of part (b) and suppose that g is the constant 1, in which case we have

$$f(x) - \frac{1}{m} = \frac{1}{h(x)} - \frac{1}{m} = \frac{m - h(x)}{mh(x)}.$$

The point here is to make sure that the denominator of the fraction is not too small. We know that the numerator is small, but if the denominator were also small, then the fraction could be big.

Since $|m|/2$ is a positive number, we can find δ_0 so that $|h(x) - m| < |m|/2$, whenever $|x - a| < \delta_0$ and $x \in S$. Then

$$|h(x)| = |m - (m - h(x))| > |m| - \frac{|m|}{2} = \frac{|m|}{2}$$

$$\text{if } |x - a| < \delta_0 \quad \text{and} \quad x \in S;$$

therefore,

$$\left| f(x) - \frac{1}{m} \right| < \frac{2|h(x) - m|}{m^2} \qquad \text{whenever } |x - a| < \delta_0 \quad \text{and} \quad x \in S.$$

Now let ϵ be a given positive number. Then $m^2\epsilon/2$ is also a positive number, and we can find δ_1 so that

$$|h(x) - m| < m^2\epsilon/2 \qquad \text{whenever} \qquad |x - a| < \delta_1 \quad \text{and} \quad x \in S.$$

Taking δ to be the smaller of δ_0 and δ_1, we have the inequality required. The three parts of the theorem are now proved.

THEOREM 2.3   *If $f(x) = x^n$, where n is any integer, then $f'(a) = na^{n-1}$, provided $a \neq 0$ if n is negative.*

Proof   The theorem is already proved if n is positive. It is obvious if $n = 0$. Let n be negative, say $n = -k$. Then

$$\frac{f(x) - f(a)}{x - a} = \frac{1}{x - a}\left(\frac{1}{x^k} - \frac{1}{a^k} \right) = \left(-\frac{x^k - a^k}{x - a} \right) \frac{1}{x^k a^k}.$$

We have seen already that the limit of the first factor is $-ka^{k-1}$ and by Theorem 2.2(c) that the limit of the second is $1/a^{2k}$. Therefore, the limit of the product is

$$-ka^{k-1-2k} = -ka^{-k-1} = na^{n-1}.$$

Exercise 3   If a is an interior point of an interval I, then

$$\lim_{\substack{x \to a \\ x \in I \cap S}} f(x) = \lim_{\substack{x \to a \\ x \in S}} f(x).$$

(That is, if one of the two limits exists, then so does the other, and they are equal.)

Theorem 2.2(c) can be improved as follows.

Exercise 4   Under the hypotheses of Theorem 2.2(c) there is an interval I with center a such that $h(x) \neq 0$ for all $x \in I \cap S$.   In this case f is defined on $I \cap S$ and

$$\lim_{\substack{x \to a \\ x \in I \cap S}} f(x) = l/m.$$

3   DERIVATIVES OF SUMS, PRODUCTS, AND QUOTIENTS

The theorems on the limits of sums, products, and quotients lead to theorems on the derivatives of sums, products, and quotients.   One additional fact is needed first, however.

THEOREM 3.1

If f is differentiable at a, then

$$\lim_{\substack{x \to a \\ x \neq a}} f(x) = f(a).$$

Proof   The number 1 is positive, so we can find a positive number δ_0 such that

$$\left| \frac{f(x) - f(a)}{x - a} - f'(a) \right| < 1 \qquad \text{whenever } |x - a| < \delta_0 \quad \text{and} \quad x \neq a.$$

Hence

$$|f(x) - f(a)| < |x - a|(|f'(a)| + 1)$$

$$\text{whenever } |x - a| < \delta_0 \quad \text{and} \quad x \neq a.$$

If ϵ is any given positive number, choose δ_1 so that

$$\delta_1(|f'(a)| + 1) < \epsilon; \qquad \text{that is, } \delta_1 < \frac{\epsilon}{|f'(a)| + 1}.$$

If δ is the smaller of δ_0 and δ_1, then we have

$$|f(x) - f(a)| < \epsilon \qquad \text{whenever } |x - a| < \delta.$$

THEOREM 3.2

Let g and h be differentiable at a.   Then $g + h$ and gh are differentiable at a, and so is g/h if $h(a) \neq 0$.   Moreover,

(a) $(g + h)' = g' + h'$,
(b) $(gh)' = g'h + gh'$,
(c) $(g/h)' = (g'h - gh')/h^2$,

where all functions and derivatives are calculated at a.

Proof   Part (a) follows directly from Theorem 2.2(a). For part (b), if $f = gh$, then

$$\frac{f(x) - f(a)}{x - a} = \frac{g(x)h(x) - g(a)h(x) + g(a)h(x) - g(a)h(a)}{x - a}$$

$$= \frac{g(x) - g(a)}{x - a} h(x) + g(a) \frac{h(x) - h(a)}{x - a}.$$

Part (b) then follows from Theorems 2.2 and 3.1.

In proving part (c) we can take account of part (b) and suppose that g is the constant 1, in which case the formula to be proved becomes

$$(1/h)' = -h'/h^2. \tag{c'}$$

If $f = 1/h$, then

$$\frac{f(x) - f(a)}{x - a} = \frac{1/h(x) - 1/h(a)}{x - a} = \frac{h(a) - h(x)}{x - a} \frac{1}{h(x)h(a)}.$$

Formula (c') follows from Theorems 2.2 and 3.1.

Note that if h is the constant α, then $h'(a) = 0$, and part (b) of the theorem gives

$$(\alpha g)' = \alpha g' \qquad \text{if } \alpha \text{ is a constant.}$$

Combining this with part (a) of the theorem, we get

COROLLARY 3.3   *If g and h are differentiable at a, and α and β are real numbers, then $\alpha g + \beta h$ is differentiable at a, and*

$$(\alpha g + \beta h)' = \alpha g' + \beta h'.$$

This was the assertion of Theorem 2.3, Chapter 1.

Remark   Theorem 3.1 says that if f is differentiable at a, then f is continuous at a.

Exercise 1   Prove Theorem 2.3 (the derivative of x^n) by using Theorem 3.2 and mathematical induction.

Exercise 2   Calculate the derivative of

$$f(x) = \frac{x^9 - 3x^6 + 12x}{x^2 + 1}.$$

Exercise 3   For a function to be differentiable at a point a it must be defined at least on some interval with a as an interior point. Is this condition met in the case of the sum, product, and quotient in Theorem 3.2?

4   CONTINUITY

The purpose of this section is to list some of the simplest properties of continuity, those that follow directly from the theorems already proved about limits.   Some of the deeper properties have very far-reaching consequences, which will be explained as they come up (for example in Chapter 3).

DEFINITION 4.1   *A function f on a set S is continuous at the point a $\in$ S if*

$$\lim_{\substack{x \to a \\ x \in S}} f(x) = f(a).$$

The function f is continuous on S, or simply continuous, if it is continuous at each point a $\in$ S.

This is the same definition that was given in Section 5 of Chapter 1, except that the earlier definition applied only to functions defined on an interval.

If the definition of limit is written out in full, then Definition 4.1 becomes

DEFINITION 4.2   *A function f on a set S is continuous at the point a $\in$ S if for every positive number ϵ there is a positive number δ such that if $|x - a| < \delta$ and $x \in S$, then*

$$|f(x) - f(a)| < \epsilon.$$

In terms of continuity the assertion of Theorem 3.1 is as follows.

THEOREM 4.3   *If f is differentiable at a, then f is continuous at a.*

Exercise 1   The hypothesis that f is differentiable at a implies that f is defined at least on some interval with center a.   Apart from this requirement, the set S on which f is defined is immaterial.

Exercise 2   Show that the function $f(x) = |x|$ is continuous but not differentiable at 0.

Exercise 3   Let $f(x) = 0$ if x is irrational and $f(x) = 1/q$ if $x = p/q$, where p is an integer, q is a positive integer, and the two have no common factor.   Show that f is continuous at each irrational point, discontinuous at each rational point, and differentiable nowhere.

There are even examples of functions that are continuous at every point but differentiable at no point, but these are not easy to construct.

Theorem 2.2 on the limits of sums, products, and quotients gives the following theorem on the continuity of sums, products, and quotients.

THEOREM 4.4   *If f and g are continuous at a, then f + g and fg are continuous at a, and so is f/g if g(a) ≠ 0.*

Here it is assumed that f and g are defined on the same set S, in which case the sum and product are defined on S and the quotient is defined on $\{x : x \in S \text{ and } g(x) \neq 0\}$.

Exercise 4   Show that there is an interval I with center a such that the quotient is defined at all points of $S \cap I$, the assumptions being those of the theorem.

Exercise 5   State and prove a theorem on the continuity of a composite function. The composite function is defined as follows. Let f be defined on S and g be defined on T. The composite function h is defined on the set $\{x : x \in T \text{ and } g(x) \in S\}$. For any point x in this set, $h(x) = f(g(x))$. If T is an interval with center a and S is an interval with center $b = g(a)$, what can be said about the set on which h is defined? What are some simple choices for f and g if $h(x) = \sqrt{\sin x}$, $h(x) = \sin \sqrt{x}$, or $h(x) = (x - 1)^6$?

5   TRIGONOMETRIC FUNCTIONS

In plane geometry an angle is an ordered pair of half-lines with a common initial point. The trigonometric functions are defined as follows. Translate and rotate the angle so that the initial point is at the origin of the coordinates of the plane and so that the first half-line coincides with the positive x axis. Then the sine of the angle is y, the cosine is x, the tangent is y/x, and so on, where (x, y) is the point where the second half-line meets the unit circle (the circle with center at the origin and radius 1).

In calculus we do not deal with functions defined on the set of ordered pairs of half-lines, but rather with functions defined on sets of real numbers.

DEFINITION 5.1   *Let θ be a real number. Let (x, y) be the point obtained by starting at the point $(0, 1)$ and traveling counterclockwise along the unit circle a distance θ if θ ≥ 0 (clockwise a distance −θ if θ < 0) (Figure 1). Then*

$$\sin \theta = y, \qquad \cos \theta = x, \qquad \tan \theta = \frac{y}{x},$$

$$\csc \theta = \frac{1}{y}, \qquad \sec \theta = \frac{1}{x}, \qquad \cot \theta = \frac{x}{y}.$$

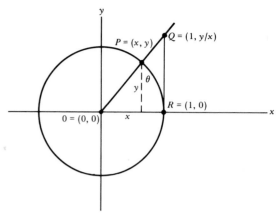

Figure 1

Exercise 1 The sine and cosine are defined for all real θ, but the others are undefined at certain exceptional values of θ. What are the exceptional values in each case?

Exercise 2 Explain the relation between Definition 5.1 and the geometric definition, and prove the familiar "addition formulas"

$$\sin(\theta + \varphi) = \sin \theta \cos \varphi + \cos \theta \sin \varphi,$$
$$\cos(\theta + \varphi) = \cos \theta \cos \varphi - \sin \theta \sin \varphi. \tag{1}$$

The calculation of the derivatives of the trigonometric functions is based on the addition formulas and on two inequalities:

$$|\sin \theta| \leq |\theta| \qquad \text{for every real } \theta, \tag{2}$$
$$\theta \leq \tan \theta \qquad \text{for } 0 \leq \theta < \pi/2. \tag{3}$$

From a geometrical point of view these inequalities are easy to prove. Inequality (2) says simply that the perpendicular distance from the point $P = (x, y)$ to the x axis is less than the distance along an arc of the unit circle. Inequality (3) says that the area of the sector $S = ORP$ is less than the area of the triangle $T = ORQ$. To see this, note that T is a right triangle with base 1 and height y/x. Therefore,

$$\text{area } T = \frac{1}{2} \cdot 1 \cdot \frac{y}{x} = \frac{1}{2} \tan \theta.$$

As for the sector, the ratio of the area of S to the area of the whole circle is equal to the ratio of the arc length θ to the arc length of the whole circle. In other words (since the radius of the circle is 1),

$$\frac{\text{area } S}{\pi} = \frac{\theta}{2\pi};$$

hence

$$\text{area } S = \frac{\theta}{2}.$$

Thus, the inequality (3) does say exactly that area $S \le$ area T (which is obviously true since $S \subset T$).

Remark   Whether this argument, or even Definition 5.1 itself, can be considered rigorous is a question that can be debated pretty hotly. On the one hand, they make use of the notions of arc length and area, which have not even been defined. These are rather complicated notions in which too-free use of intuition can lead quickly to trouble. On the other hand, we are not dealing here with the arc length of some complicated curve or the area of some complicated figure, but just with arcs and sectors of circles. In this case the geometric argument is very convincing, and, after all, the final test of rigor is whether the argument is really convincing. So the question is debatable. Presently, we shall be able to end the debate in either of two ways. One is to provide a sound general theory of arc length and area so that the above arguments no longer have to appeal to geometric intuition. The other is to take an entirely different point of view and to define the trigonometric functions by certain "infinite sums" instead of by Definition 5.1. Such a definition appears more complicated in the beginning, but in the end it is much easier to work with. For those who want to, it is all right to take the results of this section as provisional until the time (Chapters 4 and 6) when the more sophisticated methods are ready.

Now let us calculate the derivatives of the sine and cosine on the basis of the addition formulas (1), the inequalities (2) and (3), and the identity

$$\sin^2 \theta + \cos^2 \theta = 1 \qquad \text{for all real } \theta, \tag{4}$$

which comes from the fact that the point (x, y), $x = \sin \theta$, $y = \cos \theta$, is on the unit circle. Take any real number h and put $\theta = \varphi = h/2$ in the addition formula for the cosine, and then use the identity (4). The result is

$$\cos h = \cos^2 \frac{h}{2} - \sin^2 \frac{h}{2} = 1 - 2 \sin^2 \frac{h}{2};$$

hence

$$1 - \cos h = 2 \sin^2 \frac{h}{2} \qquad \text{for all real } h. \tag{5}$$

Thus, the inequality (2) gives

$$\left| \frac{1 - \cos h}{h} \right| \le \left| \frac{\sin^2(h/2)}{h/2} \right| \le \left| \frac{h}{2} \right|, \tag{6}$$

and in particular that

$$\lim_{\substack{h \to 0 \\ h \neq 0}} \frac{1 - \cos h}{h} = 0. \tag{7}$$

Inequalities (2) and (3) together show that

$$\cos h \le \frac{\sin h}{h} \le 1 \qquad \text{for } |h| < \frac{\pi}{2}. \tag{8}$$

If this is combined with (7), it shows that

$$\lim_{\substack{h \to 0 \\ h \neq 0}} \frac{\sin h}{h} = 1. \tag{9}$$

Exercise 3   Why does inequality (8) hold for $|h| < \pi/2$ and not just for $0 < h < \pi/2$?

Exercise 4   Write out the proofs of (7) and (9) with ϵ's and δ's.

THEOREM 5.2   *If* $f(x) = \sin x$, *then* $f'(x) = \cos x$. *If* $f(x) = \cos x$, *then* $f'(x) = -\sin x$.

Proof   It is clear in general that

$$f'(a) = \lim_{\substack{h \to 0 \\ h \neq 0}} \frac{f(a + h) - f(a)}{h}, \tag{10}$$

which is a more convenient formula than the original for making use of the addition formulas.

Exercise 5   Prove the obvious formula (10).

In the case of the sine we have

$$\frac{\sin(a + h) - \sin a}{h} = \frac{\sin a \cos h + \cos a \sin h - \sin a}{h}$$

$$= \sin a \frac{\cos h - 1}{h} + \cos a \frac{\sin h}{h},$$

so the result follows from formulas (7) and (9) (and the theorem on the limit of a sum or product).

Exercise 6   In the case of the cosine the proof is similar.   Carry it out.

Exercise 7   Express each of the trigonometric functions in terms of the sine and cosine, and calculate its derivative.

Exercise 8   Find the maximum of the function $f(x, y) = 3x - 2y$ on the unit circle.   [*Hint:* (x, y) is on the unit circle if and only if $x = \cos \theta$ and $y = \sin \theta$ for some θ.]

6   COMPOSITE FUNCTIONS

Another way of combining two functions is to follow one by the other.   The result is called the *composite function*.   The main theorem, called the *chain rule*, is as follows.

THEOREM 6.1   (***Chain Rule***)   *If g is differentiable at a and f is differentiable at $g(a) = b$, then the composite function $h(x) = f(g(x))$ is differentiable at a, and*

$$h'(a) = f'(b)g'(a) = f'(g(a))g'(a).$$

Example 1   The function $h(x) = (x - 1)^3$ is the composite of $f(y) = y^3$ and $g(x) = x - 1$. Therefore, $h'(a) = 3b^2 \cdot 1 = 3(a - 1)^2$.

Example 2   Let g be the function whose graph is the top half of the unit circle [the circle with center $(0, 0)$ and radius 1].   The equation of the unit circle is $x^2 + y^2 = 1$, so g satisfies the equation

$$g(x)^2 + x^2 = 1.$$

Take the derivative on both sides, considering $g(x)^2$ as the composite of g with $f(y) = y^2$.   This gives

$$2g(x)g'(x) + 2x = 0.$$

Therefore, since $g(x) = \sqrt{1 - x^2}$,

$$g'(x) = \frac{-x}{g(x)} = \frac{-x}{\sqrt{1 - x^2}}.$$

A remark is needed about the meaning of the theorem.   In order that a function be differentiable at a point a, it is necessary that the function be defined on some interval with a inside—hence (with a smaller interval if necessary) on some interval with center a.   Therefore, it is part of the hypothesis of the theorem that g is defined on an interval with center a and that f is defined on an interval with center $g(a)$.   It is part of the conclusion that the composite function h is defined on an interval with center a.

Let us examine this point.   Since f is defined on an interval with center $g(a)$, there is a positive number ϵ such that f is defined on the interval $I = \{y : |y - g(a)| < \epsilon\}$.   According to Theorem 3.1, there is a positive number δ such that g is defined on the interval $J = \{x : |x - a| < \delta\}$.   Furthermore,

$|g(x) - g(a)| < \epsilon$ whenever $x \in J$. Then the composite function h is defined on J.

Proof of the Theorem   To see what to do we write a formula that is not quite correct.

$$\frac{h(x) - h(a)}{x - a} = \frac{f(g(x)) - f(g(a))}{x - a} = \frac{f(g(x)) - f(g(a))}{g(x) - g(a)} \frac{g(x) - g(a)}{x - a}$$
$$= \frac{f(y) - f(b)}{y - b} \frac{g(x) - g(a)}{x - a},$$

where $y = g(x)$ and $b = g(a)$. As x approaches a, the limit of the second factor is $g'(a)$. The limit of y is b (by Theorem 3.1), so the limit of the first factor is $f'(b)$. Therefore, the limit of the product is $f'(b)g'(a)$.

The argument is perfectly correct, except that there may be points x with $g(x) = g(a)$, in which case it does not make sense. To avoid this trouble, define a new function $\tilde{f}$.

$$\tilde{f}(x) = \begin{cases} \dfrac{f(g(x)) - f(g(a))}{g(x) - g(a)} & \text{if } g(x) \neq g(a), \\ f'(g(a)) & \text{if } g(x) = g(a). \end{cases}$$

With this function we have the formula

$$\frac{h(x) - h(a)}{x - a} = \tilde{f}(x) \frac{g(x) - g(a)}{x - a}$$

for $g(x) - g(a)$ cancels from the top and bottom as long as it is $\neq 0$, and both sides of the formula are 0 if it is $= 0$. Now the rule that the limit of a product is the product of the limits can be used, provided that

$$\lim_{\substack{x \to a \\ x \neq a}} \tilde{f}(x) = f'(g(a)),$$

the proof of which is a good exercise for the reader.

Example 2 suggests that the chain rule can be used to calculate lots of derivatives.

THEOREM 6.2   *Let $g(x) = x^n$, where n is any rational number. Then $g'(x) = nx^{n-1}$ for any $x > 0$.*

This is the same formula that has been proved already when n is an integer.

Proof   If $n = m/k$, where m and k are integers, then

$$g(x)^k = x^m.$$

Taking the derivative of both sides at the point x, we have

$$kg(x)^{k-1}g'(x) = mx^{m-1};$$

therefore,

$$g'(x) = \frac{mx^{m-1}}{kg(x)^{k-1}} = \frac{m}{k}x^{m/k-1}.$$

Exercise 1 Find the flaw in the proof of Theorem 6.2 and in the discussion of Example 2.

7 LOGARITHMS AND EXPONENTIALS

We shall show that the function

$$L(x) = \int_1^x \frac{1}{x} \qquad \text{for } x > 0 \tag{1}$$

is the logarithm of x to a certain base and shall establish some of the properties of the logarithm function.   The basic property of logarithms is that

$$L(xy) = L(x) + L(y). \tag{2}$$

Let us prove it.

Notice first that according to Theorem 5.2 of Chapter 1,

$$L'(x) = \frac{1}{x} \qquad \text{and} \qquad L(1) = 0. \tag{3}$$

This determines L uniquely, for if two functions have the same derivative, then they must differ by a constant.   If two functions differ by a constant and take the same value at one point, then they must be identical.

Let y be fixed and consider the two functions

$$M(x) = L(xy) \qquad \text{and} \qquad N(x) = L(x) + L(y).$$

Since $L(y)$ is constant, we have

$$N'(x) = L'(x) = \frac{1}{x}.$$

By the theorem on the derivative of a composite function,

$$M'(x) = L'(xy)y = \frac{1}{xy}y = \frac{1}{x}.$$

Thus, M and N have the same derivative and clearly take the same value at $x = 1$ [that is, $L(y)$].   Hence $N = M$, and formula (2) is proved.

Formula (2) implies that

$$L(a^n) = nL(a) \tag{4}$$

for every real $a > 0$ and every rational n. First take $x = y = a$ in (2) to get $L(a^2) = 2L(a)$. Then take $x = a^2$, $y = a$, to get $L(a^3) = L(a^2) + L(a) = 3L(a)$, etc. This gives formula (4) if n is any positive integer. If n is a negative integer, write $1 = a^n a^{-n}$, to get

$$0 = L(1) = L(a^n) + L(a^{-n}) = L(a^n) - nL(a),$$

which gives formula (4) if n is any negative integer.

Finally, let n be a rational number, say $n = m/k$, $k > 0$, and let $a^n = b$. Then $a^m = b^k$, so

$$mL(a) = kL(b) \quad \text{or} \quad L(a^n) = L(b) = \frac{m}{k} L(a) = nL(a).$$

This gives formula (4) when n is any rational number.

What about the formula when n is an arbitrary real number? In this case the troublesome point is the meaning of a^n. For instance, how would you define 2^π? The best thing to do is to take formula (4) as the definition! But for this we need a theorem. (How to prove such theorems is the subject of Chapter 3. For the moment we just assume it.)

THEOREM 7.1   *For every real number y there is one and only one positive real number x such that $L(x) = y$. Moreover, the inverse function E defined by*

$$E(y) = x \quad \text{if and only if} \quad L(x) = y \tag{5}$$

is differentiable at every point.

With the aid of this theorem, a^n can be defined to be the unique number x such that $L(x) = nL(a)$.

Now, let us show that L is, in fact, a logarithm and that the inverse function E is an exponential. According to the theorem, there is a unique number e such that $L(e) = 1$. Then, according to the definition of e^x,

$$L(e^x) = xL(e) = x.$$

Hence, according to the definition of E as the inverse of L,

$$E(x) = e^x.$$

Therefore, L, which is the inverse of E, is the logarithm to the base e.

In calculus $\log x$ always means the logarithm to the base e; that is, $\log x$ means $L(x)$. (Some books use $\ln x$ instead of $\log x$.)

Exercise 1 Use the equation $L(E(x)) = x$ to show that $E'(x) = E(x) = e^x$.

Exercise 2 Show that if $A(x) = a^x$, then $A'(x) = a^x \log a$.

It is natural to question whether our definition of a^x is the right one. Exercise 2 shows that this definition does make a^x continuous, and formula (4) shows that it does give the right value when x is rational.   Therefore, Exercise 3 shows that this definition is the only one that is reasonable.

Exercise 3 If two continuous functions agree on all the rational points, then they are identical.   (*Hint:* For every real number x and every $\epsilon > 0$, there is a rational number n such that $|x - n| < \epsilon$.)

Section 1 of Chapter 3 deals with the way to prove theorems like Theorem 7.1 and Theorem 6.2.   In both cases there is a given function f.   [Here it is the function $f(x) = L(x)$; in Theorem 6.2 it is the function $f(x) = x^k$.]   The real problem is to know that for every point y there is some point x with $f(x) = y$. The fact that there is not more than one such x is immediate.   Both functions have the property that if $x_1 < x_2$, then $f(x_1) < f(x_2)$.   There is also the problem of showing that the inverse function g, defined by

$$g(y) = x \qquad \text{if and only if } f(x) = y,$$

is differentiable, but this problem is not so hard.

Exercise 4 Show that $e^x e^y = e^{x+y}$ and that $(e^x)^y = e^{xy}$.

Exercise 5 Show that if $f(x) = x^n$, $x > 0$, then $f'(x) = nx^{n-1}$, where n is an arbitrary real number.

Exercise 6 The number e is about 2.7.   Prove that $2.5 < e < 3$.

3 Deeper Properties of Continuous Functions

1   INVERSE FUNCTIONS

The problem that was left unresolved in the last two sections is one of existence: to show that for each positive real number y, there exists a positive real number x with $x^k = y$; and to show that for each real number y, there exists a positive real number x with $\log x = y$. Such problems cannot be touched without the aid of a fundamental property of the real numbers that has not appeared so far. There are many ways to state it, one of which is the following:

AXIOM
1.1
Every nonempty bounded set of real numbers has a least upper bound.

An upper bound of a set S is a number b such that $b \geq x$ for every $x \in S$. A least upper bound is one that is smaller than any other. Certainly, there cannot be more than one for each would have to be smaller than the other.

The least upper bound b of a set S is characterized by the following conditions:

A. $b \geq x$ *for every* $x \in S$.
B. *For every positive number* ϵ, *there is a point* $x \in S$ *with* $x > b - \epsilon$.

The first condition says that b is an upper bound. The second says that $b - \epsilon$ is not, no matter how small ϵ.

The greatest lower bound of a set is defined similarly. It is a lower bound that is larger than any other.

Exercise 1   Every nonempty bounded set of real numbers has a greatest lower bound.

The least upper bound of a set S is often called the *supremum* and is usually written sup S. The greatest lower bound is often called the *infimum* and is usually written inf S.

The basic theorem on the existence of a solution to an equation $f(x) = y$ is as follows.

THEOREM 1.2   *Let f be a continuous function on an interval I, and let y be any number. If there exist points a and b in I with $f(a) < y < f(b)$, then there exists a point x between a and b with $f(x) = y$.*

Proof   Suppose, to be definite, that $a < b$ and let

$$S = \{x : a < x < b \text{ and } f(x) < y\}.$$

S is nonempty, as $a \in S$, and bounded, as $S \subset [a, b]$. Therefore $c = \sup S$ exists and lies in $[a, b]$. It will be shown that $y = f(c)$.

Given $\epsilon > 0$, choose $\delta > 0$ so that if $|x - c| < \delta$, then $|f(x) - f(c)| < \epsilon$. According to property B of least upper bounds, there is a point $x \in S$ with $x > c - \delta$; according to property A, $x < c$. Consequently, $y > f(x) > f(c) - \epsilon$. Since this holds for all ϵ, $y \geq f(c)$, which requires $c < b$.

Again, given $\epsilon > 0$, choose δ as above and $< b - c$. For any point x satisfying $c < x < c + \delta$, $y < f(x) < f(c) + \epsilon$. Since this holds for all ϵ, $y \leq f(c)$. Consequently, $y = f(c)$.

The geometric version of this theorem sounds quite obvious. It says that if the graph is sometimes below a given horizontal line and sometimes above, then somewhere it must cross. The point c that is found in the proof is the last point between a and b where it does cross.

The idea of the last (or first) point at which a graph crosses a given line can be slippery. Take, for instance,

$$f(x) = x \sin \frac{1}{x} \qquad \text{on } 0 < x < 1.$$

There is certainly no first point at which the graph crosses the x axis.

Exercise 2   Give an explanation to reconcile this example with the proof of the theorem.

Exercise 3   Give a proof of the theorem that produces the first point between a and b at which the graph crosses the given horizontal line.

Exercise 4   Use the theorem to prove the existence of solutions to the equations $x^k = y$ and $\log x = y$.

The characteristic property of an interval (indeed, the best way to define an interval!) is that if it contains two points, then it must contain every point between them.

Exercise 5 Show that this definition of interval gives exactly the usual several kinds of intervals.   (You will have to use Axiom 1.1!)

In terms of this characterization of intervals, Theorem 1.2 can be restated as follows.

THEOREM 1.3   *If f is continuous on an interval I, then the set*

$$f(I) = \{y : y = f(x) \text{ for some } x \in I\}$$

is an interval.

COROLLARY 1.4   *If f is continuous and one to one on an interval I, then f is strictly increasing or strictly decreasing.*

To say that f is one to one means that $f(a) \neq f(b)$ for any two distinct points a and b.

Proof   It is not hard to see that what must be ruled out is the possibility of three points a, b, and c such that $a < b < c$, while

$$f(a) < f(b) > f(c) \quad \text{or} \quad f(a) > f(b) < f(c). \tag{1}$$

Exercise 6   Prove rigorously that this is what must be ruled out.   The first possibility is roughly that f increases for awhile, then decreases.   The second is that f decreases for awhile, then increases.

It is clear that Theorem 1.2 does rule out the possibilities envisioned in formula (1).   Take the first, for instance.   If $f(c) > f(a)$, the theorem says that there is a point x between a and b with $f(x) = f(c)$, which is impossible if f is one to one.   If $f(c) < f(a)$, then the theorem says that there is a point x between b and c with $f(x) = f(a)$, which is again impossible if f is one to one.   The second possibility is ruled out in the same way.

Before going on, let us recall some general definitions.   A function from a set X into a set Y is a rule that assigns a point of Y to each point of X.   The point assigned by the rule f to a given $x \in X$ is called the *value* of f at x and is written $f(x)$.   The set of all values of f is called the *range* of f and is written $f(X)$.

The function f is one to one if it takes distinct values at distinct points; that is, $f(x_1) \neq f(x_2)$ whenever $x_1 \neq x_2$.   If f is one to one, then a function g can be defined from $f(X)$ into X as follows: If $y \in f(X)$, then by the definition of $f(X)$ there is at least one $x \in X$ with $f(x) = y$; and by the fact that f is one to one, there is only one such x. g is the rule that assigns this point x to the point y. In other words, $g(y) = x$ if and only if $f(x) = y$.   The function g defined in this way is called the *inverse* of f.   It is a function from $f(X)$ into X.

THEOREM 1.5   *Let f be continuous on the interval I.   If f is one to one, then the inverse is continuous on $f(I)$.*

Proof   Let $b \in f(I)$, and let $\epsilon > 0$ be given.   If $b = f(a)$, then for δ we can take the smaller of the two numbers $|b - f(a - \epsilon)|$ and $|b - f(a + \epsilon)|$.   Indeed, suppose that f is increasing.   Then plainly the inverse, g, is, too; so if $f(a - \epsilon) \le y \le f(a + \epsilon)$, then $a - \epsilon \le g(y) \le a + \epsilon$.

The next question is the differentiability of the inverse function.

THEOREM 1.6   *Let f be continuous and one to one on the interval I.   If f is differentiable at a and $f'(a) \neq 0$, then the inverse, g, is differentiable at $b = f(a)$, and $g'(b) = 1/f'(a)$.*

Proof   By the definition of the inverse we have $y = f(g(y))$; so if g is differentiable at b, then Theorem 6.1 of Chapter 2 (*chain rule*) gives $1 = f'(a)g'(b)$.

To show that g is differentiable at b, note that

$$\frac{f(g(y)) - f(g(b))}{g(y) - g(b)} \frac{g(y) - g(b)}{y - b} = 1;$$

so if $x = g(y)$ and $a = g(b)$, then

$$\frac{g(y) - g(b)}{y - b} = \frac{x - a}{f(x) - f(a)}.$$

By the previous theorem, g is continuous at b; so if $y \to b$, it follows that $x \to a$.   Hence, the right-hand side approaches $1/f'(a)$.

Remark   This is the same idea as the proof of Theorem 6.1, except that g is one to one; so the problem that arose there does not arise here—that is, if $y \neq b$, then $g(y) \neq g(b)$.

Exercise 7   Show that every positive number has a unique positive kth root, and that the function $g(x) = x^{1/k}$ is continuous on $0 \le x < \infty$ and differentiable on $0 < x < \infty$.

Exercise 8 The arcsin is the inverse of the sine. More precisely, if $-1 \leq x \leq 1$, then arcsin $x = y$ if and only if $-\pi/2 \leq y \leq \pi/2$ and $\sin y = x$. Show that this makes sense, that the arcsin is differentiable, and that its derivative is $(1 - x^2)^{-1/2}$. Carry out a similar program for the arccos and the arctan.

2   UNIFORM CONTINUITY

Before turning to maxima and minima, which are also existence questions that are settled by the least-upper-bound axiom, we need a stronger version of continuity called uniform continuity. At present it may seem to be a technical notion, but in fact it is very important.

DEFINITION 2.1   *A function f on a set S is uniformly continuous on S if for every positive number ϵ there is a positive number δ such that*

$$|f(x) - f(y)| < \delta \qquad if \ x \in S, \ y \in S, \quad and \quad |x - y| < \delta. \tag{1}$$

At first glance it is hard to tell the difference from ordinary continuity. The difference is this: To say that f is continuous on S means that it is continuous at each point. This means that for each point $x \in S$ and each positive number ϵ there is a positive number δ. In short, δ is determined not by ϵ alone but by both ϵ and x. In uniform continuity δ is determined by ϵ alone.

Consider, for example,

$$f(x) = \frac{1}{x} \qquad \text{on } 0 < x < 1.$$

If we fix a positive number δ and take $|x - y| = \delta$, we have

$$|f(x) - f(y)| = \left| \frac{1}{x} - \frac{1}{y} \right| = \left| \frac{y - x}{xy} \right| \geq \frac{\delta}{x}.$$

As x moves toward 0, $|f(x) - f(y)|$ becomes as large as we please, no matter how small δ is. So the function $1/x$ is not uniformly continuous on $0 < x < 1$.

It may appear that the fact that $1/x$ is unbounded has some bearing on the matter. To some extent this is true.

THEOREM 2.2

A uniformly continuous function on a bounded set S must be bounded.

Proof

Take any ϵ, say $\epsilon = 1$, and find the corresponding δ. Take an interval I containing S and divide it in subintervals $I_1, \ldots, I_n$ of length less than δ. If $I_k \cap S$ is not empty, choose some point x_k in it.

We shall show that if M is the maximum of the numbers $|f(x_1)|, \ldots, |f(x_n)|$, then $|f(x)| \leq M + 1$ for every $x \in S$. Indeed, every $x \in S$ belongs to some I_k, hence satisfies $|x - x_k| < \delta$ for some k, and, therefore, satisfies $|f(x) - f(x_k)| < 1$ for some k.

Exercise 1

Where did we use the fact that S is bounded? Show by example that this is necessary.

Exercise 2

The function $\sin 1/x$ on $0 < x < 1$ is bounded but not uniformly continuous.

In the examples above the set S is an open interval—never a closed one. The good reason for this is the following basic theorem.

THEOREM 2.3

Every continuous function on a closed bounded interval is uniformly continuous.

Proof

We shall suppose that f is continuous but not uniformly continuous and derive a contradiction. If f is not uniformly continuous, then for some ϵ there is no δ. Thus, for every positive number r there is a pair of points x_r and y_r such that

$$|x_r - y_r| < r \quad \text{and} \quad |f(x_r) - f(y_r)| \geq \epsilon. \quad (2)$$

(Otherwise r would be a δ.) The object is to find a point c with the following property:

For each positive number δ there is a positive number r
$$\text{such that } r < \delta \text{ and } |x_r - c| < \delta. \quad (3)$$

Let us assume for the moment that we have found such a point c. Since f is continuous at c, we can find $\delta > 0$ such that

$$|f(x) - f(c)| < \epsilon/2 \quad \text{if } |x - c| < 2\delta.$$

Take this δ and take r as in (3). Then

$$|y_r - c| \leq |y_r - x_r| + |x_r - c| < r + \delta < 2\delta,$$

so $|f(y_r) - f(c)| < \epsilon/2$. But also $|x_r - c| < \delta$, so $|f(x_r) - f(c)| < \epsilon/2$. Therefore, $|f(x_r) - f(y_r)| < \epsilon$, which is in contradiction with (2).

The construction that produces a point c for which (3) holds is short but tricky. For each $s > 0$ let

$$c_s = \sup\{x_r : 0 < r < s\}, \qquad (4)$$

and then let

$$c = \inf\{c_s : s > 0\}. \qquad (5)$$

If $\delta > 0$ is given, then by the definition of the greatest lower bound we can find s so that $c_s < c + \delta$. (Otherwise $c + \delta$ would be a lower bound.) Now, if $t < s$, then plainly $c_t \leq c_s$, for c_t is the least upper bound over a smaller set. Therefore, we can suppose that $s < \delta$. By the definition of the least upper bound and the definition of c_s, we can find $r < s$ so that $x_r > c_s - \delta$. For this r we have $r < \delta$, and

$$c - \delta \leq c_s - \delta < x_r \leq c_s < c + \delta,$$

which shows that c does have the property (3).

Exercise 3   Since the theorem is false without them, the hypotheses that the interval I on which f is defined is bounded and closed ought to have been used somewhere in the proof. Where were they used?

The construction of the point c is of general value. This point is called the *limit superior* of the x_r as $r \to 0$.

Exercise 4   Let g be a bounded function defined on a set S of real numbers, and let a be a point such that for each $\delta > 0$ there is at least one point $x \in S$ with $|x - a| < \delta$. Define the limit superior of $g(x)$ as $x \to a$ and $x \in S$. [*Hint:* In the above case $g(r) = x_r$ for $r > 0$, $a = 0$, and S is the set of positive real numbers.] Define also the limit inferior of $g(x)$ as $x \to a$ and $x \in S$. Show that

$$\lim_{\substack{x \to a \\ x \in S}} g(x)$$

exists if and only if

$$\limsup_{\substack{x \to a \\ x \in S}} g(x) = \liminf_{\substack{x \to a \\ x \in S}} g(x),$$

where the lim sup and the lim inf have the obvious meanings.

Exercise 5   Why is it required in Exercise 4 that for each $\delta > 0$ there is at least one point $x \in S$ with $|x - a| < \delta$?

Exercise 6   If f is uniformly continuous on the open interval (a, b), a and b finite, then the limits

$$\lim_{\substack{x \to a \\ x > a}} f(x) \quad \text{and} \quad \lim_{\substack{x \to b \\ x < b}} f(x)$$

both exist. Consequently, $f(a)$ and $f(b)$ can be defined so that f is continuous on the closed interval $[a, b]$.

3   MAXIMA AND MINIMA

THEOREM 3.1   *A continuous function on a closed bounded interval has a maximum and a minimum.*

Proof   According to Theorem 2.3, the function f is uniformly continuous, and then, according to Theorem 2.2, it is bounded. Let

$$M = \sup\{f(x) : x \in I\}.$$

The problem is to show that there is a point $c \in I$ with $f(c) = M$. If not, then the function

$$g(x) = \frac{1}{M - f(x)}$$

is continuous on I, so by Theorems 2.2 and 2.3 it is bounded. This is absurd, for by the definition of the least upper bound we can find a point x such that

$$M \geq f(x) \geq M - \epsilon \quad \text{hence } g(x) \geq 1/\epsilon,$$

and this for any positive ϵ.

In Chapter 1 we had the following theorem.

THEOREM 3.2   *If f has a maximum or minimum at a point c, and if f is differentiable at c, then $f'(c) = 0$.*

In practice the two theorems are used together to locate maxima and minima. There is one point to watch for. In order to use Theorem 3.1 to guarantee the existence of a maximum or minimum, the interval I on which the function f is defined must be bounded and closed. That is, I must include its end points. On the other hand, a function is never differentiable at the end points of the interval on which it is defined.   Thus, end points must always be considered separately.

Example 1   Find the maximum and minimum of

$$f(x) = 3x + 2\sqrt{1 - x^2} \qquad \text{on } 0 \leq x \leq 1.$$

Since f is continuous on the closed interval, the maximum and minimum exist.   At every point x except the end points f is differentiable, and

$$f'(x) = 3 + 2 \cdot \tfrac{1}{2}(1 - x^2)^{-1/2}(-2x) = 3 - 2x(1 - x^2)^{-1/2};$$

so $f'(x) = 0$ if and only if $x = 3/\sqrt{13}$.   By Theorem 3.2 this point and the two end points are the only possibilities for the maximum and minimum.   To settle which is which we calculate the value of f at all three.   We have

$$f(0) = 2 \qquad f(1) = 3 \qquad f\left(\frac{3}{\sqrt{13}}\right) = \sqrt{13}.$$

Thus, the maximum occurs at $3/\sqrt{13}$ and is equal to $\sqrt{13}$, while the minimum occurs at 0 and is equal to 2.

Exercise 1   Justify the statements about continuity and differentiability that were made in the example above.

Exercise 2   Find the maximum and minimum of $g(x, y) = 3x - 2y$ on the unit circle. [*Hint:* If (x, y) is on the unit circle, then $x^2 + y^2 = 1$, and we have:
　　(a) *First approach.*   Solve for y is a suitable way, and reduce the problem to Example 1.
　　(b) *Second approach.*   Do not solve for y explicitly, but use the formula for the derivative of a composite function.
　　(c) *Third approach.*   Use the fact that $x = \cos\theta$, $y = \sin\theta$ when (x, y) is on the unit circle.]

Example 2   The equation of the Florida coast is $y = x^2$.   A swimmer is at the point $(3, 0)$. How far is he from shore?

The distance from the point $(3, 0)$ to the point (x, y) is $\sqrt{(x - 3)^2 + y^2}$, so the function to be minimized is

$$d(x) = \sqrt{(x - 3)^2 + x^4}.$$

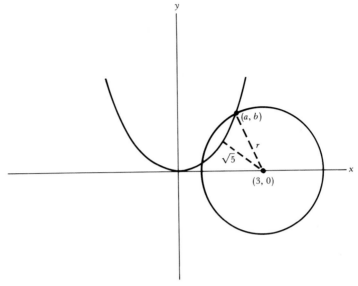

Figure 1

It simplifies matters to notice that the minimum of d and of $f(x) = d(x)^2$ must occur at the same point, for if $d(x) \geq d(a)$, then $d(x)^2 \geq d(a)^2$. Thus, the function to be minimized is

$$f(x) = (x - 3)^2 + x^4.$$

We have

$$f'(x) = 2(x - 3) + 4x^3 = 2(2x^3 + x - 3)$$
$$= 2(x - 1)(2x^2 + 2x + 3).$$

Thus, $f'(x) = 0$ if and only if $x = 1$.

This means that if the function does have a minimum, it must occur at $x = 1$, and the solution to the problem must be

$$d(1) = \sqrt{5}.$$

It can be shown that the function d does have a minimum in the following way. Choose any point (a, b) on the curve $y = x^2$, and let r be the distance from $(3, 0)$ to (a, b) (Figure 1). The points on the curve outside the circle with center $(3, 0)$ and radius r are clearly irrelevant. Their distance from $(3, 0)$ is greater than that of (a, b). In other words, if $|x - 3| > r$, then $d(x) > d(a)$; so it is sufficient to minimize d on the interval $3 - r \leq x \leq 3 + r$. Theorem 3.1 applies to this problem.

4   THE MEAN-VALUE THEOREM

The following theorem, called the *mean-value theorem*, is fundamental.

THEOREM 4.1   *Let f be continuous at every point of the closed interval I and differentiable at every interior point. If a and b are any two points of I, then there is a point ξ between them such that*

$$f(b) - f(a) = (b - a)f'(\xi).$$

A more general version is useful too and is easier to see how to prove because of the symmetry of the statement.

THEOREM 4.2   *Let f and g be continuous at every point of the closed interval I and differentiable at every interior point. If a and b are any two points of I, then there is a point ξ between them such that*

$$\big(f(b) - f(a)\big)g'(\xi) = \big(g(b) - g(a)\big)f'(\xi).$$

Theorem 4.1 is the special case in which $g(x) = x$.

Proof   Let $h(x) = \big(f(b) - f(a)\big)g(x) - \big(g(b) - g(a)\big)f(x)$ on the closed interval with end points a and b. By Theorem 3.1, h has both a maximum and a minimum. If either of these is not an end point, then it is suitable as ξ, for the equation $h'(\xi) = 0$ is just what is to be proved. Inspection shows that $h(b) = h(a)$, so the maximum and minimum cannot both occur at end points unless h is constant—in which case $h'(\xi) = 0$ for every ξ.

THEOREM 4.3   *Let f be continuous at each point of an interval I.*
    (a) If $f'(x) = 0$ at all but a finite number of points, then f is constant.
    (b) If $f'(x) \geq 0$ at all but a finite number of points, then f is increasing; that is, if $a \leq b$, then $f(a) \leq f(b)$.
    (c) If $f'(x) > 0$ at all but a finite number of points, then f is strictly increasing; that is, if $a < b$, then $f(a) < f(b)$.

Remark   Here it does not matter whether I is closed. It is taken closed in Theorems 4.1 and 4.2 in order not to exclude the end points. We shall prove part (a) of Theorem 4.3 and leave the other two parts as exercises.

Proof   The finite number of points, where $f'(x) = 0$ does not hold (and these may include points where f is not differentiable), divide I into a finite number of subintervals. Let a and b be any two points in one of these subintervals. Then the assumption of Theorem 4.1 is satisfied for the interval with end points a and b, and the theorem gives

$$f(b) - f(a) = (b - a)f'(\xi) = 0.$$

This shows that f is constant on each subinterval. But then it must be constant on the whole interval I, because it is continuous.

Part (a) of this theorem is one of the keys to the calculation of integrals. Recall that if

$$F(x) = \int_a^x f,$$

then $F'(x) = f(x)$, at least if f is continuous. Part (a) of the theorem shows that if G is any function whatever satisfying $G'(x) = f(x)$ at all but a finite number of points, then G and F differ by a constant; so

$$\int_a^b f = F(b) - F(a) = G(b) - G(a).$$

This was used in Section 5 of Chapter 1 and in Section 7 of Chapter 2.

Exercise 1   Let $f'(a) = 0$ and $f''(a) > 0$. Show that there is a positive number δ such that $f(x) \geq f(a)$ for every x with $|x - a| < \delta$. (This means that f has a "local" minimum at the point a.)

Exercise 2   Define "local" maximum. State and prove the corresponding theorem.

Exercise 3   Suppose that f and g are continuous on an interval I and that $f'(x) \leq g'(x)$ for all but a finite number of points. If $f(a) \leq g(a)$ for some point a, then $f(x) \leq g(x)$ for all points $x \geq a$. Show that $e^x \geq x^2 + 1$ for $x \geq 0$.

5   ZERO AND INFINITY

The limit of a quotient necessarily exists and is the quotient of the limits whenever the limit of the denominator is $\neq 0$. But the limit of the quotient may well exist even though the limit of the denominator is 0. One example that has been important already is the fact that

$$\lim_{\substack{x \to 0 \\ x \neq 0}} \frac{\sin x}{x} = 1.$$

Indeed, every derivative is this sort of limit.

There is a simple rule for the calculation of such limits, which says that

$$\lim_{\substack{x \to a \\ x \neq a}} \frac{f(x)}{g(x)} = \lim_{\substack{x \to a \\ x \neq a}} \frac{f'(x)}{g'(x)},$$

when both f and g have limit 0 or ∞ and the limit on the right exists. This rule is called l'Hospital's rule (after the Marquis de l'Hospital, who is revered by students everywhere for having written the first calculus book). The same rule also holds for left- and right-hand limits.

Example 1

$$\lim_{\substack{x \to 1 \\ x \neq 1}} \frac{1 - \sqrt{x}}{1 - x} = \lim_{\substack{x \to 1 \\ x \neq 1}} \frac{(-1/2)x^{-1/2}}{-1} = \frac{1}{2}.$$

Example 2

$$\lim_{\substack{x \to 0 \\ x \neq 0}} \frac{1 - \cos x}{x^2} = \lim_{\substack{x \to 0 \\ x \neq 0}} \frac{\sin x}{2x} = \frac{1}{2}.$$

Example 3

$$\lim_{\substack{x \to 0 \\ x > 0}} x \log x = \lim_{\substack{x \to 0 \\ x > 0}} \frac{\log x}{1/x}$$

$$= \lim_{\substack{x \to 0 \\ x > 0}} \frac{x^{-1}}{-x^{-2}} = 0.$$

Exercise 1   Show that

$$\lim_{\substack{x \to 0 \\ x > 0}} x^p \log x = 0$$

for every $p > 0$.

The theorem is as follows:

THEOREM 5.1

(*l'Hospital's Rule*) *Let f and g be differentiable on the open interval (a, b) with $g'(x) \neq 0$ for all x. Let f and g both have limit 0 or both have limit $\pm \infty$ as x approaches b from the left. If the limit*

$$\lim_{\substack{x \to b \\ x < b}} \frac{f'(x)}{g'(x)}$$

exists, then so does the limit

$$\lim_{\substack{x \to b \\ x < b}} \frac{f(x)}{g(x)},$$

and the two are equal.

DEFINITION
5.2

The statement

$$\lim_{\substack{x \to b \\ x < b}} f(x) = \infty$$

means that for every positive number M there is a positive number δ such that if $|x - b| < \delta$ *and* $x < b$*, then* $f(x) \geq M$.

Exercise 2 Give the definition of

$$\lim_{\substack{x \to b \\ x < b}} f(x) = -\infty.$$

The first step in the proof is to show that the function g' cannot change sign. It is always positive or always negative. If g' is continuous, then this follows from Theorem 1.2, but it is not assumed that g' is continuous.

LEMMA
5.3

Let g' *exist and be* $\neq 0$ *on* (a, b). *Then* g' *does not change sign.*

Proof

We show that if $g'(c) > 0$ and $g'(d) < 0$, then there is a point ξ between c and d with $g'(\xi) = 0$. Suppose that $c < d$, and let ξ be a point at which g is maximum on $[c, d]$. All that is necessary is to show that ξ cannot be either c or d. Since $g'(c) > 0$, it follows that $g(x) > g(c)$ for every x near c and to the right of c. Hence $\xi \neq c$. Since $g'(d) < 0$, it follows that $g(x) > g(d)$ for every x near d and to the left of d. Hence $\xi \neq d$.

Proof of Theorem
5.1

Multiplying by -1 if necessary, we can suppose that $g'(x) > 0$ for each x. Let

$$l = \lim_{\substack{x \to b \\ x < b}} \frac{f'(x)}{g'(x)},$$

and let ϵ be a given positive number. Choose $\delta > 0$ so that if $|x - b| < \delta$ and $x < b$, then

$$\left| \frac{f'(x)}{g'(x)} - l \right| < \epsilon; \qquad \text{hence } (l - \epsilon)g'(x) \leq f'(x) \leq (l + \epsilon)g'(x). \quad (1)$$

From the inequality (1) it follows that if c and d are any two points satisfying $b - \delta < c < d < b$, then

$$(l - \epsilon)(g(d) - g(c)) \leq f(d) - f(c) \leq (l + \epsilon)(g(d) - g(c)). \quad (2)$$

Indeed (for example), according to Theorem 4.3(b), the function $(l + \epsilon)g - f$ is increasing, so

$$(l + \epsilon)g(d) - f(d) \geq (l + \epsilon)g(c) - f(c),$$

which is the same as the right-hand inequality in (2).

At this point the cases where f and g have limit 0 and where f and g have limit ∞ separate. Take first the case where f and g have limit 0. In this case fix $c = x$ and let $d \to b$. The result is

$$-(l - \epsilon)g(x) \le -f(x) \le -(l + \epsilon)g(x) \qquad \text{for } b - \delta < x < b,$$

and division by $-g(x)$ gives

$$l - \epsilon \le \frac{f(x)}{g(x)} \le l + \epsilon \qquad \text{for } b - \delta < x < b,$$

as required. [Note that $g(x) < 0$, since g is strictly increasing and has limit 0 at b.]

Now suppose that f and g have limit $\pm \infty$. Of course, g must have limit $+\infty$ because of the hypothesis that $g' > 0$. Fix c in the inequality (2), take $d = x$, and divide by $g(x)$. The result is

$$(l - \epsilon)\left(1 - \frac{g(c)}{g(x)}\right) \le \frac{f(x)}{g(x)} - \frac{f(c)}{g(x)} \le (l + \epsilon)\left(1 - \frac{g(c)}{g(x)}\right). \qquad (3)$$

Since the limit of g is ∞ and c is fixed, we can choose $\delta_1 < \delta$ so that if $|x - b| < \delta_1$ and $x < b$, then

$$\left|\frac{g(c)}{g(x)}\right| < \epsilon \qquad \text{and} \qquad \left|\frac{f(c)}{g(x)}\right| < \epsilon.$$

Then we have

$$(l - \epsilon)(1 - \epsilon) - \epsilon \le \frac{f(x)}{g(x)} \le (l + \epsilon)(1 + \epsilon) + \epsilon \qquad \text{for } |x - b| < \delta_1.$$

This completes the proof.

Exercise 3   The theorem remains correct when

$$\lim_{\substack{x \to b \\ x < b}} \frac{f'(x)}{g'(x)} = \pm \infty.$$

(*Hint:* The proof is almost correct, but the expressions $l \pm \epsilon$ are nonsense.)

DEFINITION 5.4   *The statement* $\lim_{x \to \infty} f(x) = l$ *means that for every positive number ϵ there is a positive number r such that if $x > r$, then $|f(x) - l| < \epsilon$.*

Exercise 4   Give the definitions of

$$\lim_{x \to -\infty} f(x) = l, \qquad \lim_{x \to \infty} f(x) = \infty, \qquad \text{etc.,}$$

and show that Theorem 5.1 remains correct in all cases.

Exercise 5   For every $a > 1$, $p > 0$, $\lim_{x \to \infty} x^p a^{-x} = 0$.

Exercise 6   For every $p > 0$, $\lim_{x \to \infty} (\log x / x^p) = 0$.

Example 4   The function $f(x) = (1 + 1/x)^x$ is strictly increasing and $\lim_{x \to \infty} f(x) = e$.

It is equivalent to show that the function $g(y) = (1 + y)^{1/y}$ is strictly decreasing on $y > 0$ and has limit e as $y \to 0$. The obvious idea is to consider

$$h(y) = \log g(y) = \frac{\log(1 + y)}{y}.$$

It is equivalent to show that h is strictly decreasing on $y > 0$ and has limit 1 as $y \to 0$.

The fact that h has limit 1 is evident from l'Hospital's rule, so what remains is to show that $h'(y) < 0$ for $y > 0$. Now,

$$h'(y) = \frac{y - (1 + y) \log(1 + y)}{(1 + y)y^2},$$

so it is enough to prove that $k(y) = y - (1 + y) \log(1 + y) < 0$ for $y > 0$. Since $k(0) = 0$, it is enough to prove that k is decreasing, that is, that $k'(y)$ is negative. But

$$k'(y) = -\log(1 + y),$$

and this is negative because $1 + y > 1$.

4 Riemann Integration

1 AREA

It is time to reconsider the problem of area to see just what it means to say that the area under a curve exists.

Let f be a nonnegative bounded function on the interval $[a, b] = \{x : a \leq x \leq b\}$, and let A be the set under the curve $y = f(x)$,

$$A = \{(x, y) : a \leq x \leq b \text{ and } 0 \leq y \leq f(x)\}.$$

The natural approach is to approximate the set A by a union of rectangles. The way to do this is to divide the interval $[a, b]$ into small intervals, and on each small interval replace f by a constant. First we shall do this in such a way as to produce a set that contains A and, therefore, should have a larger area. Choose a finite sequence of points

$$a = x_0 < x_1 < \cdots < x_n = b.$$

For each i let

$$M_i = \sup\{f(x) : x_{i-1} \leq x \leq x_i\},$$

and let $\bar{R}_i$ be the rectangle with base $[x_{i-1}, x_i]$ and height M_i. It is clear that the set A is contained in the set

$$\bar{R} = \bigcup_{i=1}^{n} \bar{R}_i.$$

The picture looks as shown in Figure 1.

The sequence $(x_0, x_1, \ldots, x_n)$ is called a *partition* of the interval $[a, b]$. Of course, the numbers and sets defined above depend on the particular partition. If p is a partition, $\bar{R}(p)$ will denote the corresponding set $\bar{R}$. In this notation, the assertion of the last paragraph is that for every partition p,

$$A \subset \bar{R}(p). \tag{1}$$

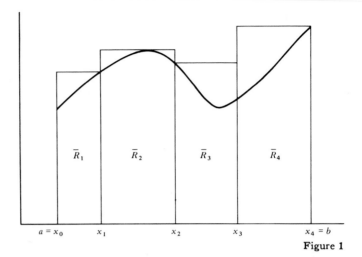

Figure 1

The area of $\bar{R}(p)$ is obviously the sum

$$\bar{S}(p) = \sum_{i=1}^{n} M_i(x_i - x_{i-1}). \tag{2}$$

Consequently, if the area of the set A does make sense, then

$$\text{area } A \leq \bar{S}(p),$$

which says that the area of A is a lower bound of all the numbers $\bar{S}(p)$. It is, therefore, $\leq$ their greatest lower bound. In other words,

$$\text{area } A \leq \bar{S} = \inf\{\bar{S}(p) : p \text{ is a partition of } [a, b]\}. \tag{3}$$

There is a similar construction to produce a union of rectangles which is contained in A. All that is necessary is to replace the least upper bound M_i by the greatest lower bound

$$m_i = \inf\{f(x) : x_{i-1} \leq x \leq x_i\}.$$

If $\underline{R}_i$ is the rectangle with base $[x_{i-1}, x_i]$ and height m_i, and

$$\underline{R}(p) = \bigcup_{i=1}^{n} \underline{R}_i,$$

then clearly $A \supset \underline{R}(p)$, so that

$$\text{area } A \geq \underline{S}(p) = \sum_{i=1}^{n} m_i(x_i - x_{i-1}); \tag{4}$$

hence

$$\text{area } A \geq \underline{S} = \sup\{\underline{S}(p): p \text{ is a partition of } [a, b]\}.$$

The two inequalities (3) and (5) give

$$\underline{S} \leq \text{area } A \leq \bar{S}. \tag{6}$$

If it happens that $\underline{S} = \bar{S}$, then there is no doubt about what the area must be—it must be the common value of $\underline{S}$ and $\bar{S}$. The idea is to start afresh and to use this as the definition of area.

DEFINITION 1.1   *The area of the set A exists if $\underline{S} = \bar{S}$. If the area does exist, then it is the common value of $\underline{S}$ and $\bar{S}$.*

The program for what must be done is fairly clear.

1. We must show that the definition makes sense in an analytical framework that is independent of the intuitive idea of area.

2. We must show that the area does exist when the function f is reasonable—say continuous.

3. We must show that the notion of area defined in this way has the properties that area ought to have.

In connection with point 2, consider the function f on $[0, 1]$, which is equal to 1 at each irrational point and to 0 at each rational point. Any subinterval contains both rational and irrational points, so each m_i is equal to 0 and each M_i is equal to 1. Hence, for any partition p, $\underline{S}(p) = 0$ and $\bar{S}(p) = 1$; consequently, $\underline{S} = 0$ and $\bar{S} = 1$. According to Definition 1.1, the area under the curve does not exist. The present theory of area and integration is designed basically for continuous functions, or at least functions that are almost continuous. In Chapter 13 there is a more general theory that does assign an area to this particular set.

Although the collected works of Riemann fill only one volume, he was one of the most profound and original mathematicians of all time.

Bernhard Riemann

2 INTEGRALS

The foregoing developments can be carried out in an analytical framework that has nothing to do with area. The expressions to consider are suggested by area, but they can be interpreted in many other interesting and useful ways.

Let f be a bounded function on the interval $[a, b]$. If p is a partition of $[a, b]$, the numbers M_i and m_i are defined as before, and the sums $\bar{S}(p)$ and $\underline{S}(p)$ are defined by

$$\bar{S}(p) = \sum_{i=1}^{n} M_i(x_i - x_{i-1}), \qquad \underline{S}(p) = \sum_{i=1}^{n} m_i(x_i - x_{i-1}). \tag{1}$$

Exercise 1 The sets $\bar{R}(p)$ and $\underline{R}(p)$ could also be defined but would not be useful. Why? (Two reasons.)

It is clear that $\underline{S}(p) \leq \bar{S}(p)$ for a single partition p, but we shall have to find a new proof of the fact that $\underline{S}(q) \leq \bar{S}(p)$ for any two partitions p and q. [The old proof depended on the interpretation of $\underline{S}(p)$ as the area of $\underline{R}(p)$, etc.] The new one is based on an investigation of what happens to $\underline{S}(p)$ and $\bar{S}(p)$ when points are added to the partition p. This investigation is important in other connections as well.

DEFINITION 2.1 *The partition r is a refinement of p, written $r \prec p$, if every point in p is also in r.*

LEMMA 2.2 *If $r \prec p$, then $\underline{S}(r) \geq \underline{S}(p)$ and $\bar{S}(r) \leq \bar{S}(p)$.*

Proof It is enough to consider the case where r is obtained from p by adding just one point y. Suppose that y is between x_{i-1} and x_i, and let

$$M_i' = \sup\{f(x) : x_{i-1} \leq x \leq y\}, \qquad M_i'' = \sup\{f(x) : y \leq x \leq x_i\}.$$

The only difference between $\bar{S}(r)$ and $\bar{S}(p)$ is that the former contains $M_i'(y - x_{i-1}) + M_i''(x_i - y)$ in place of $M_i(x_i - x_{i-1})$. It is obvious from the definitions that $M_i' \leq M_i$ and $M_i'' \leq M_i$, so

$$M_i'(y - x_{i-1}) + M_i''(x_i - y) \leq M_i(y - x_{i-1}) + M_i(x_i - y) = M_i(x_i - x_{i-1}).$$

This proves that $\bar{S}(r) \leq \bar{S}(p)$, and the other inequality is proved similarly.

LEMMA 2.3 *If p and q are any two partitions, then*

$$\underline{S}(q) \leq \bar{S}(p).$$

Proof If r is the common refinement of p and q (i.e., the partition whose points are those of p together with those of q), then Lemma 2.2 gives

$$\underline{S}(q) \leq \underline{S}(r) \leq \bar{S}(r) \leq \bar{S}(p). \tag{2}$$

Having Lemma 2.3, we can proceed as before. Fix q and take the lower bound on p to obtain $\underline{S}(q) \leq \bar{S}$. Then take the upper bound on q to obtain $\underline{S} \leq \bar{S}$. This gives

LEMMA
2.4

For any partition p,

$$\underline{S}(p) \leq \underline{S} \leq \bar{S} \leq \bar{S}(p); \qquad hence\; 0 \leq \bar{S} - \underline{S} \leq \bar{S}(p) - \underline{S}(\mathrm{p}).$$

The importance of the lemma is to show that in order to prove that $\underline{S} = \bar{S}$ it is sufficient to prove that for each positive number ϵ there is some partition p with $\bar{S}(p) - \underline{S}(p) < \epsilon$. And Lemma 2.2 shows that the situation is improved by refinement.

DEFINITION
2.5

The function f is Riemann integrable on the interval $[a, b]$ if $\underline{S} = \bar{S}$. If this is the case, the common value of $\underline{S}$ and $\bar{S}$ is called the integral of f and is written

$$\int_a^b f \qquad or \qquad \int_a^b f(x)\; dz.$$

Let us begin by showing that every continuous function is Riemann integrable.

THEOREM
2.6

If f is continuous on $[a, b]$, then f is Riemann integrable.

Proof

According to Lemma 2.4, we have to show that if $\epsilon > 0$ is given, then we can produce some partition p such that $\bar{S}(p) - \underline{S}(p) < \epsilon$. We shall show that in fact this holds for every sufficiently fine partition, the fineness being measured by the number

$$|p| = \max x_i - x_{i-1}.$$

Let $\epsilon > 0$ be given and use the uniform continuity of f to find $\delta > 0$ such that $|f(x) - f(y)| \leq \epsilon$ if $|x - y| \leq \delta$, hence such that

$$f(x) \leq f(y) + \epsilon \qquad if\; |x - y| \leq \delta. \tag{3}$$

This inequality implies that

$$M_i - m_i \leq \epsilon \qquad if\; |p| \leq \delta. \tag{4}$$

Indeed, if x and y are both in $[x_{i-1}, x_i]$, then $|x - y| \leq \delta$, so the inequality in (3) holds. First fix y and take the upper bound on x to obtain $M_i \leq f(y) + \epsilon$; hence $M_i - \epsilon \leq f(y)$. Now take the lower bound on y to obtain

$M_i - \epsilon \leq m_i$, which is just (4).   The inequality (4) gives

$$\bar{S}(p) - \underline{S}(p) = \sum_{i=1}^{n} (M_i - m_i)(x_i - x_{i-1}) \leq \epsilon \sum_{i=1}^{n} x_i - x_{i-1} = \epsilon(b - a).$$

[In order to come out with ϵ, we would of course simply start with $\epsilon/(b - a)$ instead of ϵ.]

THEOREM 2.7

If f and g are Riemann integrable on $[a, b]$ and α is a real number, then αf and $f + g$ are integrable and

$$\int_a^b \alpha f = \alpha \int_a^b f \quad and \quad \int_a^b (f + g) = \int_a^b f + \int_a^b g.$$

Exercise 2   Prove the theorem by showing that

$$\underline{S}(p; f) + \underline{S}(p; g) \leq \underline{S}(p; f + g) \leq \bar{S}(p; f + g) \leq \bar{S}(p; f) + \bar{S}(p; g).$$

THEOREM 2.8

If f is Riemann integrable on $[a, b]$, then f is Riemann integrable on each subinterval.   Conversely, let c be a point between a and b.   If f is Riemann integrable on $[a, c]$ and on $[c, b]$, then f is Riemann integrable on $[a, b]$ and

$$\int_a^b f = \int_a^c f + \int_c^b f.$$

Proof

Let c be a point between a and b.   If p' is a partition of $[a, c]$, p'' a partition of $[c, b]$, and p the partition of $[a, b]$ obtained by taking the points of p' together with those of p'', then it is obvious that

$$\bar{S}(p) = \bar{S}(p') + \bar{S}(p'') \quad and \quad \underline{S}(p) = \underline{S}(p') + \underline{S}(p''). \qquad (5)$$

Therefore,

$$\bar{S}(p) - \underline{S}(p) = \bar{S}(p') - \underline{S}(p') + \bar{S}(p'') - \underline{S}(p''). \qquad (6)$$

First suppose that f is integrable on $[a, c]$ and on $[c, b]$.   For any $\epsilon > 0$ we can choose p' and p'' so that both terms on the right of (6) are $< \epsilon$.   Then $\bar{S}(p) - \underline{S}(p) < 2\epsilon$, and since ϵ is arbitrary, it follows that f is integrable on $[a, b]$.   Prove the addition formula as an exercise by using formula (5).

Next suppose that f is integrable on $[a, b]$.   Choose a partition p of $[a, b]$ such that $\bar{S}(p) - \underline{S}(p) < \epsilon$.   Since this is only improved by refinement, it can be assumed that c is one of the points in p, in which case p decomposes into a p' and a p''.   Then formula (6) shows that $\bar{S}(p') - \underline{S}(p') < \epsilon$, which implies that f is integrable on $[a, c]$, and also shows that $\bar{S}(p'') - \underline{S}(p'') < \epsilon$, which implies that f is integrable on $[c, b]$.

If $[c, d]$ is any subinterval and f is integrable on $[a, b]$, then what has been proved shows first that f is integrable on $[a, d]$, and then that f is integrable on $[c, d]$.

THEOREM 2.9

If f is Riemann integrable on $[a, b]$ and if $m \leq f(x) \leq M$, then

$$m(b - a) \leq \int_a^b f \leq M(b - a).$$

Proof

It is clear either from the definition or from Lemma 2.2 that for each partition p we have

$$m(b - a) \leq \underline{S}(p) \leq \bar{S}(p) \leq M(b - a), \qquad (7)$$

and, consequently,

$$m(b - a) \leq \underline{S} \leq \bar{S} \leq M(b - a). \qquad (8)$$

So far the only integrable functions we know are the continuous functions. This can be improved easily to allow at least a finite number of discontinuities.

THEOREM 2.10

If f is bounded on $[a, b]$ and continuous at all but a finite number of points, then f is Riemann integrable.

Proof

Because of Theorem 2.8, it can be supposed that f is continuous at all but one of the end points, say the end point a. Let ϵ be a given positive number, let $|f(x)| \leq M$, and let $c = a + \epsilon/M$.

For any partition p' of $[a, c]$ we have

$$\bar{S}(p') - \underline{S}(p') \leq 2M(c - a) = 2\epsilon.$$

Since f is continuous on $[c, b]$, Theorem 2.6 shows that for any sufficiently fine partition p'' of $[c, b]$ we have

$$\bar{S}(p'') - \underline{S}(p'') < \epsilon.$$

If p is the partition of $[a, b]$ determined by p' and p'', then formula (6) gives $\bar{S}(p) - \underline{S}(p) < 3\epsilon$, which proves the theorem.

Theorems 2.8 and 2.9 give the properties that were used in Section 5 of Chapter 1 to prove the fundamental theorem of calculus, which expresses the relation between the integral and the derivative:

THEOREM 2.11

*(**Fundamental Theorem of Calculus**)   If f is Riemann integrable on $[a, b]$ and*

$$F(x) = \int_a^x f,$$

then F is continuous on $[a, b]$ and $F'(x) = f(x)$ at every point x where f is continuous.

Exercise 3   Look back at the proof of Theorem 5.2 of Chapter 1, and then prove this theorem.

DEFINITION 2.12   *Let f be a function on an interval I. A primitive of f is a function G that is continuous on I and satisfies $G'(x) = f(x)$ at all but a finite number of points.*

THEOREM 2.13   *Let f be bounded on $[a, b]$ and continuous at all but a finite number of points.*
   (a) *The function*
$$F(x) = \int_a^x f$$
is a primitive of f.
   (b) *If G is any primitive of f, then*
$$\int_a^b f = G(b) - G(a).$$

Proof   Part (a) comes from Theorem 2.11. Part (b) comes from this and Theorem 4.3 of Chapter 3, which gives a number α such that $G(x) = F(x) + \alpha$, and hence shows that $G(b) - G(a) = F(b) - F(a)$.

The expression $G(b) - G(a)$ recurs so frequently in the calculation of integrals that it is convenient to have a special symbol for it. The one commonly used is
$$G(x) \Big|_a^b.$$
With this notation we would write, for example,
$$\int_{-1}^{3} x^2 \, dx = \frac{x^3}{3} \Big|_{-1}^{3} = 9 - \frac{1}{3} = 8\frac{2}{3}.$$
Sometimes it is convenient to consider the integral $\int_a^b f$ when $b < a$. In this case it is defined by
$$\int_a^b f = -\int_b^a f.$$

Exercise 4   Show that $\int_a^b f + \int_b^c f = \int_a^c f$ no matter what the relative positions of a, b, and c. Show that if two of the integrals exist, so does the third.

Exercise 5   If $F(x) = \int_x^a f$, what is $F'(x)$?

Exercise 6   If $F(x) = \int_x^{x^2} f$, what is $F'(x)$?   [*Hint:* First set $G(y) = \int_a^y f$, and note that the derivative of $G(x^2)$ can be obtained from the chain rule.]

3   ELEMENTARY FUNCTIONS

The usual way to calculate an integral $\int_a^b f(x)\, dx$ is to find a primitive F and use Theorem 2.13, which says that

$$\int_a^b f(x)\, dx = F(b) - F(a),$$

although there are occasional interesting cases where the integral can be calculated for some particular a and b, while the primitive cannot.   The primitives of the elementary functions are as follows, with the symbol $\int f(x)\, dx$ standing for some primitive.

$$\int x^n\, dx = \frac{x^{n+1}}{n+1} \qquad \text{for } n \neq -1. \tag{a}$$

$$\int \frac{1}{x}\, dx = \log|x|. \tag{b}$$

$$\int \sin x\, dx = -\cos x. \tag{c}$$

$$\int \cos x\, dx = \sin x. \tag{d}$$

$$\int \tan x\, dx = -\log|\cos x|. \tag{e}$$

$$\int \cot x\, dx = \log|\sin x|. \tag{f}$$

$$\int \sec x\, dx = \log|\sec x + \tan x|. \tag{g}$$

$$\int \csc x\, dx = -\log|\csc x + \cot x|. \tag{h}$$

$$\int e^x\, dx = e^x. \tag{i}$$

$$\int \log|x|\, dx = x \log|x| - x. \tag{j}$$

The formulas are simply verified by differentiation—with a couple of precautions.   The assertion is that the formulas hold on any interval where they make sense, for primitives are only defined on intervals.   For example, it is not claimed that when n is negative, formula (a) holds on the whole line.

It holds on the interval $x > 0$ and on the interval $x < 0$, but not on any interval containing 0.

The derivative of $\log|x|$ has been discussed at points $x > 0$ (where $\log|x| = \log x$), but not at points $x < 0$. It must be shown that it is $1/x$ for $x < 0$ as well. This can be done immediately by the composite function formula.

To provide one example of the way the calculations go, consider formula (e). The function $h(x) = -\log|\cos x|$ is the composite of $f(y) = -\log|y|$ and $g(x) = \cos x$. Now $f'(y) = -1/y$ and $g'(x) = -\sin x$; therefore,

$$h'(x) = f'(y)g'(x) = \frac{\sin x}{y} = \frac{\sin x}{\cos x} = \tan x.$$

The calculation of derivatives is an orderly business. There are simple formulas for the derivatives of the elementary functions and for all their usual combinations. The combinations may be complicated, but patience and care do the job.

Calculation of integrals is more primitive. Patience and care can well go unrewarded while the prizes go to experience and cunning. The reason is that, although there are formulas for the primitives of the elementary functions, there are no formulas for combinations, except for sums and constant multiples (Theorem 2.7).

There are some general formulas, particularly for transforming one integral into another. The procedure is to use these to transform this way and that, until finally something recognizable is reached. Cunning in making the transformations and experience in recognizing the results are the important factors. These general formulas and some examples are given in the next sections.

4   CHANGE OF VARIABLE

The most effective tool for the calculation of an integral

$$\int_a^b f(x)\,dx$$

is a suitable change of variable,

$$x = \varphi(t).$$

The formula for making this change is nothing more than an integrated version of the chain rule. If F is a primitive of f and $G(t) = F(\varphi(t))$, the chain rule gives

$$G'(t) = F'(\varphi(t))\varphi'(t) = f(\varphi(t))\varphi'(t); \tag{1}$$

so if $a = \varphi(\alpha)$ and $b = \varphi(\beta)$, then Theorem 2.13 gives

$$\int_a^b f(x)\,dx = F(b) - F(a) = G(\beta) - G(\alpha) = \int_\alpha^\beta f(\varphi(t))\varphi'(t)\,dt. \qquad (2)$$

THEOREM 4.1

The formula for making the change of variable $x = \varphi(t)$ is

$$\int_a^b f(x)\,dx = \int_\alpha^\beta f(\varphi(t))\varphi'(t)\,dt, \qquad \text{where } a = \varphi(\alpha) \text{ and } b = \varphi(\beta). \qquad (3)$$

It holds under the following conditions on f and φ:

(a) *φ is continuous at every point of $[\alpha, \beta]$ and is differentiable at all but a finite number of points.*

(b) *f is continuous on $\varphi([\alpha, \beta])$.*

(c) *$f(\varphi(t))\varphi'(t)$ is continuous at all but a finite number of points of $[\alpha, \beta]$ and is bounded.*

Ordinarily the symbol $[\alpha, \beta]$ is used only when $\alpha < \beta$. Here it does not matter—$[\alpha, \beta]$ designates the closed interval with end points α and β, whatever their relative positions.

Proof

What we have to do is justify the steps that were sketched in the opening paragraph. There is just one minor problem, which will be visible in a moment.

From the results of Chapter 3 it is known that $\varphi([\alpha, \beta])$ is an interval. In fact, if m and M are the minimum and maximum of φ, then $\varphi([\alpha, \beta])$ is the interval $[m, M]$. The chain rule asserts that formula (1) holds at any point t such that φ is differentiable at t and F is differentiable at $\varphi(t)$. The minor problem is that there may be infinitely many points t with $\varphi(t) = m$ or $\varphi(t) = M$. At these points the formula does not hold because of the technicality that F is not differentiable at the end points of the interval on which it is defined. The first step is to take care of these.

LEMMA 4.2

Let f be continuous on $[m, M]$, and define

$$\tilde{f}(x) = \begin{cases} f(x) & \text{for } m \le x \le M, \\ f(m) & \text{for } x < m, \\ f(M) & \text{for } x > M. \end{cases}$$

Then $\tilde{f}$ is continuous at every real x.

Exercise 1 Prove the lemma and draw a picture.

The proof is achieved as follows. Choose any point c and set

$$\tilde{F}(x) = \int_c^x \tilde{f}.$$

By Theorems 2.6 and 2.11, $\tilde{F}(x)$ is defined for every real x and satisfies $\tilde{F}'(x) = \tilde{f}(x)$ for every real x. Here we do not have to exclude any end points. Set $G(t) = \tilde{F}(\varphi(t))$ for $\alpha \le t \le \beta$. By the chain rule we have

$$G'(t) = f(\varphi(t))\varphi'(t)$$

at every point where φ is differentiable, that is, at all but a finite number of points. Hence Theorem 2.13 gives

$$\int_a^b f(x)\,dx = \tilde{F}(b) - \tilde{F}(a) = G(\beta) - G(\alpha) = \int_\alpha^\beta f(\varphi(t))\varphi'(t)\,dt.$$

The best way to remember the change-of-variable formula is as follows:

1. *Replace x by $\varphi(t)$ and dx by $\varphi'(t)\,dt$.*
2. *Replace the limits a and b by points α and β such that $a = \varphi(\alpha)$ and $b = \varphi(\beta)$.*

This rule is one of the main reasons for writing $\int_a^b f(x)\,dx$ instead of $\int_a^b f$. It is easier to remember to replace x by $\varphi(t)$ and dx by $\varphi'(t)\,dt$ than it is to remember to replace f by the function g defined by $g(t) = f(\varphi(t))\varphi'(t)$. There are other reasons too, but for the present the dx can be considered simply a mnemonic device with no meaning whatever. It should be emphasized that

$$\int_a^b f(x)\,dx, \qquad \int_a^b f(y)\,dy, \qquad \int_a^b f(u)\,du,$$

and so on, all have exactly the same meaning.

Example 1 Calculate $\int_0^1 x\sqrt{1-x^2}\,dx$.
Make the change of variable

$$x = (1-t)^{1/2}; \qquad \text{hence } dx = -\tfrac{1}{2}(1-t)^{-1/2}\,dt.$$

Then

$$\int_0^1 x\sqrt{1-x^2}\,dx = -\tfrac{1}{2}\int_1^0 (1-t)^{1/2}t^{1/2}(1-t)^{-1/2}\,dt = \tfrac{1}{2}\int_0^1 t^{1/2}\,dt$$
$$= \tfrac{1}{3}t^{3/2}\Big|_0^1 = \tfrac{1}{3}.$$

Exercise 2 This example shows that we would not want to simplify condition (c) of Theorem 4.1 to read that φ' itself is bounded. Discuss.

Exercise 3 The problem in making a change of variable is to discover the right one. Discover the right one in the example above by working backward, that is, by putting $t = 1 - x^2$, and $dt = -2x\,dx$. Discuss why this works.

Example 2   Calculate $\int \sin^m x \cos^n x \, dx$, where m and n are nonnegative integers.

First suppose that m is odd, say $m = 2k + 1$.   Then

$$\int \sin^m x \cos^n x \, dx = \int \sin^{2k} x \cos^n x \sin x \, dx$$
$$= \int (1 - \cos^2 x)^k \cos^n x \sin x \, dx.$$

When $(1 - \cos^2 x)^k$ is multiplied out, the integral becomes a sum of terms that look like this:

$$\int \cos^r x \sin x \, dx.$$

Now it is plain what to do.   Put $t = \cos x$, in which case $dt = -\sin x \, dx$; so the integral is

$$-\int t^r \, dt = -\frac{t^{r+1}}{r+1} = -\frac{\cos^{r+1} x}{r+1}.$$

If m is even but n is odd, the procedure is entirely similar.

Now, suppose that both m and n are even, say $m = 2k$ and $n = 2l$.   In this case use the identities

$$\cos^2 x = \frac{1 + \cos 2x}{2}, \qquad \sin^2 x = \frac{1 - \cos 2x}{2}.$$

The integral becomes

$$\int \left(\frac{1 - \cos 2x}{2}\right)^k \left(\frac{1 + \cos 2x}{2}\right)^l dx,$$

which is a sum of terms that look like this:

$$\int \cos^r 2x \, dx = \frac{1}{2} \int \cos^r t \, dt \qquad \text{with } r \le k + l = \frac{m+n}{2}.$$

This is the same kind of problem as the original problem, but with an exponent r that is at most half the sum of the original exponents.   When r is odd, the first part of the discussion settles the matter.   When r is even, a repetition of the same procedure cuts the exponent in half again.

When m or n is large (and both are even), the calculation is impossibly long, but a finite number of steps does reduce the exponent to 0 or to an odd number. Explicit formulas can be given, but they do not seem particularly relevant at this stage.

Exercise 4   Calculate $\int \sin^4 x \cos^2 x \, dx$.

Exercise 5   Calculate $\int \sin^3 x \sqrt{\cos x} \, dx$, and state a general theorem about this situation.

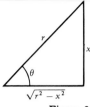

Figure 2

Example 3 $\int_{-r}^{r} \sqrt{r^2 - x^2}\, dx.$

This is the integral for the area of a semicircle.

When the integrand (function to be integrated) contains a sum or difference of squares, it is often profitable to make a trigonometric change of variable. The idea is to draw a right triangle in which the squared quantities appear on two of the sides.   In this case r goes on the hypotenuse and x on either of the other sides.   (If the integrand contained $x^2 - r^2$, then x would go on the hypotenuse and r on either of the other sides, while if it contained $x^2 + r^2$, both would go on the other sides.)

In the present case it looks as shown in Figure 2, so that $x/r = \sin \theta$. Thus, $x = r \sin \theta$ and $dx = r \cos \theta\, d\theta$.   The lower limit is $\alpha = -\pi/2$ (or any other point α with $\sin \alpha = -1$?), and the upper limit is $\beta = \pi/2$.   Inspection of the triangle shows that $\sqrt{r^2 - x^2}/r = \cos \theta$, so the integral becomes

$$r^2 \int_{-\pi/2}^{\pi/2} \cos^2 \theta\, d\theta = r^2 \int_{-\pi/2}^{\pi/2} \frac{1 + \cos 2\theta}{2}\, d\theta$$

$$= \frac{r^2}{2}\left(\theta + \frac{\sin 2\theta}{2}\right)\Bigg|_{-\pi/2}^{\pi/2} = \frac{\pi r^2}{2}. \tag{4}$$

What about the limits of integration?   We have taken $\alpha = -\pi/2$ and $\beta = \pi/2$, but it would appear from Theorem 4.1 that we can take α to be any point such that $\sin \alpha = -1$, and β to be any point such that $\sin \beta = 1$.   If we take, for instance, $\alpha = -\pi/2$, $\beta = 5\pi/2$, then we seem to get $3\pi r^2/2$, which is not the answer above.

Exercise 6   Where is the trouble?

5   INTEGRATION BY PARTS

The formula for integration by parts is nothing more than an integrated version of the formula for the derivative of a product: $(fg)' = f'g + fg'$.   If this is integrated from a to b the result is

$$\int_a^b f'g \, dx + \int_a^b fg' \, dx = fg \Big|_a^b. \tag{1}$$

The formula is useful when the integrand can be written as a product in such a way that one factor can be integrated and the other differentiated with a net effect that is good.

Example 1 Calculate $\int_a^b x \sin x \, dx$.

If x is differentiated and $\sin x$ is integrated, the effect is clearly good. Therefore, let $g(x) = x$ and $f(x) = -\cos x$. Then

$$\begin{aligned}
\int_a^b x \sin x \, dx = \int_a^b f'g \, dx &= -\int_a^b fg' \, dx + fg \Big|_a^b \\
&= \int_a^b \cos x \, dx - x \cos x \Big|_a^b \\
&= \sin b - b \cos b - \sin a + a \cos a.
\end{aligned}$$

Another notation is sometimes easier to remember. With the convention that if u is a function of x, then $du = u' \, dx$, formula (1) reads

$$\int_a^b u \, dv + \int_a^b v \, du = uv \Big|_a^b. \tag{2}$$

Example 2 Calculate $\int_1^3 \log x \, dx$.

Clearly, it will be helpful to differentiate $\log x$ and to integrate 1. Therefore, let $u = \log x$ and $dv = dx$. Then $du = 1/x \, dx$ and $v = x$. Then

$$\begin{aligned}
\int_1^3 \log x \, dx = \int_1^3 u \, dv \\
= -\int_1^3 v \, du + uv \Big|_1^3 &- \int_1^3 dx + x \log x \Big|_1^3 \\
= -2 + 3 \log 3.
\end{aligned}$$

Example 3 Calculate $\int_0^x e^t \cos t \, dt$.

This time the trick is to integrate by parts twice. Things do not look so good after the first one, but they look much better after the second.

$$\begin{aligned}
\int_0^x e^t \cos t \, dt &= -\int_0^x e^t \sin t \, dt + e^x \sin x \\
&= -\int_0^x e^t \cos t \, dt + e^x \cos x - 1 + e^x \sin x,
\end{aligned}$$

from which it follows that

$$\int_0^x e^t \cos t \, dt = \frac{e^x \cos x + e^x \sin x - 1}{2}.$$

The relevant theorem, which results directly from the definition of a primitive and the formula for the derivative of a product, is as follows:

THEOREM 5.1   *Let f and g be continuous on [a, b] and differentiable at all but a finite number of points.   Let f′ and g′ be bounded and continuous at all but a finite number of points.   Then*

$$\int_a^b f'g \, dx + \int_a^b fg' \, dx = fg \Big|_a^b.$$

Exercise 1   Write out the proof of Theorem 5.1 in full detail.

6   RIEMANN SUMS

If f is integrable, then for any partition p

$$\underline{S}(p) \leq \underline{S} = \int_a^b f(x) \, dx = \bar{S} \leq \bar{S}(p). \tag{1}$$

Moreover, for *certain* partitions p, the sums $\underline{S}(p)$ and $\bar{S}(p)$ approximate the integral as well as we please.   We shall show now that these sums approximate the integral for *every* sufficiently fine partition.   In view of formula (1) it is enough to prove the following Theorem.

THEOREM 6.1   *Let f be integrable.   For every positive number ϵ there is a positive number δ such that if p is any partition with $|p| < \delta$, then*

$$\bar{S}(p) - \underline{S}(p) < \epsilon.$$

Proof   Let $\epsilon > 0$ be given.   Since f is integrable, there is some partition q such that $\bar{S}(q) - \underline{S}(q) < \epsilon$.   Let N be the number of points in q, and let δ_0 be the length of the smallest interval.   Let $M = \sup\{f(x) : a \leq x \leq b\}$.   We shall prove the following assertion.

LEMMA 6.2   *If p is any partition with $|p| \leq \delta < \delta_0$, then*

$$\bar{S}(p) - \underline{S}(p) < \epsilon + 4MN\delta.$$

This will prove the theorem, for M and N are fixed numbers, and we can take δ as small as we please.

Proof of the Lemma   Let r be the common refinement of p and q. Since r is a refinement of q, we have

$$\bar{S}(r) - \underline{S}(r) \le \bar{S}(q) - \underline{S}(q) < \epsilon. \tag{2}$$

On the other hand, the sums for r and p can be compared. The terms in $\bar{S}(r)$ and $\bar{S}(p)$ are the same with the following exceptions. If y is a point of r that is not in p, then the two adjacent points x_{i-1} and x_i must belong to p. This is the effect of taking $\delta < \delta_0$. $\bar{S}(p)$ contains the term $M_i(x_i - x_{i-1})$, while in place of it $\bar{S}(r)$ contains the sum $M_i'(y - x_{i-1}) + M_i''(x_i - y)$, where M_i' and M_i'' are the suprema over the two smaller intervals $[x_{i-1}, y]$ and $[y, x_i]$. The error committed in replacing the single term by the sum of the two is clearly at most $2M(x_i - x_{i-1}) < 2M\delta$. This error is committed at worst at the N points of q, so the total error is at most $2MN\delta$. Thus, we have

$$\bar{S}(p) < \bar{S}(r) + 2MN\delta,$$

and, similarly,

$$\underline{S}(p) > \underline{S}(r) - 2MN\delta.$$

Combining these with (2), we get the lemma.

There are other sums associated with a partition p that lie between $\underline{S}(p)$ and $\bar{S}(p)$, and hence also approximate the integral. For each i, choose any point ξ_i between x_{i-1} and x_i and let

$$S(p) = \sum_{i=1}^{n} f(\xi_i)(x_i - x_{i-1}). \tag{3}$$

[The sums $\bar{S}(p)$ and $\underline{S}(p)$ are obtained by choosing ξ_i at the maximum or minimum of f on $[x_{i-1}, x_i]$—at least if f is continuous so that the maximum and minimum exist.]

Exercise 1   For the sake of logical notation, the definition of partition should be altered so as to include the points $\xi_1, \ldots, \xi_n$ as well as the points $x_0, \ldots, x_n$. State the definition correctly.

It is clear that for every partition p,

$$\underline{S}(p) \le S(p) \le \bar{S}(p), \tag{4}$$

so Theorem 6.1 has the following consequence.

THEOREM 6.3   *Let f be integrable. For every positive number ϵ there is a positive number δ such that if p is any partition with $|p| < \delta$, then*

$$\left| S(p) - \int_a^b f(x)\ dx \right| < \epsilon.$$

**Exercise 2
(Converse of the
Theorem)** Let f be any function on $[a, b]$ and let L be a number. If for every $\epsilon > 0$ there is a $\delta > 0$ such that if $|p| < \delta$, then $|S(p) - L| < \epsilon$, then f is integrable, and the integral is L.

The statement of Exercise 2 could be taken as the definition of integrability and of the integral. Whether it would be a more natural definition than the one given depends on the problem one has in mind at the time. In the problem of area, the original definition is the natural one, for the sums $\underline{S}(p)$ and $\bar{S}(p)$ are the ones that trap the area in between. In the problem of arc length, which is discussed in Section 7, these two have no special role, but certain $S(p)$ do. In problems of volume (Section 9), $\underline{S}(p)$ and $\bar{S}(p)$ do play a role in special cases but not in the general case.

The big advantage of the original definition is that it provides some definite numbers $\underline{S}$ and $\bar{S}$ to work with. The alternative definition gives no clue as to what the number L is.

The sums $S(p)$ are usually called the *Riemann sums* for the integral. The sums $\bar{S}(p)$ and $\underline{S}(p)$ are called either the *upper and lower Riemann sums* or the *upper and lower Darboux sums*.

7 ARC LENGTH

There is a nice integral formula for the length of the curve $y = f(x)$, $a \leq x \leq b$. The natural way to define the length is to approximate the curve by inscribed polygons.

Choose points $a = x_0 < x_1 < \cdots < x_n = b$, let $y_i = f(x_i)$, and let $A_i = (x_i, y_i)$. The polygon $P = (A_0, A_1, \ldots, A_n)$ is the sequence of line segments $A_0A_1, A_1A_2, \ldots, A_{n-1}A_n$. Its length is the sum of the lengths of the segments, which is

$$l(P) = \sum_{i=1}^{n} \sqrt{(x_i - x_{i-1})^2 + (y_i - y_{i-1})^2}. \tag{1}$$

The polygon P is an approximation to the curve $y = f(x)$ as the number

$$|P| = \max (x_i - x_{i-1}) \tag{2}$$

goes to 0, so it is natural to make the following definition:

**DEFINITION
7.1** *The length of the curve $y = f(x)$, $a \leq x \leq b$, is the limit of $l(P)$ as $|P|$ approaches 0. This means that the length is L if for each positive number ϵ there is a positive number δ such that $|l(P) - L| < \epsilon$ whenever $|P| < \delta$.*

Let us make use of the mean-value theorem in formula (1). There is a point ξ_i between x_{i-1} and x_i such that

$$y_i - y_{i-1} = f(x_i) - f(x_{i-1}) = f'(\xi_i)(x_i - x_{i-1}).$$

Therefore,

$$l(P) = \sum_{i=1}^{n} \sqrt{1 + f'(\xi_i)^2}\,(x_i - x_{i-1}), \tag{3}$$

which is nothing but the sum $S(p)$ for the function $\sqrt{1 + f'^2}$ and the partition p composed of the sequences $(x_0, \ldots, x_n)$ and $(\xi_1, \ldots, \xi_n)$. Theorem 6.3 has the following consequence.

THEOREM 7.2   *Let f be continuous on $[a, b]$ and differentiable at each interior point. If $\sqrt{1 + f'^2}$ is integrable, then the length of the curve $y = f(x)$ exists and is equal to $\int_b^a \sqrt{1 + f'^2}\,dx$.*

The hypothesis that f is continuous on $[a, b]$ and differentiable at each interior point is what is needed to apply the mean-value theorem.

Example   Find the arc length of the unit circle $y = \sqrt{1 - x^2}$ between $x = a$ and $x = b$.

In this case $f'(x) = -x(1 - x^2)^{-1/2}$, so $1 + f'^2 = (1 - x^2)^{-1}$; therefore,

$$\text{arc length} = \int_a^b \frac{dx}{\sqrt{1 - x^2}} = \arcsin b - \arcsin a. \tag{4}$$

Taking, for instance, $a = 0$ and $b = 1$, we find that the length of a quarter-circle is

$$\arcsin 1 - \arcsin 0 = \frac{\pi}{2} - 0 = \frac{\pi}{2}.$$

This is perfectly correct, but it is not justified by Theorem 7.2. The integrand is not bounded near 1, so it is not Riemann integrable.

The problem here lies not with the curve, but with the equation $y = f(x)$ that represents it. The same quarter-circle is also represented by the equation $x = g(y) = \sqrt{1 - y^2}$, but the same problem arises. In the first case the derivative f' is unbounded because the tangent line to the circle at $(1, 0)$ is parallel to the y axis. In the second case the derivative g' is unbounded because the tangent line at $(0, 1)$ is parallel to the x axis. The problem arises because the x and y axes are forced to play a special role.

A better way to write the equation of a curve is to express both x and y as functions of a third variable t:

$$x = x(t), \qquad v = y(t), \qquad a \le t \le b. \tag{5}$$

It is convenient to think of t as the time and of the point $(x(t), y(t))$ as the position of a moving object at time t. During the time interval $[a, b]$ the object traces out a curve in the plane. The equations (5) are called *parametric equations of the curve*, and t is called the *parameter*.

Exercise 1 Show that $x = \cos t$, $y = \sin t$, $0 \leq t \leq 2\pi$, are parametric equations of the circle $x^2 + y^2 = 1$.

Exercise 2 Show that $x = a \cos t$, $y = b \sin t$, $0 \leq t \leq 2\pi$, are parametric equations of the ellipse $x^2/a^2 + y^2/b^2 = 1$.

The examples in the exercises show a second advantage of the parametric equations. They describe the full circle and the full ellipse. The equation $y = f(x)$ can describe either the top half or the bottom half, but not both at once. On the other hand, any curve in the form $y = f(x)$ can be put immediately in parametric form by setting

$$x = t, \qquad y = f(t).$$

The natural way to approximate a parametric curve

$$x = x(t), \qquad y = y(t), \qquad a \leq t \leq b,$$

is to partition the time interval $[a, b]$. If $p = (t_0, \ldots, t_n)$ is such a partition and $A_i = (x(t_i), y(t_i))$, then the sequence of segments $A_0A_1, A_1A_2, \ldots, A_{n-1}A_n$ is an inscribed polygon whose length is

$$l(p) = \sum_{i=1}^{n} \sqrt{(x(t_i) - x(t_{i-1}))^2 + (y(t_i) - y(t_{i-1}))^2}. \tag{6}$$

The big square root is just the distance from A_{i-1} to A_i, or the length of the segment $A_{i-1}A_i$.

DEFINITION 7.3 *The length of the curve $x = x(t)$, $y = y(t)$, $a \leq t \leq b$, is the limit of $l(p)$ as $|p| \to 0$. This means that the length is L if for each positive number ϵ there is a positive number δ such that $|l(p) - L| < \epsilon$ whenever $|p| < \delta$.*

Again, there is a nice integral formula for the length, which results from using the mean-value theorem in formula (6) to show that $l(p)$ is almost a Riemann sum. Choose a point ξ_i between t_{i-1} and t_i so that

$$x(t_i) - x(t_{i-1}) = x'(\xi_i)(t_i - t_{i-1}),$$

and choose a point η_i between t_{i-1} and t_i so that

$$y(t_i) - y(t_{i-1}) = y'(\eta_i)(t_i - t_{i-1}).$$

Substitute these into formula (6) to get

$$l(p) = \sum_{i=1}^{n} \sqrt{x'(\xi_i)^2 + y'(\eta_i)^2} \, (t_i - t_{i-1}). \tag{7}$$

This is very much like a Riemann sum for the function

$$\sqrt{x'(t)^2 + y'(t)^2}.$$

Indeed, it would be a Riemann sum if it were true that $\eta_i = \xi_i$, so it suggests the formula

$$L = \int_a^b \sqrt{x'(t)^2 + y'(t)^2} \, dt. \tag{8}$$

THEOREM
7.4

The length of the curve $x = x(t)$, $y = y(t)$, $a \le t \le b$, is given by formula (8) if the derivatives x' and y' exist on (a, b) and are uniformly continuous.

The idea is to compare $l(p)$ as it appears in formula (7) with the Riemann sum

$$S(p) = \sum_{i=1}^{n} \sqrt{x'(t_i)^2 + y'(t_i)^2} \, (t_i - t_{i-1}).$$

Exercise 3   To make this comparison you will need the inequality

$$\sqrt{a^2 + b^2} - \sqrt{c^2 + d^2} \le \sqrt{(a - c)^2 + (b - d)^2}. \tag{9}$$

Prove this inequality by considering the triangle with vertices (a, b), (c, d), and $(0, 0)$ and using the geometric theorem that the sum of any two sides of a triangle is greater than the third.

Exercise 4   Use inequality (9) to show that

$$|l(p) - S(p)| \le \sum_{i=1}^{n} \sqrt{(x'(t_i) - x'(\xi_i))^2 + (y'(t_i) - y'(\eta_i))^2} \, (t_i - t_{i-1}). \tag{10}$$

Proof of Theorem 7.4   Let $\epsilon > 0$ be given, and use the uniform continuity of x' and y' to find $\delta > 0$ such that

$$|x'(t) - x'(s)| < \epsilon \quad \text{and} \quad |y'(t) - y'(s)| < \epsilon \qquad \text{if } |t - s| < \delta.$$

If p is any partition with $|p| < \delta$, then formula (10) gives

$$|l(p) - S(p)| \le \sum_{i=1}^{n} \sqrt{\epsilon^2 + \epsilon^2} \, (t_i - t_{i-1}) = \sqrt{2}\epsilon(b - a).$$

This proves the theorem because $S(p)$ differs from the integral in (7) by as little as we please when $|p|$ is small.

Remark   The assumption in Theorem 7.4 that x' and y' are uniformly continuous on (a, b) is certainly stronger than is desirable. It rules out polygons! A more reasonable assumption would be that x' and y' are uniformly continuous on each open interval of a partition of $[a, b]$. To handle this type of thing it is necessary to make an analysis of arc length along the lines of our analysis of the Riemann integral. In particular it should be shown that if c is a point between a and b, then the arc length over $[a, b]$ exists if and only if the arc lengths over $[a, c]$ and $[c, b]$ exist, in which case it is the sum. The crucial feature of arc length that makes these things relatively easy to prove is the following theorem.

THEOREM 7.5   *The arc length of the curve $x = x(t)$, $y = y(t)$, $a \leq t \leq b$, is the sup of $l(p)$ over all partitions p (in the sense that if either exists, so does the other and the two are equal).*

We shall not prove this theorem at present or the facts above, but why not try it for yourselves? (There are proofs in Section 6 of Chapter 8.)

Exercise 5   Let C be the circular arc described by $x = r \cos \theta, y = r \sin \theta, a \leq \theta \leq b$, where $0 < b - a \leq 2\pi$. Show that the length of C is $r(b - a)$ and that the area of the corresponding circular sector is $r^2(b - a)/2$. Do this both by integration and by geometry (similar triangles, etc.).

8   POLAR COORDINATES

Some curves and figures in the plane are described more simply in what are called polar coordinates than in the rectangular coordinates that we have been using. A point $(x, y) \neq (0, 0)$ in the plane is completely determined by its distance from the origin and the counterclockwise angle from the positive x axis to the half-line starting at the origin and going through (x, y). These two numbers are the polar coordinates of the point (x, y). More precisely, let r be the distance from (x, y) to the origin (that is, $r = \sqrt{x^2 + y^2}$), and let θ be the arc length (counterclockwise) along the unit circle from the point $(1, 0)$ to the point $(x/r, y/r)$. By the definition of the sine and cosine we have

$$x = r \cos \theta \quad \text{and} \quad y = r \sin \theta.$$

DEFINITION 8.1   *The numbers r and θ are called polar coordinates of the point (x, y) if*

$$x = r \cos \theta \quad \text{and} \quad y = r \sin \theta. \tag{1}$$

The argument above shows that each point $(x, y) \neq (0, 0)$ has exactly one set of polar coordinates r and θ such that $r > 0$ and $0 \leq \theta < 2\pi$.

Exercise 1   Discuss the polar coordinates of the point $(0, 0)$.

Exercise 2   The pairs (r, θ) and (ρ, φ) are polar coordinates of the same point $\neq (0, 0)$ if and only if either

(a) $\rho = r$ and $\varphi = \theta + 2k\pi$ for some integer k, or
(b) $\rho = -r$ and $\varphi = \theta + (2k + 1)\pi$ for some integer k.

Example 1   Draw the curve whose equation in polar coordinates is $r = \sin\theta$, $0 \leq \theta < 2\pi$. This is just the circle with center at $(0, \frac{1}{2})$ and radius $\frac{1}{2}$.   To see this, multiply through by r to get $r^2 = r\sin\theta$, or $x^2 + y^2 - y = 0$, or $x^2 + (y - \frac{1}{2})^2 - \frac{1}{4} = 0$. As θ goes from 0 to π, the point (r, θ) runs around the circle once in the counterclockwise direction.   As θ goes from π to 2π, the point runs around again in the counterclockwise direction, for in this case r is negative.

Example 2   Draw the curve whose equation is $r = \sin 2\theta$, $0 \leq \theta < 2\pi$.   It looks as shown in Figure 3.   As θ goes from 0 to $\pi/2$, the point (r, θ) runs around loop 1 in the counterclockwise direction.   As θ goes from $\pi/2$ to π, (r, θ) runs counterclockwise around loop 2.   Note that r is negative!

Exercise 3   Analyze the examples in detail.

Exercise 4   In Example 2 it appears that the curve has two tangent lines at the origin—the x and y axes.   Try to discuss this question

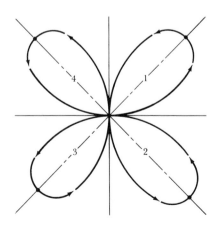

Figure 3

Exercise 5   Find equations in rectangular coordinates for the curve in Example 2.

The reasoning of Section 1 leads to a simple formula for the area of a set of the form

$$A = \{(r, \theta) : a \leq \theta \leq b \text{ and } 0 \leq r \leq f(\theta)\}, \tag{2}$$

where f is a given positive function on $[a, b]$ and $b - a \leq 2\pi$.

Exercise 6   Why assume that $b - a \leq 2\pi$?

As in Section 1, let $a = \theta_0 < \theta_1 < \cdots < \theta_n = b$ be a partition of $[a, b]$, let M_i and m_i be the sup and inf of f on $[\theta_{i-1}, \theta_i]$, and let

$$\bar{R}_i = \{(r, \theta) : \theta_{i-1} \leq \theta \leq \theta_i \text{ and } 0 \leq r \leq M_i\},$$
$$\underline{R}_i = \{(r, \theta) : \theta_{i-1} \leq \theta \leq \theta_i \text{ and } 0 \leq r \leq m_i\}.$$

It is clear that

$$\cup \underline{R}_i \subset A \subset \cup \bar{R}_i,$$

so

$$\Sigma \text{ area } \underline{R}_i \leq \text{ area } A \leq \Sigma \text{ area } \bar{R}_i. \tag{3}$$

In this case the $\bar{R}_i$ and $\underline{R}_i$ are not rectangles but circular sectors. The picture looks as shown in Figure 4.

By Exercise 5 of Section 7, we have

$$\text{area } \bar{R}_i = \frac{M_i^2(\theta_i - \theta_{i-1})}{2}, \qquad \text{area } \underline{R}_i = \frac{m_i^2(\theta_i - \theta_{i-1})}{2}, \tag{4}$$

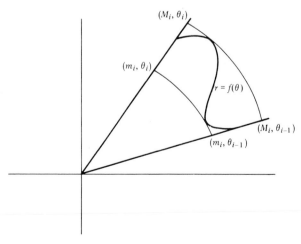

Figure 4

so the sums in inequality (3) are nothing but the lower and upper Riemann sums for the function $f(\theta)^2/2$. This gives the following theorem.

THEOREM 8.2

Let f be nonnegative and Riemann integrable on $[a, b]$, $0 \leq b - a \leq 2\pi$. The area of the set

$$A = \{(r, \theta) : a \leq \theta \leq b \text{ and } 0 \leq r \leq f(\theta)\}$$

is the integral

$$\int_a^b \frac{f(\theta)^2}{2} \, d\theta.$$

Exercise 7   Calculate the area of one of the loops in Example 2.

Any curve with the equation $r = f(\theta)$, $a \leq \theta \leq b$, in polar coordinates can be put immediately in parametric form. Because of the relations $x = r \cos \theta$, $y = r \sin \theta$, the parametric equations are

$$x = f(\theta) \cos \theta, \quad y = f(\theta) \sin \theta, \qquad a \leq \theta \leq b.$$

Exercise 8   Write parametric equations for the curve in Example 2 and find its length.

Exercise 9   As captain of the Coast Guard Cutter U.S.S. Polar Coordinates, you are chasing a rumrunner off the foggy coast of San Francisco. The fog lifts for a moment, you spot him 4 miles due north, and then the fog comes down. From past experience with this fellow you know that he will take off on a straight line at full speed, which is 10 mph. The U.S.S. Polar Coordinates will do 30 mph. What plan should you follow to catch him?

9   VOLUME

There is also a nice integral formula for the volume of a set in three dimensions. Let V be such a set. For each x_0, let $V(x_0)$ be the section of V by the plane $x = x_0$; that is,

$$V(x_0) = \{(y, z) : (x_0, y, z) \in V\},$$

and let $A(x_0)$ be the area of $V(x_0)$. [$V(x_0)$ is a set in the plane, so this makes sense.]

THEOREM 9.1

If V_a^b is the part of V between the planes $x = a$ and $x = b$, $a \leq b$, then (under mild restrictions on V)

$$\text{vol. } V_a^b = \int_a^b A(x) \, dx$$

Example 1   Find the volume of a right circular cone with height h and base of radius r.

Put the vertex of the cone at the origin and the axis along the positive x axis. The first problem is to calculate $A(x)$. Now, $V(x)$ is a circle, so the problem is to find its radius. The way to do this is to look at the section of V cut by the (y, z) plane. This is just a triangle, and consideration of similar triangles shows that the radius of the circle $V(x)$ is $(r/h)x$. Therefore, $A(x) = (\pi r^2/h^2)x^2$, and

$$\text{vol.} = \int_0^h \frac{\pi r^2}{h^2} x^2 \, dx = \tfrac{1}{3}\pi r^2 h.$$

Let us see what can be done about a proof of the theorem. Let p be a partition, and consider the Riemann sum

$$S(p) = \sum_{i=1}^{n} A(\xi_i)(x_i - x_{i-1}).$$

We would like to interpret this sum as the volume of a set that approximates the given set V.

For each i, let $R(\xi_i)$ be the cylinder whose base is $V(\xi_i)$ and whose height is $x_i - x_{i-1}$ (Figure 5)—more precisely, let

$$R(\xi_i) = \{(x, y, z) : (y, z) \in V(\xi_i) \text{ and } x_{i-1} \leq x \leq x_i\}.$$

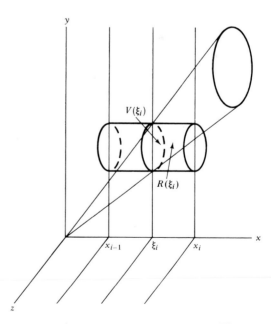

Figure 5

If it is assumed that the volume of a cylinder is the area of the base times the height (which seems fair enough), then $A(\xi_i)(x_i - x_{i-1})$ is the volume of $R(\xi_i)$; therefore, $S(p)$ is the volume of

$$R(p) = \bigcup_{i=1}^{n} R(\xi_i).$$

This set "approaches" the set V as $|p| \to 0$, so $S(p)$ ought to approach the volume of V, and the theorem ought to be true.

The obvious idea is to try to justify the argument by using upper and lower sums $\bar{S}(p)$ and $\underline{S}(p)$ as in the case of area. But there is a problem. The set inclusion that is needed [corresponding to formulas (1) and (4) of Section 1] is

$$\underline{R}(p) \subset V \subset \bar{R}(p), \tag{1}$$

which is simply not true. The fact that one set in the plane has a smaller area than another does not at all imply that it is included in the other. (See Exercise 2.) The theorem itself is true (subject to a very mild restriction on the set V, which is always satisfied in practice), and we shall use it freely; but we cannot prove it fully without a sound definition of volume, which will appear in Chapter 13.

There are, however, some interesting cases in which formula (1) is true. One is the solid of revolution. The set V is a solid of revolution about the x axis if each of the sections $V(x)$ is a circle with center $(0, 0)$. It is plain that if one such circle has a smaller area than another, then it is contained in the other. In this case formula (1) is obvious, and the proof of Theorem 9.1 is complete, except for the assumption that the volume of a cylinder is the area of the base times the height (which is also elementary, since the cylinders in question are right circular cylinders!). One example of the solid of revolution is the cone of Example 1. An example that is not a solid of revolution, but where the same argument works, is the following.

Exercise 1 Find the volume of the pyramid with height h and base a square of side s.

Exercise 2 Let C be a circle of radius r in the plane $x = h$ (arbitrary center), and let V be the cone obtained by joining each point of C to the origin. Find the volume of V.

Exercise 3 Let D be the doughnut obtained by revolving the circle $x^2 + (y - 2)^2 \leq 1$ about the x axis. Find the volume of D by looking at D as the difference of two solids of revolution. What is the equation of the surface that bounds D?

Chapter 13 contains a more general theory of integration that puts the notion of volume on a perfectly sound footing.

10   IMPROPER INTEGRALS

Reconsideration of the arc length of a quarter-circle suggests an extension of our notion of integral.   We have seen in Section 7 that the length of the curve $y = \sqrt{1 - x^2}$ on $0 \leq x \leq b$ is the number

$$L(b) = \int_0^b \frac{dx}{\sqrt{1 - x^2}} = \arcsin b \qquad \text{for } 0 \leq b < 1.$$

If b approaches 1, then the limit of $L(b)$ is $\pi/2$, which is exactly the length of the curve on $0 \leq x \leq 1$.

This suggests that perhaps we should maintain that the integral

$$\int_0^1 \frac{dx}{\sqrt{1 - x^2}}$$

exists, even though the integrand is unbounded and is not Riemann integrable in the original sense.   Such an integral is called an improper integral, and the definition is as follows.

DEFINITION 10.1   *Let f be integrable on $[a, b]$ for each $b < c$.   If the limit*

$$\lim_{\substack{b \to c \\ b < c}} \int_a^b f(x)\, dx$$

exists, it is called the improper integral of f from a to c.   Similarly, if f is integrable on $[b, c]$ for each $b > a$ and the limit

$$\lim_{\substack{b \to a \\ b < a}} \int_b^c f(x)\, dx$$

exists, it is called the improper integral of f from a to c.   In both cases the integral is written $\int_a^c f(x)\, dx$ as usual.

Exercise 1   The first part of the definition makes sense when $c = \infty$, the second part when $a = -\infty$.   Write it out in full in both cases.

Exercise 2
$$\int_0^1 \frac{dx}{\sqrt{1 - x^2}} = \frac{\pi}{2}.$$

Exercise 3
$$\int_1^\infty \frac{dx}{x^2} = 1.$$

Neither of the integrals

$$\int_{-1}^{1} \frac{dx}{\sqrt{|x|}} \quad \text{nor} \quad \int_{0}^{1} \frac{2x-1}{\sqrt{x^2-x}} \, dx \tag{1}$$

falls under the hypotheses envisioned in Definition 10.1. The definition allows only one trouble spot, and that at one of the end points. The first integral in (1) has only one trouble spot, but it is not an end point. The second causes trouble at both end points. It is necessary to broaden the definition.

DEFINITION 10.2

Let f be defined on $[a, b]$ *except at a finite number of points. The improper integral of f from a to b exists if there exist points* $a = a_0 < a_1 < \cdots < a_n = b$ *such that each of the integrals*

$$\int_{a_{i-1}}^{a_i} f(x) \, dx \tag{2}$$

exists, either as an ordinary integral or as an improper integral in the sense of Definition 10.1.

It is important to understand that the definition requires each of the integrals in (2) to exist. In some cases there is another way of looking at it that seems equally reasonable, but is not correct. For instance, it is tempting to say that

$$\int_{-1}^{1} \frac{dx}{x} = \lim_{a \to 0} \left\{ \int_{-1}^{-a} \frac{dx}{x} + \int_{a}^{1} \frac{dx}{x} \right\} = \lim_{a \to 0} 0 = 0,$$

whereas, in fact, the integral does not exist.

Exercise 4   Prove that this improper integral does not exist.

Exercise 5   Decide whether the improper integrals in formula (1) exist.

Exercise 6   For what values of p do the improper integrals

$$\int_{0}^{1} x^p \, dx \quad \int_{1}^{\infty} x^p \, dx \quad \int_{0}^{\infty} x^p \, dx$$

exist?

Exercise 7   Write down the integral for the arc length of the curve $y = x \sin(1/x)$, $0 < x < 1$. Does the improper integral exist? Does the arc length exist?

It is not difficult to develop a theory of improper integrals, but neither is it very interesting.   We shall rely on the examination of individual cases and on the reader's common sense.   (As a matter of fact, there are fascinating problems about improper integrals, but they are not the kind of problems we have been facing so far.)   As a rule, when we speak of integrable functions or of integrals existing, we shall have in mind the original definitions and not improper integrals, unless it is plain from the context that these are included.

5 | Taylor's Formula

1   TAYLOR'S FORMULA

Taylor's formula deals with the approximation of arbitrary differentiable functions by polynomials. There are many kinds of approximation in mathematics, each with particular advantages designed for particular problems. The one studied here is designed to approximate very well in a small neighborhood of a given point. It is achieved by matching the function and its derivatives to a certain order at the point. Note that the tangent line is exactly this sort of thing. It is obtained by matching the function and the derivative at the point. The first question is how to manage the additional matching.

THEOREM 1.1   *Suppose that the derivatives $f, f', \ldots, f^m$ all exist at the point a. Then the polynomial*

$$P(x) = \sum_{k=0}^{m} \frac{f^k(a)}{k!} (x - a)^k$$

satisfies $P^k(a) = f^k(a)$ for $k \leq m$ and is the only polynomial of degree $\leq m$ that does. This polynomial is called the Taylor polynomial of f of degree m at a, and is written $T_a^m f$.

The proof of the theorem rests on the following lemma.

LEMMA 1.2   *If $Q(x) = \sum_{k=0}^{n} a_k(x - a)^k$, then $a_k = Q^k(a)/k!$.*

Proof   (By induction on the number m)   We have

$$Q'(x) = \sum_{k=1}^{m} k a_k(x - a)^{k-1},$$

so the induction hypothesis gives that

$$ka_k = \frac{(Q')^{k-1}(a)}{(k-1)!} = \frac{Q^k(a)}{(k-1)!}.$$

This gives the assertion of the lemma if $k > 0$. The assertion is obvious if $k = 0$. (Simply set $x = a$.)

The serious question is how well the Taylor polynomial does approximate f.

THEOREM 1.3

(*Taylor's Formula*)   Let $f, f', \ldots, f^{m+1}$ all exist on an open interval I. If a and x are any two points of I, then there is a point ξ between them such that

$$f(x) = T_a^m f(x) + \frac{f^{m+1}(\xi)}{(m+1)!}(x-a)^{m+1}. \tag{1}$$

Note that for $m = 0$, this formula is nothing but the mean-value theorem. Therefore, the theorem is true for $m = 0$, and we are in a position to use induction. The induction is based on the mean-value theorem and on the fact that

$$(T_a^m f)' = T_a^{m-1} f'. \tag{2}$$

Exercise 1   Prove formula (2).

Proof of the Theorem

We shall apply the mean-value theorem in the strong form 4.2 of Chapter 3 to the function

$$g(x) = f(x) - T_a^m f(x)$$

and the function $h(x) = (x-a)^{m+1}$. It says that for any point x there is a point y between a and x such that

$$\frac{f(x) - T_a^m f(x)}{(x-a)^{m+1}} = \frac{g(x) - g(a)}{h(x) - h(a)} = \frac{g'(y)}{h'(y)} = \frac{g'(y)}{(m+1)(y-a)^m}.$$

Formula (2) shows that

$$g'(y) = f'(y) - T_a^{m-1} f'(y).$$

If we assume as an induction hypothesis that the theorem holds for $m - 1$, then we find that there is point ξ between a and y such that

$$f'(y) - T_a^{m-1} f'(y) = \frac{f^{m+1}(\xi)}{m!}(y-a)^m.$$

The last three formulas give just what is required.

Taylor's formula provides the means for interesting calculations. To calculate the value of a function f at a point x, we look for a point a at which we can calculate the value of f and of all derivatives. Then we can calculate $T_a^m f(x)$, and the idea is to use Taylor's formula to evaluate the error, or remainder, $f(x) - T_a^m f(x)$. Because of the factor $(x - a)^{m+1}$, the closer we can take a to x, the easier it will be to show that the error is small.

Example 1   Calculate e to within 0.01.

If $f(x) = e^x$, then $f^k(x) = e^x$ for all k. The one point at which we know the value of e^x is $x = 0$, so we write Taylor's formula with $a = 0$, which is

$$e^x = \sum_{k=0}^{m} \frac{x^k}{k!} + \frac{e^\xi}{(m+1)!} x^{m+1}. \tag{3}$$

We do not know what ξ is, but we do know that it is between 0 and x. In the present case we are interested in $x = 1$, so (Exercise 6, Section 7 of Chapter 2)

$$e^\xi < e^1 = e < 3.$$

Therefore, the error term that cannot be evaluated explicitly is in any event less than $3/(m+1)!$, and we have only to choose m so that this is less than 0.01. It $m = 5$, then

$$\frac{3}{(m+1)!} = \frac{3}{6!} = \frac{3}{720} < 0.005.$$

Therefore, to within 0.005 we have

$$e = 1 + 1 + \tfrac{1}{2} + \tfrac{1}{6} + \tfrac{1}{2}\tfrac{1}{120} = \tfrac{326}{120} = 2.71666 \ldots ,$$

so

$$e = 2.72 \qquad \text{to within } 0.01. \tag{4}$$

Example 2   Calculate $\sin 36°$ to within 0.001. If $f(x) = \sin x$, then

$$f^k(x) = \begin{cases} \pm \cos x & \text{if } k \text{ is odd,} \\ \pm \sin x & \text{if } k \text{ is even.} \end{cases}$$

Since both the sin and the cos are at most 1 in absolute value, we have the following evaluation for the error term:

$$\left| \frac{f^{m+1}(\xi)}{(m+1)!} (x-a)^{m+1} \right| \leq \frac{|x-a|^{m+1}}{(m+1)!}.$$

We must choose a point a for which we know both $\sin a$ and $\cos a$, and it will be helpful to choose a as close to $36°$ as possible. The obvious choice is $a = 30° = \pi/6$, in which case $x - a = 6° = \pi/30$. Now, we must choose m so that

$$\frac{1}{(m+1)!} \left(\frac{\pi}{30}\right)^{m+1} < 0.001.$$

This will be true for $m = 2$. Therefore, to within 0.001,

$$\sin 36° = f\left(\frac{\pi}{6}\right) + f'\left(\frac{\pi}{6}\right) \frac{\pi}{30} + \frac{f''(\pi/6)}{2} \left(\frac{\pi}{30}\right)^2$$

$$= \frac{1}{2} + \frac{\pi}{30 \sqrt{3}} - \frac{\pi^2}{3600}.$$

Exercise 2　What value of m is needed in order to do the above calculation with $a = 0$ instead of $a = \pi/6$?

2　EQUIVALENT FORMULAS

Taylor's formula, or at least others equivalent to it, can sometimes be obtained by tricks. For instance, let

$$S = 1 + y + y^2 + \cdots + y^m = \sum_{k=0}^{m} y^k,$$

$$yS = y + y^2 + \cdots + y^{m+1}.$$

Then $S - yS = 1 - y^{m+1}$, so

$$S = \frac{1 - y^{m+1}}{1 - y} = \frac{1}{1 - y} - \frac{y^{m+1}}{1 - y};$$

therefore,

$$\frac{1}{1 - y} = \sum_{k=0}^{m} y^k + \frac{y^{m+1}}{1 - y} \qquad \text{for } y \neq 1. \qquad (1)$$

This is a little better than Taylor's formula, since the error term is quite explicit and does not involve an unknown point ξ.

Substitution of $y = -x^2$ in formula (1) gives

$$\frac{1}{1 + x^2} = \sum_{k=0}^{m} (-1)^k x^{2k} + \frac{(-1)^{m+1} x^{2m+2}}{1 + x^2}, \qquad (2)$$

and again the same comment applies. Integration of (2) gives a formula for the arctan:

$$\arctan z = \int_0^z \frac{dx}{1+x^2}$$

$$= \sum_{k=0}^m \frac{(-1)^k x^{2k+1}}{2k+1} + (-1)^{m+1} \int_0^z \frac{x^{2m+2}}{1+x^2}\, dx. \qquad (3)$$

Exercise 1   Calculate π to within 0.01. See what value of m will be needed if you use the fact that $\pi/4 = \arctan 1$—and then think of a better idea.

Formulas (1), (2), and (3) are not visibly Taylor's formula, since the error, or remainder, terms do not have quite the form envisioned in Taylor's formula. However, they serve the same purpose, and, indeed, the polynomial parts are exactly the Taylor polynomials. This is no accident.

THEOREM 2.1   *If $f, f', \ldots, f^m$ all exist at a, then for every positive number ϵ there is a positive number δ such that*

$$|f(x) - T_a^m f(x)| \le \epsilon |x - a|^m \qquad if\ |x - |a < \delta. \qquad (4)$$

Moreover, $T_a^m f$ is the only polynomial of degree $\le m$ for which this is true.

Proof   Formula (4) is established by induction. The case $m = 1$ is nothing more than the definition of the derivative. The induction hypothesis and formula (2) of Section 1 show that if $g = f - T_a^m f$, then $|g'(x)| \le \epsilon |x - a|^{m-1}$ if $|x - a| < \delta$. Since $g(a) = 0$, integration from a to x gives the required result.

Exercise 2   If $m = 2$, the integration is not legitimate, since nothing guarantees that g' is integrable. Complete the argument in this case.

In order to prove the other half of the theorem, let P be any polynomial satisfying (4), let $Q = T_a^m f - P$, and use the following lemma.

LEMMA 2.2   *If a polynomial Q of degree $\le m$ satisfies*

$$\lim_{\substack{x \to a \\ x \ne a}} \frac{Q(x)}{(x-a)^m} = 0, \qquad (5)$$

then $Q = 0$.

Proof   Let

$$Q(x) = \sum_{k=0}^{m} a_k(x-a)^k. \tag{6}$$

If a_j is the first nonvanishing coefficient, then

$$Q(x) = (x-a)^j \sum_{k=j}^{m} a_k(x-a)^{k-j} = (x-a)^j R(x),$$

where R is a polynomial with $R(a) = a_j \neq 0$. It is clear that condition (5) cannot possibly be satisfied. Therefore, there cannot be a first non-vanishing coefficient; so all coefficients are 0, and Q is identically 0.

Remark   Both here and in the proof of Theorem 1.1 we have used the fact that a polynomial of degree $\leq m$ can be put in the form (6). A polynomial of degree $\leq m$ is defined to be a function for which there are numbers $\alpha_0, \ldots, \alpha_m$ such that

$$Q(x) = \sum_{k=0}^{m} \alpha_k x^k.$$

To show that such a function can be put in the form (6) we have only to write $x = (x - a) + a$, substitute in this formula, and multiply out.

Exercise 3   Show that Theorem 2.1 implies that the polynomials in formulas (1), (2), and (3) are the Taylor polynomials. Use this result to calculate $f^k(0)$ in each case.

From Theorem 2.1 we can deduce the formula for the kth derivative of a product of two functions. It is easy to see that

$$T_a^m f(x)\, T_a^m g(x) = \sum_{k=0}^{m} c_k(x-a)^k + P(x)(x-a)^{m+1},$$

where P is a polynomial and

$$c_k = \sum_{i=0}^{k} \frac{f^i(a)}{i!} \frac{g^{k-i}(a)}{(k-i)!}.$$

Now, $f = T_a^m f + R$ and $g = T_a^m g + S$, where R and S have the property that for every positive number ϵ there is a positive number δ such that $|R(x)| \leq \epsilon|x-a|^m$ and $|S(x)| \leq \epsilon|x-a|^m$ if $|x-a| < \delta$. Therefore,

$$f(x)g(x) = \sum_{k=0}^{m} c_k(x-a)^k + P(x)(x-a)^{m+1} + T_a^m f(x)S(x)$$

$$+ T_a^m g(x)R(x) + R(x)S(x),$$

which shows that for every positive number ϵ there is a positive number δ such that

$$\left| f(x)g(x) - \sum_{k=0}^{m} c_k(x-a)^k \right| \leq \epsilon |x-a|^m \qquad \text{for } |x-a| < \delta.$$

By Theorem 2.1, this polynomial is the Taylor polynomial of fg, and we have the following theorem.

THEOREM 2.3

$$(fg)^k = \sum_{i=0}^{k} \frac{k!}{i!(k-i)!} f^i g^{k-i}.$$

Exercise 4   Prove Theorem 2.3 by induction.

Exercise 5   Start with the identity

$$f(x) = f(a) + \int_a^x f'(t)\, dt$$

and integrate by parts to obtain

$$f(x) = f(a) + f'(a)(x-a) + \int_a^x f''(t)(x-t)\, dt.$$

Continue to integrate by parts to obtain

$$f(x) = T_a^m f(x) + \frac{1}{m!} \int_a^x f^{m+1}(t)(x-t)^m\, dt. \qquad (7)$$

This is called Taylor's formula with the remainder in integral form. State carefully the hypotheses that are needed to carry through the proof.

Exercise 6   Suppose that f and p are continuous on $[a, b]$ and that $p \geq 0$. Show that there is a point ξ between a and b such that

$$\int_a^b f(t)p(t)\, dt = f(\xi) \int_a^b p(t)\, dt.$$

This is called the *mean-value theorem for integrals*.

Exercise 7   Use formula (7) and the mean-value theorem for integrals to derive the original form of Taylor's formula—but under slightly worse hypotheses.

3   LOCAL MAXIMA AND MINIMA

If a function f has a maximum or minimum at a point a, and if $f'(a)$ exists, then $f'(a) = 0$. On the other hand, $f'(a)$ may well be 0 without there being a maximum or minimum at a.   What we want now is a supplementary condition to add to the condition $f'(a) = 0$ that will guarantee a maximum or minimum. The supplementary condition will bear on the higher derivatives of f at a. Since these depend only on the behavior of f near a, the notions of maximum and minimum must be modified.

DEFINITION 3.1

A function f on a set S has a local minimum at the point $a \in S$ if there is a positive number δ such that

$$f(x) \geq f(a) \qquad if \ x \in S \ and \ |x - a| < \delta.$$

THEOREM 3.2

Suppose that $f', \ \ldots \ , f^{m-1}$ are all 0 at a and that f^m exists but is not 0 at a.
(a) If m is odd, then f has neither a local maximum nor a local minimum at a.
(b) If m is even, then f has one or the other—a minimum if $f^m(a) > 0$, and a maximum if $f^m(a) < 0$.

Example 1 $f(x) = x^3$ has neither a local maximum nor a local minimum at $x = 0$, because the first nonvanishing derivative is the third, which is odd.

Example 2 $f(x) = x^4$ has a local minimum at $x = 0$ because the first nonvanishing derivative is the fourth, which is even, and its value is 24, which is positive.

Proof of the Theorem

Write out the inequality in Theorem 2.1 with

$$\epsilon = \frac{1}{2} \frac{|f^m(a)|}{m!}.$$

Since $f', \ \ldots \ , f^{m-1}$ are all 0 at a, it reads

$$\left| f(x) - f(a) - \frac{f^m(a)}{m!} (x - a)^m \right| \leq \frac{1}{2} \frac{|f^m(a)|}{m!} |x - a|^m$$
$$\text{if } |x - a| < \delta. \quad (1)$$

This implies that $f(x) - f(a)$ has the same sign as $[f^m(a)/m!](x - a)^m$. If m is odd, the sign changes as x crosses from one side of a to the other, so there can be neither a maximum nor a minimum.   If m is even, the sign remains the same, so there must be one or the other.   For example, if $f^m(a) > 0$, then $f(x) - f(a)$ is positive for all x near a; hence $f(x) > f(a)$ for all x near a, and f has a strict local minimum.

Theorem 3.2 does cover most cases, but not all.   It may happen, for instance, that $f', \ \ldots \ , f^{m-1}$ are all 0 at a, while f^m does not exist.   In this case anything can happen.

Exercise 1 Give examples.

It may also happen that $f^k(a) = 0$ for every k.   Consider the function

$$f(x) = e^{-1/x^2} \quad \text{for } x \neq 0, \qquad f(0) = 0.$$

This function has derivatives of every order at every point, including 0, and all derivatives vanish at 0.   To establish these properties, carry out the following exercises.

Exercise 2   For each positive integer k,

$$f^k(x) = P(x)\, x^{-3k} e^{-1/x^2}, \; P \text{ a polynomial.}$$

Exercise 3   For each positive integer m,

$$\lim_{\substack{x \to 0 \\ x \neq 0}} x^{-m} e^{-1/x^2} = 0.$$

Exercise 2 is handled immediately by induction.   Exercise 3 reduces to

$$\lim_{y \to \infty} y^m e^{-y^2} = 0,$$

which can be handled by l'Hospital's rule.

The function f obviously has a minimum at 0.   The function $-f$ has the same properties and has a maximum.   The function

$$g(x) = \begin{cases} f(x) & \text{if } x \geq 0, \\ -f(x) & \text{if } x < 0, \end{cases}$$

also has the same properties and has neither.

6 | Sequences and Series

1 SEQUENCES AND SERIES

DEFINITION 1.1

A sequence in a set S is a function from the positive integers into S. If s is a sequence, it is customary to write s_n for $s(n)$ and $\{s_n\}$ for s.

Eventually we shall discuss sequences in a variety of sets S, but for the present the important one is the set of real numbers. In this case the limit at ∞ is defined as it was in Section 5 of Chapter 3.

DEFINITION 1.2

If $\{s_n\}$ is a sequence of real numbers, then $\lim_{n \to \infty} s_n = l$, or $s_n \to l$, if for every positive number ϵ there is a positive integer n_0 such that if $n > n_0$, then $|s_n - l| < \epsilon$. If the limit exists, the sequence is said to converge. If it does not, the sequence is said to diverge.

Example 1 The sequence $\{(1 + (1/n))^n\}$ is strictly increasing and has the limit e.

To say that $\{s_n\}$ is strictly increasing means that $s_{n+1} > s_n$ for every n. To say it is increasing means that $s_{n+1} \geq s_n$ for every n. The assertion of the example follows from Example 4 at the end of Section 5 of Chapter 3.

Before taking up other examples, let us prove an inequality that gives a pretty good idea how fast $n!$ increases.

LEMMA 1.3

$$n! > (n/e)^n \text{ for } n \geq 1.$$

The lemma is proved by induction. Assume that it holds for n, and multiply both sides by $n + 1$ to get

$$(n + 1)! > \left(\frac{n}{e}\right)^n (n + 1).$$

Therefore, it is sufficient to show that

$$\left(\frac{n}{e}\right)^n (n+1) > \left(\frac{n+1}{e}\right)^{n+1} = \left(\frac{n+1}{e}\right)^n \frac{n+1}{e},$$

or, in other words, that

$$e > \left(\frac{n+1}{n}\right)^n = \left(1 + \frac{1}{n}\right)^n.$$

This follows from Example 1.

Example 2   For each real x, $x^n/n! \to 0$.

This follows from the lemma, for

$$\left|\frac{x^n}{n!}\right| \leq \left|\left(\frac{ex}{n}\right)^n\right| \leq \left|\frac{ex}{n}\right| \qquad \text{if } n \geq |ex|,$$

and the term on the right is certainly as small as we please when n is large.

Example 3   For each positive integer n and each real x, let

$$s_n(x) = \sum_{k=0}^{n} \frac{x^k}{k!}.$$

For each real x, we have $s_n(x) \to e^x$.

It is not surprising that what is involved here is Taylor's formula, for s_n is the Taylor polynomial for e^x.   Taylor's formula gives

$$e^x - s_n(x) = \frac{e^\xi x^{n+1}}{(n+1)!} \qquad \text{with } |\xi| < |x|.$$

The result follows from Example 2.

This example suggests that it should be profitable to define a notion of "infinite sum."

DEFINITION 1.4

Let $\{a_k\}$ be a sequence of real numbers.   The sequence $\{s_n\}$ defined by

$$s_n = \sum_{k=0}^{n} a_k$$

is called the series associated with the sequence $\{a_k\}$.   The limit of $\{s_n\}$ is called the sum of the series and is written $\sum_{k=0}^{\infty} a_k$, provided, of course, the limit exists.   If the limit does exist, the series converges; if not, it diverges.

In terms of this definition the result of Example 3 is that

$$e^x = \sum_{k=0}^{\infty} \frac{x^k}{k!} \qquad \text{for all real } x. \tag{1}$$

Exercise 1 Establish the following formulas:

$$\sin x = \sum_{k=0}^{\infty} (-1)^k \frac{x^{2k+1}}{(2k+1)!} \qquad \text{for all real } x. \tag{2}$$

$$\cos x = \sum_{k=1}^{\infty} (-1)^k \frac{x^{2k}}{(2k)!} \qquad \text{for all real } x. \tag{3}$$

$$\frac{1}{1-x} = \sum_{k=0}^{\infty} x^k \qquad \text{for } |x| < 1. \tag{4}$$

$$\log(1-x) = -\sum_{k=1}^{\infty} \frac{x^k}{k} \qquad \text{for } |x| < 1. \tag{5}$$

$$\arctan x = \sum_{k=0}^{\infty} (-1)^k \frac{x^{2k+1}}{2k+1} \qquad \text{for } |x| < 1. \tag{6}$$

These series are called Taylor series. The proofs of the formulas call for an evaluation of the error, or remainder, term in Taylor's formula—or one of its equivalents from Section 2 of the last chapter.

In general, if f is a function which has derivatives of all orders at a point a, then the Taylor series of f at the point a is the series

$$\sum_{k=0}^{\infty} \frac{f^k(a)}{k!} (x-a)^k.$$

There are a couple of points to be wary of here. First, the series may not converge for any value of x (except $x = a$). Second, even if it does converge, the sum may not be $f(x)$. For the time being at least, the way to show that the series does converge and that its sum is $f(x)$ is to estimate the remainder in Taylor's formula. One nasty example is the function

$$f(x) = e^{-1/x^2} \qquad \text{for } x \neq 0, \quad f(0) = 0$$

considered at the end of the last section. It was shown in the exercises there that $f^k(0) = 0$ for every k. Therefore, the Taylor series at 0 is the series of 0's for every x. This certainly converges for every x and has the sum 0—which is not at all equal to $f(x)$.

Some ambiguity in the notations may have been noticed. According to Definition 1.1, a sequence $\{a_k\}$ should start with the term a_1, while in Definition 1.4, and in all the Taylor series, it starts with a_0. Obviously, the starting point does not matter.

The same symbol $\sum_{k=0}^{\infty} a_k$ has been used both for the sum of the series and for the series itself. According to the definition, it should stand for the sum, and some other symbol, such as Σa_k, should be used for the series itself. We shall make use of this convention but not very systematically. The ambiguity does exist throughout the mathematical literature and must be borne.

The number a_n is called the nth term of the sequence $\{a_n\}$. It is also called the nth term of the series Σa_n, which is another ambiguity. It ought to be the number

$$s_n = \sum_{k=0}^{n} a_k,$$

which is called the nth term of the series. The latter is called the nth partial sum.

From the definition of a series as a sequence of partial sums, it is plain that any theorem about convergence of sequences leads to a corresponding theorem about convergence of series. On the other hand, theorems about series also lead to theorems about sequences, for if s_n is the nth partial sum of the series Σa_k, then $a_n = s_n - s_{n-1}$; so the sequence $\{a_k\}$ can be recovered from the series Σa_k. A second way to recover a given sequence $\{a_k\}$ from a series, although not the corresponding series of partial sums, is the formula

$$a_n = a_1 + \sum_{k=2}^{n} (a_k - a_{k-1}).$$

Thus, the study of sequences is more or less equivalent to the study of series. Often, however, a theorem that is interesting and natural for one of the two becomes very awkward when it is rephrased for the other.

2   INCREASING SEQUENCES AND POSITIVE SERIES

With nothing but the definition at hand, there would be serious difficulties standing in the way of a satisfactory theory of convergence. To decide whether a given sequence converges, one must know ahead of time the limit to which it converges! For any given number l Definition 1.2 gives the test to determine whether $\lim_{k\to\infty} s_k = l$—which is useless without knowing what number l to test. (Test them all?)

Consider, for example, the series

$$\sum_{k=0}^{\infty} \frac{1}{k!}. \tag{1}$$

Taylor's formula shows that the series does converge and that the sum is e. Looking just at the series, however, and not at Taylor's formula, there would be no way to guess that the sum is e and no way to prove that the series converges.

For the series

$$\sum_{k=1}^{\infty} \frac{1}{k^2} \tag{2}$$

there is no way to guess the sum and no way to prove that the series converges.

What are needed are criteria of convergence that bear only on the sequence or series itself.

THEOREM 2.1	*Every convergent sequence is bounded.*
THEOREM 2.2	*Every bounded increasing sequence converges to its least upper bound.*
Proofs	Suppose that $\lim_{n\to\infty} s_n = l$. Take any positive ϵ, say $\epsilon = 1$, and find a corresponding integer n_0 such that

$$|s_n - l| < 1; \qquad \text{hence } |s_n| < |l| + 1 \qquad \text{if } n > n_0.$$

Then $M = \max(|s_1|, |s_2|, \ldots, |s_{n_0}|, |l| + 1)$ is a bound.

Suppose that $\{s_n\}$ is increasing (that is, $s_{n+1} \geq s_n$) and bounded, and let l be the least upper bound. Let ϵ be a given positive number. Then $l - \epsilon$ is no longer an upper bound, so there exists n_0 with $s_{n_0} > l - \epsilon$. If $n > n_0$, then

$$l \geq s_n \geq s_{n_0} > l - \epsilon.$$

These two theorems have a wide range of application, coming partly from the fact that a series with nonnegative terms always has increasing partial sums.

Exercise 1 Show that $\Sigma_{k=1}^{\infty} 1/k$ diverges (that is, does not converge) by showing that

$$s_n = \bar{S}\left(p; \frac{1}{x}\right) \geq \int_1^n \frac{dx}{x} = \log n,$$

where p is the partition $(1, 2, \ldots, n)$ of the interval $[1, n]$. Draw a picture.

Exercise 2   Show that $\Sigma_{k=1}^{\infty} 1/k^2$ converges by showing that

$$s_n - s_1 = S\left(p; \frac{1}{x^2}\right) \leq \int_1^n \frac{dx}{x^2} = 1 - \frac{1}{n}.$$

Exercise 3   For what values of p does the series $\Sigma_{k=1}^{\infty} k^{-p}$ converge?

Exercise 4   What about the series $\Sigma_{k=2}^{\infty} 1/k \log k$?

Exercise 5   If $0 \leq a_k \leq b_k$ holds for all large k, and $\Sigma_{k=1}^{\infty} b_k$ converges, then $\Sigma_{k=1}^{\infty} a_k$ converges.

The result of the last exercise is called the *comparison test*. Probably it is the most valuable convergence test there is—but plainly the value depends on having a large stock of series to use in comparison.

Exercise 6   Show by the comparison test that $\Sigma_{k=0}^{\infty} 1/k!$ and $\Sigma_{k=0}^{\infty} 2^{-k}$ converge.

3   CAUCHY SEQUENCES

Theorems 2.1 and 2.2 tell the whole story for increasing sequences and for series with positive terms. Together they give a simple, necessary, and sufficient condition for convergence. Obviously, they also do the job for decreasing sequences and series with negative terms. But they do nothing whatever for general sequences and series.

The problem is to get one's fingers on the limit. In the case of an increasing sequence this is accomplished easily by taking the least upper bound. In the general case it is more complicated.

**DEFINITION
3.1**

If $\{s_n\}$ is a bounded sequence, let

$$b_n = \sup\{s_k : k \geq n\} \qquad and \qquad b = \inf\{b_n\}.$$

The number b is called the limit superior of the sequence $\{s_n\}$ and is written lim sup s_n.

The point is that every bounded sequence has a limit superior, and if the sequence happens to have a limit, then it must be the limit superior. This notion appeared already in formula (4), Section 2 of Chapter 3, at which point the following lemma was proved.

LEMMA 3.2

Let $b = \limsup s_n$. For every positive number δ there is a positive integer k such that

$$k > \frac{1}{\delta} \quad and \quad |b - s_k| < \delta.$$

The main theorem on the convergence of general sequences is as follows:

THEOREM 3.3

The sequence $\{s_n\}$ converges if and only if it has the following property: For every positive number ϵ there is a positive integer n_0 such that if $n > n_0$ and $m > n_0$, then $|s_n - s_m| < \epsilon$.

A sequence with this property is called a *Cauchy sequence* (after a Frenchman named Cauchy).   Two easy parts of the theorem are left as exercises.

Exercise 1   Every convergent sequence is Cauchy.

Exercise 2   Every Cauchy sequence is bounded.

Proof of the Theorem

What remains to be proved is that every Cauchy sequence converges to its limit superior.   Let $b = \limsup s_n$, and let $\epsilon > 0$ be given.   First choose n_0 so that if $n > n_0$ and $m > n_0$, then $|s_n - s_m| < \epsilon$.   Take any positive number δ smaller than $1/n_0$ and smaller than ϵ, and choose k as in the lemma.   If $n > n_0$, then we have

$$|b - s_n| \leq |b - s_k| + |s_k - s_n| < \epsilon + \epsilon = 2\epsilon.$$

Indeed, $|b - s_k| < \delta < \epsilon$, while $|s_k - s_n| < \epsilon$ follows from the fact that $k > n_0$.

The theorem looks harmless enough, possibly not even interesting.   This is deceptive.   For instance, it allows the results on series with positive terms to be brought to bear on general series.

THEOREM 3.4

If $\sum_{k=1}^{\infty} |a_k|$ converges, then so does $\sum_{k=1}^{\infty} a_k$.

Proof

Let s_n be the nth partial sum of the first series and t_n be the nth partial sum of the second.   For $n > m$ we have

$$|t_n - t_m| = \left| \sum_{k=m+1}^{n} a_k \right| \leq \sum_{k=m+1}^{n} |a_k| = s_n - s_m,$$

from which it is plain that if $\{s_n\}$ is Cauchy, then so is $\{t_n\}$.

A series such that $\sum_{k=1}^{\infty} |a_k|$ converges is said to converge absolutely. It is pretty hard to prove that a series converges without proving that it converges absolutely, but there are examples. One, according to the following theorem and Exercise 1 of the last section, is the series $\sum_{k=1}^{\infty} (-1)^k (1/k)$.

THEOREM 3.5

If $\{a_k\}$ is decreasing and $\lim_{k \to \infty} a_k = 0$, then $\sum_{k=1}^{\infty} (-1)^k a_k$ converges.

Proof

What can be shown is that if $n > m$, then

$$|s_n - s_m| \leq a_{m+1},\tag{1}$$

which proves the theorem because of the assumption that $a_m \to 0$.

To see that formula (1) holds, write

$$(-1)^{m+1}(s_n - s_m) = a_{m+1} - a_{m+2} + a_{m+3} - a_{m+4} + \cdots a_n.$$

Group the terms in pairs, the first two together, the next two together, and so on. Since the sequence is decreasing, each pair is nonnegative, and if there is a term left over that cannot be paired, it too is nonnegative. Therefore,

$$(-1)^{m+1}(s_n - s_m) \geq 0.$$

Now group in pairs, but leave out the first term a_{m+1}. This time each pair is nonpositive, and if there is a term left over, it, too, is nonpositive. Therefore,

$$(-1)^{m+1}(s_n - s_m) \leq a_{m+1}.$$

The two inequalities together give (1).

Exercise 3 Show that the improper integral $\int_0^\infty (\sin x/x)\, dx$ exists. [*Hint:* Let $a_k = |\int_{k\pi}^{(k+1)\pi} (\sin x/x)\, dx|$ and use Theorem 3.5.]

Exercise 4 Define the limit inferior (lim inf) of a bounded sequence, and prove Lemma 3.2 for the lim inf.

Exercise 5 If $\sum a_k$ converges, then $a_k \to 0$.

Exercise 6 If $\sum a_k$ and $\sum b_k$ converge, or converge absolutely, then so do $\sum(a_k + b_k)$ and $\sum \alpha a_k$ (α real).

In many respects infinite sums behave much like finite sums, but in some respects you have to watch out. In a finite sum, for example, the order of the terms is immaterial—$3 + 2 = 2 + 3$. In an infinite sum this is not always the case. To state the theorem, the following definition is needed.

DEFINITION 3.6   *A permutation of a set X is a one-to-one function from X onto itself.*

THEOREM 3.7   *If Σa_k converges absolutely, then for every permutation ν of the positive integers the series $\Sigma a_{\nu(k)}$ converges absolutely to the same sum.*

So far so good, but see how strange things are for series that do not converge absolutely.

THEOREM 3.8   *If the series Σa_k converges, but does not converge absolutely, then for any number s whatever there is a permutation ν of the positive integers such that $\Sigma a_{\nu(k)} = s$.*

Proof of Theorem 3.7   Let $s = \Sigma a_k$ and let $\epsilon > 0$ be given.   Choose n_0 so that

$$\sum_{k=n_0}^{n} |a_k| < \epsilon \qquad \text{if } n > n_0. \tag{2}$$

Exercise 7   Show that (2) implies that $|s - \Sigma_{k=1}^{n_0} a_k| \le \epsilon$.

Now let ν be any permutation, and let

$$m_0 = \max \{\nu^{-1}(1), \ \ldots \ , \nu^{-1}(n_0)\}.$$

If $m > m_0$, then we have

$$\left| s - \sum_{k=1}^{m} a_{\nu(k)} \right| \le \left| s - \sum_{k=1}^{n_0} a_k \right| + \sum_{k=n_0}^{n} |a_k|,$$

where $n = \max\{\nu(1), \ \ldots \ , \nu(m)\}$.   Formula (2) and Exercise 7 show that the right side is at most 2ϵ and prove the theorem.

Exercise 8   Theorem 3.8 is not as mysterious as it looks.   Develop a proof along the following lines: Let $\{b_k\}$ be the sequence of nonnegative terms in the sequence $\{a_k\}$ and let $\{c_k\}$ be the sequence of negative terms (both picked out in order).   Show that both series Σb_k and Σc_k diverge.   Now pick out b's and c's according to the following plan: Pick just enough b's to get a sum $>s$, then just enough c's to get a total sum $<s$, then just enough b's to get a total sum $>s$, and so on.   Use Exercise 5 to show that the resulting series converges to s.

In view of Theorem 3.7, it is natural to wonder whether the definition of absolute convergence cannot be made in a way that is independent of the order of the terms.   Consider first the case where each $a_k \ge 0$.   For each finite set F of positive integers, let s_F be the sum of the a_k's with $k \in F$, and let $s = \sup s_F$, the upper bound being taken over all finite sets F.

Exercise 9 $s = \Sigma a_k$ if each $a_k \geq 0$ (in the sense that the equality holds if either is finite).

Exercise 10   Formulate a condition on the s_F's that is equivalent to absolute convergence (a_k not necessarily ≥ 0, of course).   (In proving that your condition works, you will have to use Theorem 3.8.)

4   SEQUENCES OF FUNCTIONS

In the case of sequences and series of functions a new problem comes up: to deduce properties of the limit from known properties of the individual terms. For instance, if each term is continuous, is the limit continuous?   If each term is differentiable, is the limit differentiable and can we differentiate term by term? If each term is integrable, is the limit integrable and can we integrate term by term?   Consider the Taylor series for the sine:

$$\sin x = \sum_{k=0}^{\infty} (-1)^k \frac{x^{2k+1}}{(2k+1)!}.$$

If we differentiate term by term we obtain the series

$$\sum_{k=1}^{\infty} (-1)^k \frac{x^{2k}}{(2k)!},$$

which we recognize as the series for the cosine.   In this case, differentiation term by term is all right.

Example 1   Let $f_n(x) = (1 - x^2)^n$ for $-1 \leq x \leq 1$.   For each $x \neq 0$, we have $f_n(x) \to 0$, and for $x = 0$ we have $f_n(0) = 1$ for all n.   Thus $\{f_n(x)\}$ converges for each x, $-1 \leq x \leq 1$, and each term is perfectly differentiable; but the limit is not even continuous at 0.

Exercise 1   Show that integration term by term is all right in Example 1.

Example 2   Define f_n on $[0, 1]$ so that on $[0, 1/n]$ the graph of f_n is the isosceles triangle with height $2n$, and so that $f_n = 0$ on $[1/n, 1]$ (Figure 1).

It is plain that $f_n(x) \to 0$ for every x, while

$$\int_0^1 f_n(x) \, dx = 1.$$

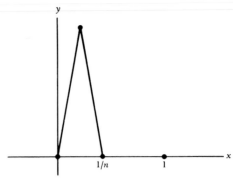

Figure 1

This is a case where we cannot integrate term by term.

Exercise 2   Write the equation for f_n and verify the assertions above.

    The examples show that pointwise convergence of a sequence of functions (that is, convergence at each point) is insufficient to imply very much about the limit function, and certainly is insufficient to allow either integration or differentiation term by term.   What is needed is a stronger kind of convergence.

DEFINITION 4.1   *The sequence $\{f_n\}$ of real-valued functions on the set I converges uniformly to the function f if for each positive number ϵ there is a positive integer n_0 such that*

$$|f(x) - f_n(x)| < \epsilon \qquad \text{for all } x \in I \text{ and all } n > n_0.$$

Exercise 3   Explain the distinction between pointwise convergence and uniform convergence. Compare it with the distinction between continuity and uniform continuity.

Exercise 4   Show that the sequences in Examples 1 and 2 do not converge uniformly.   In both cases, however, show that the convergence is uniform on any closed subinterval that does not contain 0.

Exercise 5   Define the notions of pointwise Cauchy and uniformly Cauchy sequences of functions.

Exercise 6   Every uniformly convergent sequence is pointwise convergent, and every uniformly Cauchy sequence is pointwise Cauchy.

Exercise 7   If I is a finite set, then every pointwise convergent (or Cauchy) sequence is uniformly convergent (or Cauchy).

Exercise 8   Define uniform convergence of a series of functions. Show that the Taylor series for $\sin x$ converges uniformly on each interval $[-r, r]$, but not on the whole real line.

THEOREM 4.2   *Every pointwise Cauchy sequence converges pointwise. Every uniformly Cauchy sequence converges uniformly.*

Proof   If $\{f_n\}$ is pointwise Cauchy, then for each $x \in I$ the sequence $\{f_n(x)\}$ is a Cauchy sequence of real numbers. Let $f(x)$ be the limit, which exists by Theorem 3.3. Then f is a real-valued function on I, and $f_n \to f$ pointwise.

Let $\{f_n\}$ be a uniformly Cauchy sequence, and let f be the pointwise limit, which exists because of Exercise 3 and the first part of the theorem. It will be shown that $f_n \to f$ uniformly. Let $\epsilon > 0$ be given and choose n_0 so that if $n > n_0$ and $m > n_0$, then

$$|f_n(x) - f_m(x)| < \epsilon \qquad \text{for all } x \in I.$$

This is possible because $\{f_n\}$ is uniformly Cauchy. The n_0 so determined is the one we are looking for. Indeed, let x be any point of I. Since $f_n(x) \to f(x)$, we can choose $m > n_0$ so that

$$|f_m(x) - f(x)| < \epsilon.$$

Then, if $n > n_0$, we have

$$|f_n(x) - f(x)| \leq |f_n(x) - f_m(x)| + |f_m(x) - f(x)| < 2\epsilon.$$

THEOREM 4.3   *Let I be a set of real numbers. If $f_n \to f$ uniformly and each f_n is continuous, then f is continuous.*

Proof   Let $\epsilon > 0$ and $a \in I$ be given. First choose n_0 so that if $n > n_0$, then

$$|f_n(x) - f(x)| < \epsilon \qquad \text{for all } x \in I. \tag{1}$$

Take any fixed $n > n_0$, and use the fact that f_n is continuous at a to find $\delta > 0$ so that

$$|f_n(x) - f_n(a)| < \epsilon \qquad \text{if } |x - a| < \delta. \tag{2}$$

If $|x - a| < \delta$, then we have

$$|f(x) - f(a)| \leq |f(x) - f_n(x)| + |f_n(x) - f_n(a)| + |f_n(a) - f(a)|.$$

Each of the three terms is less than ϵ—the first and last because of (1) and the second because of (2).

Exercise 9   Show that if each f_n is uniformly continuous, then f is uniformly continuous.

Example 1 shows that a pointwise limit of continuous functions is not necessarily continuous.

THEOREM 4.4   *Let $I = [a, b]$. If $f_n \to f$ uniformly and each f_n is integrable, then f is integrable, and*

$$\int_a^b f_n(x) \, dx \to \int_a^b f(x) \, dx.$$

Proof   Let $\epsilon > 0$ be given and choose n_0 so that if $n > n_0$,

$$f_n(x) - \epsilon \le f(x) \le f_n(x) + \epsilon \qquad \text{for } a \le x \le b. \tag{3}$$

If f is known to be integrable (for instance, if each f_n is continuous, then Theorem 4.3 shows that f is integrable), then we can simply integrate inequality (3) to get

$$\int_a^b f_n(x) \, dx - \epsilon(b - a) \le \int_a^b f(x) \, dx \le \int_a^b f_n(x) \, dx + \epsilon(b - a) \tag{4}$$

for $n > n_0$, which proves the theorem.

Since f is not known to be integrable, we have to go back to partitions. Inequality (3) gives

$$\bar{S}(p; f) \le \bar{S}(p; f_n + \epsilon) = \bar{S}(p; f_n) + \epsilon(b - a)$$

for every partition p; hence

$$\bar{S}(f) \le \bar{S}(f_n) + \epsilon(b - a).$$

With a similar inequality for $\underline{S}$, we have

$$\underline{S}(f_n) - \epsilon(b - a) \le \underline{S}(f) \le \bar{S}(f) \le \bar{S}(f_n) + \epsilon(b - a).$$

Since f_n is integrable, we get $\bar{S}(f) - \underline{S}(f) \le 2\epsilon(b - a)$, which shows that f is integrable. (So the initial part of the proof applies.)

Example 2 shows that the theorem is false with pointwise convergence instead of uniform convergence. (However, we shall prove a much better theorem in Chapter 13, which shows that pointwise convergence is almost good enough.) Example 1 shows that the corresponding theorem on differentiation is false. In this case much stronger hypotheses are needed—effectively uniform convergence of the derivatives.

THEOREM
4.5

Let I be an open interval, and assume that
(a) *Each f_n is differentiable at every point of I and f_n' is continuous.*
(b) *$\{f_n'\}$ converges uniformly on each closed subinterval.*
(c) *For some $a \in I$, $\{f_n(a)\}$ converges.*
*Then $\{f_n\}$ converges uniformly on each closed subinterval to a limit f,
and $f_n' \to f'$ (uniformly on each closed subinterval).*

Proof

Let $f_n' \to g$ and let $f_n(a) \to b$. By Theorem 4.4,

$$\int_a^x g(t)\, dt = \lim_{n \to \infty} \int_a^x f_n'(t)\, dt = \lim_{n \to \infty} \left(f_n(x) - f_n(a) \right)$$
$$= \lim_{n \to \infty} f_n(x) - b.$$

This shows that $f(x) = \lim f_n(x)$ exists for every x and that

$$f(x) = b + \int_a^x g(t)\, dt.$$

It follows from the fundamental theorem of calculus that $f'(x) = g(x)$ at every point where g is continuous—i.e., at every point, by virtue of Theorem 4.3.

Exercise 10   The proof used the fact that if $s_n \to s$ and $t_n \to t$, then $s_n + t_n \to s + t$.   Prove this and point out where it was used.

Exercise 11   Why does $f_n \to f$ uniformly on each closed subinterval?

Exercise 12   Instead of (c), assume that $f_n(x) \to f(x)$ for every $x \in I$. Now prove the theorem without using integration.   [*Hint:* By the mean-value theorem we have

$$\frac{f_n(x) - f_n(a)}{x - a} = f_n'(\xi_n) = g(a) + g(\xi_n) - g(a) + f_n'(\xi_n) - g(\xi_n);$$

hence

$$\left| \frac{f_n(x) - f_n(a)}{x - a} - g(a) \right| \le |g(\xi_n) - g(a)| + |f_n'(\xi_n) - g(\xi_n)|.$$

The second term is small by the uniform convergence and the first term is small by the continuity of g.]

The above results apply equally well to series. The usual way to show that a series of functions converges uniformly is to find a sequence of numbers M_n such that

$$|f_n(x)| \le M_n \qquad \text{for all } x, \text{ and } \Sigma M_n \text{ converges.} \tag{5}$$

THEOREM
4.6

(*Weierstrass M Test*)   *If* (5) *holds for* $n > n_0$, *then the series* Σf_n *converges uniformly and absolutely.*

Exercise 13   Prove the theorem.   (*Hint:* Use Theorem 4.2 and look back at the proof of Theorem 3.4.)

Exercise 14   As an introduction to Section 5, show that each of the Taylor series in formulas (1) through (6) of Section 1 converges uniformly on any closed subinterval of the interval indicated and that the differentiated series do the same.

5   POWER SERIES

The results of Section 4 are especially pretty for the special series called *power series*.   A power series with center a is a series $\Sigma a_k(x - a)^k$.   The Taylor series of a function is always a power series—and, as a matter of fact, every power series is the Taylor series of a function, but this will be proved only when the series converges for at least one point $x \neq a$.   The principal theorem is as follows.

THEOREM
5.1

Let $f(x) = \Sigma_{k=0}^{\infty} a_k(x - a)^k$.   *Then*
  (a) *There is an* $r \geq 0$ (*possibly* ∞) *such that the series converges for* $|x - a| < r$ *and diverges for* $|x - a| > r$.
  (b) *The convergence is absolute and uniform on each closed subinterval of* $|x - a| < r$.
  (c) *The series can be differentiated term by term; that is,*

$$f'(x) = \sum_{k=1}^{\infty} k a_k(x - a)^{k-1} \qquad \textit{for } |x - a| < r,$$

(*and this series also diverges for* $|x - a| > r$).

It should be emphasized that the series is what is given, and the function f is defined to be the sum of the series at whatever points x the sum exists.   The theorem then asserts that this set of points must be an interval with center a. The number r is called the *radius of convergence* of the series.   It may be 0, in which case the series converges for no point $x \neq a$.

According to part (c), the derivative of f exists on $|x - a| < r$ and is obtained by differentiating term by term just as if the sum were finite.   Moreover, the radius of convergence of the differentiated series is precisely the same number r.   Since the differentiated series is again a power series, the theorem can be applied to conclude that f'' exists on the same interval $|x - a| < r$ and is obtained by differentiating term by term, and again to conclude that f^3 exists, and so on.

COROLLARY 5.2

If $f(x) = \sum_{k=0}^{\infty} a_k(x - a)^k$ on $|x - a| < r$, $r \neq 0$, then f has derivatives of all orders on $|x - a| < r$, and

$$a_k = \frac{f^k(a)}{k!}.$$

This corollary shows that the coefficients in a power series are uniquely determined by the sum f—the series has to be the Taylor series of f. To prove the last part, use part (c) of the theorem to differentiate m times and get

$$f^m(x) = \sum_{k=m}^{\infty} k(k - 1) \cdots (k - m + 1)a_k(x - a)^{k-m}.$$

Now put $x = a$. The only nonzero term is the one with $k = m$, and the formula gives $f^m(a) = m!a_m$.

Before taking up the general theorem, we shall discuss the particular series Σy^k. The general theorem can be reduced to this particular case by the theorems of the last section. In Section 2 of Chapter 5 the following formula was established:

$$\frac{1}{1 - y} = \sum_{k=1}^{n} y^k + \frac{y^{n+1}}{1 - y}. \tag{1}$$

Differentiation gives

$$\frac{1}{(1 - y)^2} = \sum_{k=1}^{n} ky^k + \frac{y^n(n + 1 - ny)}{(1 - y)^2}. \tag{2}$$

LEMMA 5.3

The series Σy^k converges if $|y| < 1$ and diverges if $|y| > 1$. The series Σky^k converges if $|y| < 1$ and diverges if $|y| > 1$.

The first part of the lemma is plain from formula (1). The second part follows from formula (2), but is not so plain. It depends on the fact that

$$\lim_{n \to \infty} ny^n = 0 \qquad \text{if } |y| < 1. \tag{3}$$

Exercise 1   Formula (3) was already established in Exercise 4, Section 5 of Chapter 3. If you do not remember it, do it again with l'Hospital's rule. Use this to complete the proof of the lemma.

It is quite surprising that the general Theorem 5.1 reduces to this special case, and also that there is a simple formula for the number *r*.

THEOREM 5.4

The radius of convergence of the series $\Sigma a_k(x - a)^k$ is the number r defined by

$$\frac{1}{r} = \lim \sup (|a_k|)^{1/k}, \tag{4}$$

and $r = 0$ if $\{(|a_k|)^{1/k}\}$ is unbounded.

Proof of Theorems 5.1 and 5.4

Suppose that the series converges for some $x \neq a$. According to Exercise 5, Section 3, $a_k(x - a)^k \to 0$; so, in particular, there is a positive integer k_0 such that if $k > k_0$, then $|a_k(x - a)^k| < 1$. Therefore,

$$(|a_k|)^{1/k} < \frac{1}{|x - a|} \qquad \text{if } k > k_0,$$

which implies that

$$\frac{1}{r} = \lim \sup (|a_k|)^{1/k} \leq \frac{1}{|x - a|},$$

and hence that $|x - a| \leq r$. This proves half of part (a) of Theorem 5.1

To prove part (b) and with it the other half of (a), let r' be any positive number smaller than r, and let r'' be any number between them. Since $1/r'' > 1/r$, it follows from the definition of r that there is a positive integer k_0 such that

$$(|a_k|)^{1/k} < \frac{1}{r''} \qquad \text{if } k > k_0.$$

Therefore,

$$|a_k| < \left(\frac{1}{r''}\right)^k \qquad \text{if } k > k_0.$$

Consequently,

$$\left.\begin{aligned}
|a_k(x - a)^k| &\leq \left(\frac{r'}{r''}\right)^k = y^k \\
|ka_k(x - a)^k| &\leq ky^k
\end{aligned}\right\} \text{ if } k > k_0 \text{ and } |x - a| < r'. \tag{5}$$

The lemma and Theorem 4.6 show that both series $\Sigma a_k(x - a)^k$ and $\Sigma ka_k(x - a)^k$ converge absolutely and uniformly on $|x - a| \leq r'$. Then Theorem 4.5 shows that the term-by-term differentiation in (c) is all right. The only thing that remains is to show that the differentiated series cannot actually have a larger radius of convergence. If it did we could use Theorem 4.4 to integrate term by term and get a larger radius of convergence for the original.

Exercise 2   Write out the proof of this last statement. [You will have to use part (b) of the present theorem!]

Usually the easiest way to find the radius of convergence of a power series is to use the following theorem (called the ratio test):

THEOREM
5.5

(*Ratio Test*) *If* $b_k > 0$, *then the series* Σb_k *converges if*

$$\limsup \frac{b_{k+1}}{b_k} < 1$$

and diverges if

$$\liminf \frac{b_{k+1}}{b_k} > 1.$$

Example   The Taylor series for e^x is the series

$$\sum \frac{x^k}{k!}.$$

With x fixed, take $b_k = |x^k/k!|$.   Then

$$\lim \frac{b_{k+1}}{b_k} = \lim \frac{|x|}{k+1} = 0.$$

Therefore, the series converges absolutely for each x.

Exercise 3   Apply the theorem to find the radius of convergence of the several Taylor series in Section 1.

Proof of the Theorem   If

$$\limsup \frac{b_{k+1}}{b_k} < 1,$$

then there is a number $y < 1$ such that

$$\limsup \frac{b_{k+1}}{b_k} < y,$$

and there is a positive integer k_0 such that

$$b_{k+1} < y b_k \qquad \text{if } k \geq k_0.$$

Hence

$$b_{k_0+1} < yb_{k_0},\ b_{k_0+2} < yb_{k_0+1} < y^2b_{k_0},\ b_{k_0+3} < yb_{k_0+2} < y^3b_{k_0},\ \ldots.$$

In general,

$$b_{k_0+p} < y^p b_{k_0} \qquad \text{for every } p.$$

The series converges by comparison with Σy^p, $y < 1$.

Exercise 4   Prove the other half of the theorem by using a similar argument to show that the sequence $\{b_k\}$ is unbounded (hence does not approach 0).

The theorem is effective on power series because of the cancellation between the powers of $x - a$ in one term and the next. Usually it is not much good on other series. Too often the limits in question turn out to be 1.

Note that the theorem can show that a Taylor series converges, but it cannot show that the Taylor series converges to the function. This is the topic of the next section.

6   ANALYTIC FUNCTIONS

The point of departure in the last section was the series. We established various criteria of convergence and properties of the sum. The point of departure here is the function. We begin with a function and ask when the Taylor series not only converges, but converges to the function. Because of Corollary 5.2, we consider only functions that possess derivatives of all orders on some interval with center at the point in question.

DEFINITION 6.1   *A function f is of class C^∞ at a point a if there is a positive number r such that f is defined and possesses derivatives of all orders on the interval $|x - a| < r$.*

DEFINITION 6.2   *A function f is analytic at a point a if there exist a positive number r and a power series $\Sigma a_k(x - a)^k$ such that*

$$f(x) = \sum_{k=0}^{\infty} a_k(x - a)^k \qquad \text{for } |x - a| < r. \tag{1}$$

It is implicit, of course, that f is defined on $|x - a| < r$ and that the series converges on $|x - a| < r$. It follows from Corollary 5.2 that if f is analytic at a, then f is C^∞, and the series in (1) must be the Taylor series. The function

$$f(x) = e^{-1/x^2}, \qquad f(0) = 0,$$

is one which is C^∞ at every point, but is not analytic at 0. Its Taylor series at 0 is identically 0.

Note that if f is of class C^∞ at a, then it is also of class C^∞ at every point x sufficiently close to a—indeed, at every point x with $|x - a| < r$, with r as in Definition 6.1. On the other hand, suppose that f is analytic at a. It follows from Corollary 5.2 that f is C^∞ at every point x with $|x - a| < r$, with r as in Definition 6.2. It is true, but not implied by Corollary 5.2, that f is analytic at every such point.

It is obvious that the sum and product of C^∞ functions is C^∞ and almost obvious that the quotient is C^∞ if the denominator is $\neq 0$ at the point a. It is obvious that the sum of analytic functions is analytic, but it is not at all obvious that the product or quotient is. Both this question and the one raised in the last paragraph can be settled by the following theorem.

THEOREM 6.3   *Let f be of class C^∞ at the point a. Then f is analytic at a if and only if there exist positive numbers δ and M such that*

$$|f^k(x)| \leq M^k k! \qquad \text{for } |x - a| < \delta. \tag{2}$$

Proof   Suppose first that f satisfies condition (2). We shall prove that f is analytic by using Taylor's formula. The remainder term in Taylor's formula is

$$\left| \frac{f^{n+1}(\xi)}{(n + 1)!} (x - a)^{n-1} \right| \leq M^{n+1} |x - a|^{n+1} \qquad \text{for } |x - a| < \delta.$$

The term on the right goes to 0 if $M|x - a| < 1$. Therefore, the Taylor series of f converges to f if $|x - a| < r$, where

$$r = \min\left(\frac{1}{M}, \delta \right).$$

Suppose that f is analytic and that

$$f(x) = \sum_{k=0}^{\infty} a_k(x - a)^k \qquad \text{for } |x - a| < r, \tag{3}$$

and, therefore, that

$$f^m(x) = \sum_{k=m}^{\infty} k(k - 1) \cdots (k - m + 1)a_k(x - a)^{k-m}$$
$$\text{for } |x - a| < r. \tag{4}$$

As in Section 5, take any $r' < r$ and then r'' between the two. Since the series (3) converges for $|x - a| = r''$, the sequence $\{a_k(r'')^k\}$ goes to 0;

so there is a constant A such that

$$|a_k(r'')^k| \leq A \qquad \text{or} \qquad |a_k| \leq A(r'')^{-k} \qquad \text{for all } k.$$

If $|x - a| \leq r'$, then by (4)

$$|f^m(x)| \leq \sum_{k=m}^{\infty} k(k-1) \cdots (k-m+1)A(r'')^{-k}(r')^{k-m}.$$

If we put $y = r'/r''$, this gives

$$|f^m(x)| \leq A(r')^{-m} \sum_{k=m}^{\infty} k(k-1) \cdots (k-m+1)y^k.$$

The sum on the right is known. It is just the mth derivative of the function $g(y) = 1/1 - y$, which is $m!(1-y)^{-m-1}$. (Differentiate the Taylor series of g.) Therefore,

$$|f^m(x)| \leq A(r')^{-m}(1-y)^{-m-1}m! \leq M^m m! \qquad \text{for } |x-a| \leq r', \quad (5)$$

if we choose M large enough so that

$$M^m \geq A(r')^{-m}(1-y)^{-m-1}. \qquad (6)$$

Exercise 1   If $A \geq 1$, which plainly can be assumed, then (6) holds if

$$M = \frac{A}{r'(1-y)^2}.$$

THEOREM 6.4   *If $f(x) = \sum_{k=0}^{\infty} a_k(x-a)^k$ for $|x-a| < r$, then f is analytic at each point of the interval $|x-a| < r$.*

Proof   Given a point x with $|x-a| < r$, choose r' with $|x-a| < r' < r$. Then use inequality (5) and the theorem.

THEOREM 6.5   *If f and g are analytic at a, then so is fg. If*

$$f(x) = \sum_{k=0}^{\infty} a_k(x-a)^k \qquad g(x) = \sum_{l=0}^{\infty} b_l(x-a)^l \qquad \text{for } |x-a| < r,$$

then

$$f(x)g(x) = \sum_{n=0}^{\infty} c_n(x-a)^n \qquad \text{where } c_n = \sum_{k=0}^{n} a_k b_{n-k}.$$

Note that on the one hand, c_n is what it must be by virtue of Theorem 2.3 of Chapter 5 on the nth derivative of the product fg. On the other hand, it is just what is obtained by multiplying the series together term by term and collecting together all the terms with a given exponent n.

Proof

Take any $r' < r$ and use (5) to find M such that

$$|f^m(x)| \le M^m m! \quad \text{and} \quad |g^m(x)| \le M^m m! \quad \text{for } |x - a| \le r'.$$

Theorem 2.3 of Chapter 5 gives

$$|(fg)^k(x)| = \left| \sum_{i=0}^{k} \frac{k!}{i!(k-i)!} f^i(x) g^{k-i}(x) \right| \le M^k(k+1)!$$

$$\text{for } |x - a| \le r'.$$

The remainder in Taylor's formula for fg is

$$\left| \frac{(fg)^{n+1}(\xi)}{(n+1)!} (x-a)^{n+1} \right| \le (n+2)M^{n+1}|x-a|^{n+1} \quad \text{for } |x-a| \le r'.$$

Several times we have seen that this converges to 0 if $M|x - a| < 1$, which shows that the Taylor series of fg converges to fg on $|x - a| < \delta$, with $\delta = \min\left(\dfrac{1}{M}, r'\right)$.

This brings up an interesting point. On the one hand, the theorem shows that fg is analytic at every point x with $|x - a| < r$. On the other hand, the Taylor series of fg converges at every point x with $|x - a| < r$.

Exercise 2 Prove the last statement.

This suggests a general theorem.

THEOREM 6.6 *If f is analytic at every point of $|x - a| < r$ and the Taylor series converges at every point of $|x - a| < r$, then the Taylor series converges to f at every point of $|x - a| < r$.*

Exercise 3 Prove Theorem 6.6 by using Theorem 6.7 below.

THEOREM 6.7 *Let f be analytic at each point of an open interval I. If f vanishes identically on some open subinterval, then f vanishes identically on I.*

Proof   Choose a point c in the subinterval and let $b = \sup\{y : f(x) = 0$ for $c < x < y\}$. What we have to show is that b is the right-hand end point of I. Then the same proof will show that $a = \inf\{y : f(x) = 0$ for $a < x < c\}$ is the left-hand end point, and the theorem will be proved.

If b is not the right-hand end point of I, then f is analytic at b, so

$$f(x) = \sum_{k=0}^{\infty} \frac{f^k(b)}{k!}(x - b)^k \qquad \text{for } |x - b| < \delta$$

for some positive δ. On the other hand, since f vanishes identically between c and b, it follows that all derivatives of f vanish at b, whence $f(x) = 0$ for $b \leq x < b + \delta$, which is a contradiction.

This whole business is rather slippery. Theorem 6.6 suggests a stronger theorem—if f is analytic at every point of $|x - a| < r$, then the Taylor series of f at a converges to f at every point of $|x - a| < r$—which is false. It is not hard to see (and follows from the next theorem) that

$$f(x) = \frac{1}{1 + x^2}$$

is analytic at every point. On the other hand, its Taylor series

$$\Sigma(-1)^k x^{2k}$$

at $a = 0$ has the radius of convergence 1.

The explanation lies in the fact that power series should be considered in the complex domain, not in the real. Considering complex numbers x, as well as real ones, we see that f behaves badly as $x \to \pm i$. The distance from $\pm i$ to $a = 0$ is 1, and this is why the radius of convergence is 1. With the right theorems in the complex domain, the last several theorems, and the next one as well, become trivial.

THEOREM   *If f is analytic at a and $f(a) \neq 0$, then $1/f$ is analytic at a.*
6.8

Proof   This theorem is included partly for the sake of the result and partly for the sake of the proof, which is typical of a number of proofs in analytic function theory.

Let

$$f(x) = \sum_{k=0}^{\infty} a_k(x - a)^k.$$

It is clearly permissible to suppose that $a_0 = f(a) = 1$. (Prove this.) What is needed is a function g, analytic on some interval with center a, such that $f(x)g(x) = 1$ on some interval with center a. To get started, suppose that

$$g(x) = \sum_{k=0}^{\infty} b_k(x - a)^k \qquad (7)$$

is such a function. According to Theorem 6.5, it must be true that

$$\sum_{k=0}^{n} a_k b_{n-k} = \begin{cases} 1 & \text{if } n = 0, \\ 0 & \text{if } n > 0, \end{cases}$$

or, equivalently (taking account of the fact that $a_0 = 1$), that

$$b_0 = 1 \quad \text{and} \quad b_n = -\sum_{k=1}^{n} a_k b_{n-k}. \qquad (8)$$

Now start afresh and use formula (8) to define the sequence $\{b_n\}$. This is an inductive definition. It determines b_0 first. And once $b_0, \ldots, b_{n-1}$ are determined, it determines b_n. By virtue of Theorem 6.5, it is now entirely a question of proving that the series (7), with coefficients determined by formula (8), converges for some $x \neq a$. By Theorem 5.4, this is equivalent to proving that for some number N

$$|b_n| \leq N^n \qquad \text{for all } n. \qquad (9)$$

Since the series for f does converge for some $x \neq a$, there is a number M such that

$$|a_n| \leq M^n \qquad \text{for all } n. \qquad (10)$$

Let us try to prove formula (9) with $N = 2M$. By induction and (8) we have [if (9) holds for integers $<n$]

$$|b_n| \leq \sum_{k=1}^{n} M^k N^{n-k} = N^n \sum_{k=1}^{n} \frac{1}{2^k} \leq N^n,$$

which proves (9), and hence the theorem.

In somewhat the same vein, there is a theorem on composite functions. It is stated as follows, but will not be proved, since there is no advantage in laboring things that become obvious once the right (complex) point of view is adopted.

THEOREM 6.9

If g is analytic at a, and f is analytic at $b = g(a)$, then the composite function $h(x) = f(g(x))$ is analytic at a.

7 EXAMPLES

The exponential and trigonometric functions can be defined easily by power series. This avoids some of the sticky points about arc length, for instance, but it requires fairly substantial knowledge of series.

Take first the exponential. Define

$$E(x) = \sum_{k=0}^{\infty} \frac{x^k}{k!}. \tag{1}$$

The ratio test (Theorem 5.5) shows that the series converges for every x. Differentiation term by term shows that

$$E'(x) = E(x). \tag{2}$$

The basic formula for the exponential is

$$E(a)E(b) = E(a + b). \tag{3}$$

To prove it, let $f(x) = E(ax)$ and $g(x) = E(bx)$. By Theorem 6.5,

$$f(x)g(x) = \sum_{n=0}^{\infty} c_n x^n, \quad \text{where } c_n = \sum_{k=0}^{n} \frac{a^k b^{n-k}}{k'(n-k)!} = \frac{(a+b)^n}{n!},$$

so $E(ax)E(bx) = E((a + b)x)$, which gives (3) when $x = 1$.

If $x > 0$, then each term in the series is >0, so $E(x) > 0$. On the other hand, (3) shows that $E(-x) = 1/E(x)$, so $E(x) > 0$ for all x. Hence $E'(x) = E(x) > 0$ for all x, and E is strictly increasing.

Again, if $x > 0$, then each term in the series is >0, so the sum is larger than any one term; hence $E(x) > x$, and

$$\lim_{x \to \infty} E(x) = \infty.$$

The fact that $E(-x) = 1/E(x)$ gives

$$\lim_{x \to -\infty} E(x) = 0.$$

Consequently, E has a differentiable inverse L defined on $0 < y < \infty$ by

$$L(y) = x \quad \text{if and only if } E(x) = y.$$

From (2) and the rule for differentiating composite functions it follows that

$$L'(y) = \frac{1}{y}.$$

This should be enough to show how the development goes.

Exercise 1   Define e to be $E(1)$ and show that $E(x) = e^x$ when x is rational.

Remark 1   The function L is actually analytic at each point $y > 0$. The relevant theorem is the following:

THEOREM 7.1   *Let f be analytic at a, and let $f'(a) \neq 0$. Then f has an inverse which is defined on some interval with center $b = f(a)$ and is analytic at b.*

Note that the existence and differentiability of the inverse are known already. Since f is analytic, the derivative f' must exist on some interval with center a. Furthermore, it must be $\neq 0$, since it is continuous and $f'(a) \neq 0$. This implies that f is either strictly increasing or strictly decreasing, so the inverse exists. The differentiability follows from Section 1 of Chapter 3. Note also that the condition $f'(a) \neq 0$ is necessary for the existence of a differentiable inverse (differentiate the composite)—therefore, certainly for the existence of an analytic inverse.

The fact that the inverse is analytic is another one of those problems that is easy from the complex point of view, and a nuisance at present. It can be solved along lines similar to those in Theorem 6.8, but we shall not do it.

Remark 2   In proving formula (3) it would be more natural simply to multiply together the series for $E(a)$ and $E(b)$. So far we have considered the multiplication of power series but not of general series.

THEOREM 7.2   *If $a = \sum_{k=0}^{\infty} a_k$ and $b = \sum_{k=0}^{\infty} b_k$ and the series converge absolutely, then*

$$ab = \sum_{n=0}^{\infty} c_n \qquad where \ c_n = \sum_{k=0}^{n} a_k b_{n-k}.$$

Proof   Let

$$f(x) = \sum_{k=0}^{\infty} a_k x^k \qquad g(x) = \sum_{k=0}^{\infty} b_k x^k \qquad h(x) = \sum_{n=0}^{\infty} c_k x^n.$$

According to Theorem 6.5, we do have that $h(x) = f(x)g(x)$ for $|x| < 1$, and the whole point is to justify putting $x = 1$. First it must be shown that h is defined at $x = 1$.

Exercise 2   Show that the series Σc_n converges absolutely. (*Hint:* This is easy.)

The theorem will be proved if it can be shown that f, g, and h are continuous at $x = 1$:

THEOREM
7.3

If $f(x) = \sum_{k=0}^{\infty} a_k(x-a)^k$ converges absolutely for $|x-a| = r$, then the convergence is uniform on $|x-a| \leq r$, so f is continuous on $|x-a| \leq r$.

Exercise 3   Prove the theorem.   (*Hint:* This also is easy.)

Now let us turn to the trigonometric functions.   Let

$$S(x) = \sum_{k=0}^{\infty} (-1)^k \frac{x^{2k+1}}{(2k+1)!},$$

$$C(x) = \sum_{k=0}^{\infty} (-1)^k \frac{x^{2k}}{(2k)!}. \tag{4}$$

The ratio test shows that both series converge for every x.   Differentiation term by term shows that

$$S'(x) = C(x) \qquad C'(x) = -S(x). \tag{5}$$

Setting $x = 0$ in the series, we get

$$S(0) = 0 \qquad C(0) = 1. \tag{6}$$

Consider the function $f(x) = S(x)^2 + C(x)^2$.   We have

$$f' = 2SS' + 2CC' = 2SC - 2CS = 0.$$

Therefore, f is constant, and by (6) the constant must be 1.   Hence we have the identity

$$S(x)^2 + C(x)^2 = 1 \qquad \text{for all } x. \tag{7}$$

It follows that

$$|S(x)| \leq 1 \quad \text{and} \quad |C(x)| \leq 1 \qquad \text{for all } x. \tag{8}$$

Consider the Taylor series for S at an arbitrary point a and for $x = a + h$:

$$S(a+h) = \sum_{k=0}^{\infty} \frac{S^k(a)}{k!} h^k.$$

This is valid for every a and every h, for by virtue of (8) and (5) the remainder in Taylor's formula is

$$\left| \frac{S^{n+1}(\xi)}{(n+1)!} h^{n+1} \right| \leq \left| \frac{h^{n+1}}{(n+1)!} \right| \to 0.$$

According to (5),

$$S^{2k+1}(x) = (-1)^k C(x) \qquad S^{2k}(x) = (-1)^k S(x). \tag{9}$$

Substituting these in the Taylor series, we find that

$$S(a + h) = \sum_{k=0}^{\infty} \frac{(-1)^k C(a) h^{2k+1}}{(2k+1)!} + \sum_{k=0}^{\infty} \frac{(-1)^k S(a) h^{2k}}{(2k)!}.$$

Factoring out $C(a)$ and $S(a)$ and looking at the two series, we observe the identity (addition formula for the sine)

$$S(a + h) = S(a)C(h) + C(a)S(h). \tag{10}$$

Differentiate with respect to h to get

$$C(a + h) = S'(a + h) = C(a)C(h) - S(a)S(h). \tag{11}$$

Define the number π by

$$\frac{\pi}{2} = \inf\{x : x > 0 \text{ and } C(x) = 0\}. \tag{12}$$

To do this we must show that there exists some $x > 0$ with $C(x) = 0$. If not, then $C(x) > 0$ for all $x > 0$. Now, (11) and (7) give

$$C(2x) = C(x)^2 - S(x)^2 = 2C(x)^2 - 1,$$

so if $C(x) > 0$ for all $x > 0$, then it follows that

$$C(x) > \frac{1}{\sqrt{2}}; \quad \text{hence } S(x) \geq \frac{1}{\sqrt{2}} x,$$

which contradicts (8).

Therefore, (12) makes sense and defines π. From the definition it follows that

$$C\left(\frac{\pi}{2}\right) = 0 \quad \text{and} \quad C(x) > 0 \quad \text{for } 0 \leq x < \frac{\pi}{2}. \tag{13}$$

From this it follows that $S(\pi/2) = 1$, and that S is strictly increasing on $0 \leq x \leq \pi/2$. Hence $S > 0$ on this interval, and $C' = -S$ is negative. Therefore, C is strictly decreasing. The behavior on the interval $\pi/2 < x < \pi$ can be deduced from this and (10) and (11), which give

$$S\left(x + \frac{\pi}{2}\right) = C(x), \quad C\left(x + \frac{\pi}{2}\right) = -S(x). \tag{14}$$

In particular, $S(\pi) = 0$ and $C(\pi) = -1$; then

$$S(x + \pi) = -S(x), \quad C(x + \pi) = -C(x), \tag{15}$$

and, finally,

$$S(x + 2\pi) = S(x), \quad C(x + 2\pi) = C(x). \tag{16}$$

As for the behavior for negative x, since the series for S has only odd exponents and the series for C has only even ones, we have

$$S(-x) = -S(x), \qquad C(-x) = C(x). \tag{17}$$

In particular, S is strictly increasing on $-\pi/2 \le x \le \pi/2$ and has a non-zero derivative. So it has an inverse function A, and A is also differentiable. Since $S(A(x)) = x$, differentiation gives $S'(A(x))A'(x) = 1$. We have to calculate $S'(A(x)) = C(A(x))$, which is easy because

$$S(A(x))^2 + C(A(x))^2 = 1 \qquad \text{or} \qquad x^2 + C(A(x))^2 = 1.$$

$C \ge 0$ on this interval, so $C(A(x)) = \sqrt{1 - x^2}$. Thus,

$$A'(x) = \frac{1}{\sqrt{1 - x^2}}. \tag{18}$$

Exercise 4   Get back to the original definition of the sine and cosine by calculating a suitable arc length (by the formula of Section 7 of Chapter 4, of course).

Exercise 5   For any two numbers a and b with $a^2 + b^2 = 1$, there is exactly one point x with

$$\cos x = a, \qquad \sin x = b, \qquad 0 \le x < 2\pi.$$

[We now write $\sin x$ and $\cos x$ for $S(x)$ and $C(x)$. This exercise must be done entirely from the present point of view. No geometric intuition!]

Exercise 6   If $\sin y = \sin x$ and $\cos y = \cos x$, then x and y differ by an integer multiple of 2π.

8   WEIERSTRASS APPROXIMATION THEOREM

The power series that we have been studying provide one means of approximating functions by polynomials (the Taylor polynomials). This kind of approximation is very special. It works only for functions that are analytic and, in particular, C^∞. Now we shall look at another approximation that works for all continuous functions.

THEOREM 8.1   (*Weierstrass Approximation Theorem*) *If f is continuous on an interval I, there is a sequence $\{f_n\}$ of polynomials such that $f_n \to f$ uniformly on every bounded closed subinterval.*

Proof   To begin with (and this is the meat of the proof, the rest is easy) we shall suppose that f is continuous on the whole line and vanishes identically for

$|x| \geq \frac{1}{2}$. In this case we set

$$p_n(x) = c_n(1 - x^2)^n,$$

where c_n is chosen so that

$$\int_{-1}^{1} p_n(x) \, dx = 1. \tag{1}$$

Then we define

$$f_n(x) = \int_{-\infty}^{\infty} f(y) p_n(x - y) \, dy = \int_{-\infty}^{\infty} f(x - y) p_n(y) \, dy. \tag{2}$$

Note that the integrals are not really improper because f vanishes identically for $|y| \geq \frac{1}{2}$. To go from the first integral to the second, simply make the change of variable $y = x - z$.

**LEMMA
8.2**

Each f_n is a polynomial, and $f_n \to f$ uniformly on $|x| \leq \frac{1}{2}$.

From the first part of formula (2) it is plain that f_n is a polynomial. Just multiply out $(1 - (x - y)^2)^n$ and remove the powers of x from the integral.

If $|x| \leq \frac{1}{2}$, then the second part of formula (2) gives

$$f_n(x) = \int_{-1}^{1} f(x - y) p_n(y) \, dy, \tag{3}$$

for $f(x - y)$ vanishes unless $|x - y| \leq \frac{1}{2}$, hence unless $|y| \leq 1$. This formula and (1) show that

$$f(x) - f_n(x) = \int_{-1}^{1} [f(x) - f(x - y)] p_n(y) \, dy. \tag{4}$$

Let $\epsilon > 0$ be given, let M be the maximum of $|f|$, and choose $\delta > 0$ so that if $|y| < \delta$, then $|f(x) - f(x - y)| < \epsilon$. Then formula (4) gives

$$|f(x) - f_n(x)| \leq \int_{|y| < \delta} \epsilon p_n(y) \, dy + \int_{\delta \leq |y| \leq 1} M p_n(y) \, dy$$

$$\leq \epsilon + M \int_{\delta < |y| \leq 1} p_n(y) \, dy. \tag{5}$$

Therefore, the whole problem is to show that

$$\int_{\delta \leq |y| \leq 1} p_n(y) \, dy \to 0 \qquad \text{for each } \delta > 0. \tag{6}$$

Take any r, $0 < r < 1$. By the definition of c_n we have

$$\frac{1}{c_n} = \int_{-1}^{1} (1 - x^2)^n \, dx \geq \int_{-r}^{r} (1 - r^2)^n \, dx = 2r(1 - r^2)^n,$$

so

$$c_n \leq \frac{1}{2r(1 - r^2)^n}.$$

Therefore,

$$\int_{\delta \le |y| \le 1} p_n(y) \, dy \le \frac{1}{2r(1-r^2)^n} \int_{-1}^{1} (1-\delta^2)^n \, dy = \frac{(1-\delta^2)^n}{r(1-r^2)^n}.$$

If we fix $r < \delta$ and let $n \to \infty$, we see that (6) holds.

Now we have the lemma, and we shall use it to prove the theorem. Suppose first that f vanishes for $|x| \ge r$. Then $F(x) = f(2rx)$ vanishes for $|x| \ge \frac{1}{2}$. For each $\epsilon > 0$ the lemma provides a polynomial P_ϵ such that $|P_\epsilon(x) - F(x)| < \epsilon$ for $|x| \le \frac{1}{2}$. Then $Q_\epsilon(x) = P_\epsilon(x/2r)$ is a polynomial that satisfies $|Q_\epsilon(x) - f(x)| < \epsilon$ for $|x| \le r$.

Finally, let f be continuous on the open interval (a, b). For each positive integer n, choose a continuous function φ_n on (a, b) that is 1 on $[a + 1/n, b - 1/n]$ and vanishes identically near a and b. By what has been proved there is a polynomial f_n such that $|f_n(x) - \varphi_n(x)f(x)| < 1/n$ on (a, b), and the sequence $\{f_n\}$ clearly does the job of approximating uniformly on every bounded closed subinterval.

Exercise 1 How do you finish up the argument if $a = -\infty$ or $b = \infty$?

Exercise 2 How do you finish up the argument if the initial interval is not open but closed at one or the other end point?

Exercise 3 Draw pictures of the polynomials p_n and discuss why it is reasonable to expect that the f_n in Lemma 8.2 should approximate f.

In Section 12 of Chapter 7 we shall give a substantial generalization of the Weierstrass approximation theorem that can be used to find approximations by other kinds of functions than polynomials.

PART II

7 Metric Spaces

1 THE SPACE $\mathbf{R}^n$

DEFINITION 1.1

The real n-dimensional space, called $\mathbf{R}^n$, is the set of all n tuples of real numbers.

Geometrically, the one-dimensional space $\mathbf{R}^1$ is the line, the two-dimensional space $\mathbf{R}^2$ is the plane, and the three-dimensional space $\mathbf{R}^3$ is the three-dimensional space. In each case the identification of the geometric object with the set of real numbers, pairs of real numbers, or triples of real numbers presupposes that a coordinate system is given.

There are three natural algebraic operations on the space $\mathbf{R}^n$.

1. *Addition: If $x = (x_1, \ldots, x_n)$ and $y = (y_1, \ldots, y_n)$, then*
$$x + y = (x_1 + y_1, \ldots, x_n + y_n).$$

2. *Scalar multiplication: If $x = (x_1, \ldots, x_n)$ and α is a real number, then*
$$\alpha x = (\alpha x_1, \ldots, \alpha x_n).$$

3. *Inner product: If $x = (x_1, \ldots, x_n)$ and $y = (y_1, \ldots, y_n)$, then*
$$\langle x, y \rangle = \sum_{k=1}^{n} x_k y_k.$$

These operations are natural in the sense that on the one hand they suggest themselves to some extent, and on the other hand (what is far more important) they have a geometric significance.

Exercise 1 Show that if x and y are points in the plane, then $x + y$ is the fourth vertex of the parallelogram of which the other three vertices are x, y, and $0 = (0, 0)$. Find a geometric interpretation of αx.

Exercise 2 Show that if x and y are points in the plane, then $\langle x, y \rangle = |x| \, |y| \cos \theta$, where $|x|$ is the distance from x to 0 and θ is the angle determined by the half-lines $0x$ and $0y$.

The same geometrical interpretations can be established in the three-dimensional space, but the calculations are more complicated. For spaces of dimension greater than three, the meaning of the phrase "geometrical significance" will have to be made clear.

There is a common belief that spaces of dimension greater than three are illusions to which most mathematicians and occasional physicists like Einstein are subject. This is not quite correct.

Consider the problem of describing the motion of the earth and the moon around the sun. If coordinates are chosen in the three-dimensional space with the origin at the sun, then the position of the earth is described by three coordinates and so is the position of the moon. The two together are described by six coordinates, that is, by a point in $\mathbf{R}^6$. The motion of the two bodies is described by a "curve" in $\mathbf{R}^6$.

Exercise 3 Discuss the description of a box full of gas (e.g., a boiler full of steam) and its behavior over an interval of time.

Implausible as it sounds, this is one way these things are analyzed.

Before going on, we record the following simple properties of the inner product, which can be established by inspection.

THEOREM 1.2

(a) $\langle x, x \rangle \geq 0$.
(b) $\langle x, y \rangle = \langle y, x \rangle$.
(c) $\langle \alpha x, y \rangle = \alpha \langle x, y \rangle$.
(d) $\langle x + y, z \rangle = \langle x, z \rangle + \langle y, z \rangle$.

The inner product actually satisfies a stronger condition than (a), which is

$$\langle x, x \rangle > 0 \text{ unless } x = 0 = (0, \ldots, 0). \tag{a'}$$

Later on we shall come upon some inner products which satisfy (a) but not (a').

The inner product in $\mathbf{R}^n$ is not a product in the usual sense of the word unless $n = 1$, for if $x \in \mathbf{R}^n$ and $y \in \mathbf{R}^n$, then the inner product $\langle x, y \rangle$ lies not in $\mathbf{R}^n$ but in $\mathbf{R}^1$. In $\mathbf{R}^2$ there does exist a very important natural product of the usual kind.

DEFINITION 1.3   *If $x, y \in \mathbf{R}^2$, then the product xy is defined by*

$$xy = (x_1y_1 - x_2y_2,\ x_2y_1 + x_1y_2).$$

Exercise 4   Check that the usual rules of arithmetic hold for this product, that is, that $xy = yx$, $x(yz) = (xy)z$, and $x(y + z) = xy + xz$.

Exercise 5   Consider $\mathbf{R}^1$ as a subset of $\mathbf{R}^2$ by identifying the real number α with the point $(\alpha, 0)$ in $\mathbf{R}^2$.   (This just identifies the real line with the x_1 axis in $\mathbf{R}^2$.)   Show that this identification is consistent with all four operations on $\mathbf{R}^2$.   (If α and β are real, you can form the sum in $\mathbf{R}^1$ and then identify it with a point in $\mathbf{R}^2$. On the other hand, you can identify α and β individually with points in $\mathbf{R}^2$ and then form the sum in $\mathbf{R}^2$.   The problem is to show that the two procedures yield the same result, and so on.)

Exercise 6   More generally, consider $\mathbf{R}^n$ as a subset of $\mathbf{R}^{n+1}$ by identifying the point $x = (x_1, \ldots, x_n)$ with the point $(x_1, \ldots, x_n, 0)$ in $\mathbf{R}^{n+1}$.   Show that this identification is consistent with the three operations (a), (b), and (c).

DEFINITION 1.4   *In situations where the multiplication on $\mathbf{R}^2$ plays a role, $\mathbf{R}^2$ is usually called the complex plane or the complex number system, and the points of $\mathbf{R}^2$ are called complex numbers.*

The complex number $(0, 1)$ is usually called i.   According to the definition of multiplication and the convention of Exercise 5, we have

$$i^2 = (-1, 0) = -1. \tag{1}$$

If $x = (x_1, x_2)$ is any point of $\mathbf{R}^2$, then by Exercise 5 we have

$$x = (x_1, x_2) = (x_1, 0) + (0, x_2) = x_1 + x_2 i.$$

Thus, we get the following result.

THEOREM 1.5   *Every complex number z can be expressed uniquely in the form*

$$z = x + yi, \text{ where } x \text{ and } y \text{ are real and } i^2 = -1. \tag{2}$$

Exercise 7   Prove the uniqueness.

This is the usual form in which complex numbers are written, rather than with subscripts $x_1 + x_2 i$ or as pairs (x_1, x_2).   The numbers x and yi are called the *real* and *imaginary* parts of z and are written Re z and Im z.

Exercise 8   The multiplication formula of Definition 1.3 seems mysterious.   Show that it follows from formula (2) and the usual rules of arithmetic.   [Thus, formula (2) is what should be remembered, not Definition 1.3.]

DEFINITION 1.6   *The conjugate of the complex number $z = x + yi$, x and y real, is the complex number*

$$\bar{z} = x - yi.$$

Geometrically, $\bar{z}$ is just the reflection of z across the x axis.

Exercise 9   Show that

$$\overline{z + w} = \bar{z} + \bar{w}, \quad \overline{zw} = \bar{z}\bar{w}, \quad |\bar{z}| = |z|, \quad \text{and} \quad z\bar{z} = |z|^2.$$

(Recall that if $z \in \mathbf{R}^2$, then $|z|$ is the distance from z to 0; that is, if $z = x + yi$, then $|z|^2 = x^2 + y^2$.)

Exercise 10   Use the fact that $z\bar{z} = |z|^2$ to show that every nonzero complex number has a reciprocal.   Calculate $(2 - 3i)/(3 + i)$; that is, put it in the form $x + yi$ with x and y real.

Remark   It can be shown (not easily) that if $n > 2$, there is no multiplication on $\mathbf{R}^n$ with the properties listed in Exercises 4, 5, and 10.

The complex numbers of absolute value 1 are precisely the ones on the unit circle, and hence precisely the ones of the form $(\cos \theta, \sin \theta) = \cos \theta + i \sin \theta$.   This representation is unique if θ is restricted to lie in the interval $[0, 2\pi)$.   If z is any complex number $\neq 0$, then $z/|z|$ has absolute value 1; so z can be written uniquely in the form

$$z = r(\cos \theta + i \sin \theta), \qquad r > 0, \quad 0 \leq \theta < 2\pi. \qquad (3)$$

The number r is just $|z|$.   The number θ is called the *principal value of the argument of z*.

Exercise 11   If $z = r(\cos \theta + i \sin \theta), r \geq 0$, then $r = |z|$.   If $z \neq 0$, then θ differs from the principal value of the argument of z by an integer multiple of 2π.   (Any such θ is called an argument of z.)

Exercise 12   If $z = r(\cos \theta + i \sin \theta)$ and $w = s(\cos \varphi + i \sin \varphi)$, then

$$zw = rs(\cos(\theta + \varphi) + i \sin (\theta + \varphi));$$

hence

$$z^k = r^k(\cos k\theta + i \sin k\theta), \qquad k \text{ an integer.}$$

Exercise 13   Find the three solutions to the equation $z^3 = i$.

Exercise 14   The inner product in $\mathbf{R}^2$ is expressed in terms of the product by the formula

$$\langle z, w \rangle = \text{Re } z\bar{w}, \tag{4}$$

and from this it follows that

$$\langle iz, z \rangle = 0. \tag{5}$$

Also deduce formula (5) from Exercises 2 and 12.   (Exercise 12 shows that iz is obtained from z by a rotation through a counterclockwise angle of 90°.)

2   ABSOLUTE VALUE IN R^n

It is possible to define an "absolute value" in $\mathbf{R}^n$, which plays a role very much like that of the usual absolute value of a real number.   In $\mathbf{R}^2$ and $\mathbf{R}^3$ the geometric interpretation is that the absolute value is the distance from the point to the origin.

DEFINITION 2.1   *If $x = (x_1, \ldots, x_n)$, then the absolute value of x is the number*

$$|x| = \sqrt{\langle x, x \rangle} = \sqrt{\sum_{k=1}^{n} (x_k)^2}.$$

The basic properties of the absolute value are as follows.

THEOREM 2.2
(a) $|x| \geq 0$, *and* $|x| = 0$ *only if* $x = 0$.
(b) $|\alpha x| = |\alpha| \, |x|$ *if α is a real number.*
(c) $|x + y| \leq |x| + |y|$.

As indicated already, the point 0 in $\mathbf{R}^n$ is the point $(0, \ldots, 0)$.
Properties (a) and (b) are obvious, but property (c) is not.   It is based on the following theorem, which is famous in its own right and is called the *Cauchy–Schwarz inequality*.   Cauchy is the same fellow that appeared on the scene with Cauchy sequences.   Schwarz is another fellow.   The Russians call it the Buniakowski inequality.

THEOREM 2.3   *(Cauchy–Schwarz Inequality)*   *If $\langle x, y \rangle$ is an inner product and $|x| = \sqrt{\langle x, x \rangle}$, then*

$$|\langle x, y \rangle| \leq |x| \, |y|.$$

The theorem is put this way so that it can be used not only for the inner product and absolute value in $\mathbf{R}^n$, but for any inner product satisfying the conditions (a) through (d) of Theorem 1.2.

Cauchy–Schwarz

Proof   If α is any real number, then according to (b), (c), and (d) of Theorem 1.2, we have

$$\langle x + \alpha y, x + \alpha y \rangle = \langle x, x \rangle + \langle x, \alpha y \rangle + \langle \alpha y, x \rangle + \langle \alpha y, \alpha y \rangle$$
$$= |x|^2 + 2\alpha \langle x, y \rangle + \alpha^2 |y|^2.$$

Therefore, according to (a),

$$|x|^2 + 2\alpha \langle x, y \rangle + \alpha^2 |y|^2 \geq 0 \qquad \text{for all real } \alpha. \tag{1}$$

If we assume that $\langle y, y \rangle \neq 0$ and take $\alpha = -\langle x, y \rangle / \langle y, y \rangle$ in (1), then we get

$$|x|^2 - \frac{\langle x, y \rangle^2}{|y|^2} \geq 0,$$

which is just what is needed to prove the theorem.

Exercise 1   Go back to formula (1) to show that if $\langle y, y \rangle = 0$, then $\langle x, y \rangle = 0$, so that the theorem holds in this case, too.

Exercise 2   The mysterious value of α that was used in the above proof is simply the one that minimizes the function

$$f(\alpha) = |x|^2 + 2\alpha \langle x, y \rangle + \alpha^2 |y|^2.$$

In order to establish condition (c) in Theorem 2.2 we use the Cauchy–Schwarz inequality as follows:

$$|x + y|^2 = \langle x + y, x + y \rangle = \langle x, x \rangle + 2 \langle x, y \rangle + \langle y, y \rangle$$
$$\leq |x|^2 + 2|x|\,|y| + |y|^2 = (|x| + |y|)^2.$$

Then take the square root on both sides.

3   METRIC SPACES

The best way to discuss convergence and continuity in higher-dimensional spaces is to do it abstractly. There are two advantages—generality and simplicity. The results apply not only to the plane and the three-dimensional space, which are the guiding examples, but to $\mathbf{R}^n$ and many other interesting situations. And even in the three-dimensional space, the abstraction tends to make the notation simpler and the ideas clearer. What is needed is a notion of distance.

DEFINITION 3.1

A metric space is a set X on which there is a distance subject to the following conditions:

(a) $d(x, x) = 0$, and $d(x, y) > 0$ if $x \neq y$.
(b) $d(y, x) = d(x, y)$.
(c) $d(x, z) \leq d(x, y) + d(y, z)$ *(triangle inequality).*

Of course, $d(x, y)$ denotes the distance from x to y. It is remarkable that such simple conditions are adequate to develop the elementary properties of convergence and continuity.

These conditions are geometrically obvious in the case of the line, the plane, or the three-dimensional space. The last one, for instance, can be read as saying that the distance from a point x straight to a point z is at most as long as the distance around by way of a point y.

THEOREM 3.2

$\mathbf{R}^n$ is a metric space if the distance is defined by

$$d(x, y) = |x - y|.$$

Proof

Each of the conditions (a), (b), and (c) follows from the corresponding one in Theorem 2.2.

It is plain how to define convergent sequences and continuous functions on metric spaces.

DEFINITION 3.3

$x_n \to x$ in the metric space X if for each positive number ϵ there is a positive integer n_0 such that if $n > n_0$, then $d(x_n, x) < \epsilon$.

DEFINITION 3.4

A function f from a metric space X to a metric space Y is continuous at a point $a \in X$ if for each positive number ϵ there is a positive number δ such that if $d(x, a) < \delta$, then $d\big(f(x), f(a)\big) < \epsilon$. f is continuous on X, or simply continuous, if it is continuous at every point.

THEOREM 3.5

The function f is continuous at the point a if and only if it has the following property: If $x_n \to a$, then $f(x_n) \to f(a)$.

Proof   Let f be continuous at a, let $x_n \to a$, and let $\epsilon > 0$ be given.   Choose δ in accordance with the definition, and choose n_0 so that if $n > n_0$, then $d(x_n, a) < \delta$; hence $d(f(x_n), f(a)) < \epsilon$.   Since this can be done for every positive ϵ, it follows that $f(x_n) \to f(a)$.

Now suppose that f is not continuous at a.   Then there is some positive ϵ for which there is no δ.   In particular, for each positive integer n, there is a point x_n satisfying

$$d(x_n, a) < \frac{1}{n} \quad \text{and} \quad d(f(x_n), f(a)) \geq \epsilon.$$

(Otherwise $1/n$ would be a δ!)   It is plain that $x_n \to a$, but $f(x_n) \nrightarrow f(a)$.

Exercise 1   Show that a sequence cannot converge to two different points.

Exercise 2   Define a Cauchy sequence in a metric space.

Exercise 3   Define a uniformly continuous function.

4   FUNCTION SPACES

The best models to keep in mind when thinking about metric spaces are the plane and the three-dimensional space.   It is important, too, to realize that there are metric spaces quite different from these.   The following ones are very important in their own right.

Let I be any set, and let $\mathcal{B}(I)$ be the set of all bounded real-valued functions on I.   An "absolute value" can be defined on $\mathcal{B}(I)$ by the formula

$$\|x\| = \sup\{|x(t)| : t \in I\}, \tag{1}$$

and then a distance by the formula

$$D(x, y) = \|x - y\|. \tag{2}$$

Exercise 1   Show that the absolute value just defined satisfies the three conditions in Theorem 2.2, and then that the distance satisfies the conditions in Definition 3.1.

Convergence in the space $\mathcal{B}(I)$ is an old friend (a wolf?) in new clothing.   Inspection of the definition shows that it is nothing but uniform convergence of functions.   In these terms Theorem 4.2 of Chapter 6 reads

THEOREM 4.1   *In the space $\mathcal{B}(I)$ every Cauchy sequence converges.*

It is clear that any subset of a metric space is a metric space. The distance is already defined! Let $I = [a, b]$ be a closed bounded interval, and let $\mathfrak{R}(I)$ and $\mathfrak{C}(I)$ denote the Riemann integrable functions and the continuous functions on I. Both are subsets of $\mathfrak{B}(I)$, so both are metric spaces. Combining Theorem 4.1 with Theorem 4.3 of Chapter 6 we get

THEOREM 4.2 *In the space $\mathfrak{C}(I)$ every Cauchy sequence converges.*

Combining Theorem 4.1 and Theorem 4.4 of Chapter 6 we get

THEOREM 4.3 *In the space $\mathfrak{R}(I)$ every Cauchy sequence converges. Moreover, the function*

$$S(x) = \int_a^b x(t)\ dt$$

is continuous.

In the second part of the theorem we are looking at S as a function from the metric space $\mathfrak{R}(I)$ to the metric space $\mathbf{R}^1$ (the line) and are using Definition 3.4. The assertion depends on the sequential characterization of continuity given in Theorem 3.5.

DEFINITION 4.4 *A metric space is complete if every Cauchy sequence converges.*

Exercise 2 Show that an open interval is not complete.

Exercise 3 Show that the rational numbers are not complete.

There is an entirely different way to define a metric or distance on the space $\mathfrak{C}(I)$ that is also very important. First, an "inner product" is defined by the formula

$$\langle x, y \rangle = \int_a^b x(t)y(t)\ dt, \tag{3}$$

and then an absolute value by the usual formula

$$|x| = \sqrt{\langle x, x \rangle}, \tag{4}$$

and finally a distance by the usual formula

$$d(x, y) = |x - y|. \tag{5}$$

Exercise 4 Show that this inner product and absolute value satisfy the conditions in Theorems 1.2 and 2.2. Hence the Cauchy–Schwarz inequality holds, and the distance is indeed a distance in the sense of Definition 3.1.

Exercise 5 With this distance the space $\mathfrak{C}(I)$ is not complete.

The space $\mathcal{C}(I)$ with this distance resembles the spaces $\mathbf{R}^n$ very closely in many respects. It is considered their closest "infinite-dimensional" analog. However, the fact that it is not complete is a serious disadvantage. This can be remedied by the addition of certain discontinuous functions which are "limits" of Cauchy sequences that do not have continuous limits. One might think, for example, of adding all Riemann integrable functions. These have to be added, but still the space [now $\mathcal{R}(I)$] is not complete. More complicated functions than the Riemann integrable ones must be used.

5   EQUIVALENT METRICS

We have just seen an example of two metrics on the space $\mathcal{C}(I)$ which are quite different. Not only are the actual numbers $D(x, y)$ and $d(x, y)$ different for given points x and y, but the Cauchy sequences, the convergent sequences, the continuous functions, and so on, are quite different in the two cases—as can be seen from the fact that $\mathcal{C}(I)$ is complete with the metric D, but is not complete with the metric d. Now we shall look at the opposite situation to see when two metrics, although they may be different, must produce the same Cauchy sequences, the same convergent sequences, the same continuous functions, and so on.

DEFINITION 5.1

Two metrics d and D on a set X are equivalent if there exist positive numbers m and M such that

$$mD(x, y) \leq d(x, y) \leq MD(x, y) \quad \text{for all } x, y \in X.$$

Two absolute values are equivalent if there exist positive numbers m and M such that

$$m\|x\| \leq |x| \leq M\|x\| \quad \text{for all } x \in X.$$

It is plain that if two absolute values are equivalent, then the corresponding metrics are equivalent. It is also plain that if two metrics are equivalent, then they do produce the same Cauchy sequences, the same convergent sequences, and the same continuous functions. (The converse of this remark is almost, but not quite, correct. Can you give an example on $X = [0, 1]$?)

Consider $\mathbf{R}^n$ with its initial absolute value defined by

$$|x| = \sqrt{\langle x, x \rangle} = \sqrt{\sum_{k=1}^{n} x_k^2} \tag{1}$$

and with a new one defined by

$$\|x\| = \max\{|x_k| : k = 1, \ldots, n\}. \tag{2}$$

Exercise 1   Verify that $\|x\|$ is an absolute value and that

$$\|x\| \leq |x| \leq \sqrt{n}\,\|x\|. \tag{3}$$

Now, let us consider $\mathbf{R}^n$ from a different point of view. An n-tuple of real numbers is (by definition, in fact!) a real-valued function on the set

$$I_n = \{1, 2, \ldots, n\}.$$

In general, an n-tuple in a set X is a function from I_n into X. The notation is like the notation for sequences. If x is an n-tuple, then x_k is usually written in place of $x(k)$.

From this point of view, $\mathbf{R}^n$ is exactly the same set as $\mathcal{B}(I_n)$. Moreover, the addition and scalar multiplication in $\mathbf{R}^n$ are the same as those in $\mathcal{B}(I_n)$.

Exercise 2   Check the last statement.

The absolute value defined in formula (2) is just the absolute value on $\mathcal{B}(I_n)$. All of which gives a theorem.

THEOREM 5.2 $\mathbf{R}^n$ *is complete.*

Proof   By Theorem 4.1, $\mathcal{B}(I_n)$ is complete with the metric (2), and by formula (3) the metrics (1) and (2) are equivalent; so they have both the same Cauchy sequences and the same convergent sequences.

The fact of the matter, and we shall prove it in Section 10, is that any two absolute values on $\mathbf{R}^n$ are equivalent. This is useful to know, although in specific cases the equivalence is usually easy to prove, just as it was above.

Exercise 3   Show that

$$\|x\|_1 = \sum_{k=1}^{n} |x_k|$$

is an absolute value on $\mathbf{R}^n$, and that it is equivalent to the two already defined.

The notion of equivalent metrics certainly is not necessary just to prove that $\mathbf{R}^n$ is complete. It is a notion that is needed later, and it does give the nice proof above. However, a proof can also be made along the following lines. It will simplify the notation (avoid double subscripts) to write the n-tuples as functions on I_n.

Exercise 4   A sequence $\{x_k\}$ in $\mathbf{R}^n$ is Cauchy or convergent if and only if each of the sequences $\{x_k(i)\}$, $i = 1, \ldots, n$, has the same property.

Exercise 5   Use Exercise 4 and the fact that $\mathbf{R}^1$ is complete to show that $\mathbf{R}^n$ is complete.

6   OPEN AND CLOSED SETS

By analogy with everyday terminology in $\mathbf{R}^3$, balls and spheres are defined as follows:

DEFINITION 6.1

In a metric space X the open ball with center a and radius $r > 0$ is the set

$$B(a; r) = \{x : d(x, a) < r\}.$$

The closed ball with center a and radius r is the set

$$\bar{B}(a; r) = \{x : d(x, a) \leq r\}.$$

The sphere with center a and radius r is the set

$$S(a; r) = \{x : d(x, a) = r\}.$$

It is always assumed, unless the contrary is stated explicitly, that the radius of a ball or sphere is positive.

In $\mathbf{R}^1$ the open and closed balls with center a are the open and closed intervals with center a.

DEFINITION 6.2

A set $G \subset X$ is open in X if for every point $a \in G$ there is some ball with center a that is contained in G. A set $F \subset X$ is closed in X if its complement $X - F$ is open in X.

Another way to state the definition is that a set $G \subset X$ is open in X if for every point $a \in G$ there is a positive number δ such that if $x \in X$ and $d(x, a) < \delta$, then $x \in G$.

Usually a set is called simply open or closed, rather than open or closed in X, but the latter is always understood. A set is never open or closed on its own, but only open or closed in some given metric space X. For example, an open interval in $\mathbf{R}^1$ is an open set in $\mathbf{R}^1$, but is not an open set in $\mathbf{R}^2$ when we consider $\mathbf{R}^1 \subset \mathbf{R}^2$ in the usual way.

Exercise 1   In any metric space X, the empty set and X itself are both open—and hence both closed.

THEOREM 6.3

The open ball is an open set. The closed ball and the sphere are closed sets.

Proof   Let b be any point of the open ball $B(a, r)$, and let $\delta = r - d(b, a)$—which is positive by definition of the open ball. If $d(x, b) < \delta$, then

$$d(x, a) \leq d(x, b) + d(b, a) < r.$$

Thus, $B(b, \delta) \subset B(a, r)$, and $B(a, r)$ is open.

Let b be a point in the complement of the closed ball $\bar{B}(a, r)$. Then $\delta = d(b, a) - r$ is positive, and if $d(x, b) < \delta$, then, since $d(a, b) \leq d(a, x) + d(x, b)$, we have

$$d(a, x) \geq d(a, b) - d(x, b) > r.$$

Thus, $B(b, \delta)$ is contained in the complement of $\bar{B}(a, r)$, which shows that the complement is open, and hence that $\bar{B}(a, r)$ itself is closed.

Exercise 2   Do the case of the sphere yourself.

The theorem relating continuous functions to open and closed sets is the following.

THEOREM 6.4   *A function f from X to Y is continuous if and only if it has the property that $f^{-1}(G)$ is open in X whenever G is open in Y.*

When G is any set in Y, the set $f^{-1}(G)$ is the set in X defined by

$$f^{-1}(G) = \{x : x \in X \text{ and } f(x) \in G\}.$$

Proof   Suppose that f is continuous, let G be an open set in Y, and let a be any point in $f^{-1}(G)$. Then $b = f(a) \in G$, and since G is open, there is a positive number ϵ such that if $d(y, b) < \epsilon$, then $y \in G$. Use the continuity of f to find a positive number δ such that if $d(x, a) < \delta$, then $d(f(x), b) < \epsilon$; hence $f(x) \in G$; hence $x \in f^{-1}(G)$. Then $B(a; \delta) \subset f^{-1}(G)$, so $f^{-1}(G)$ is open.

Suppose that f has the property indicated in the theorem, let a be any point of X, and let $\epsilon > 0$ be given. By Theorem 6.3, $G = B(f(a); \epsilon)$ is open, and, therefore, so is $f^{-1}(G)$. Hence, there is a $\delta > 0$ such that if $d(x, a) < \delta$, then $x \in f^{-1}(G)$; hence $f(x) \in G = B(f(a); \epsilon)$; hence $d(f(x), f(a)) < \epsilon$. This shows that f is continuous at the arbitrary point a.

The same kind of theorem holds for closed sets.

THEOREM 6.5   *A function f from X to Y is continuous if and only if it has the property that $f^{-1}(F)$ is closed in X whenever F is closed in Y.*

Proof   This follows from Theorem 6.4 and the simple identity

$$f^{-1}(Y - F) = X - f^{-1}(F).$$

Remark   This theorem can be used to prove that the closed ball and the sphere are closed sets in the following way.   First we show that the function $f(x) = d(x, a)$ is continuous from X to the real numbers.   The triangle inequality gives $d(x, a) \leq d(x, y) + d(y, a)$; hence

$$d(x, a) - d(y, a) \leq d(x, y).$$

Interchange of x and y gives

$$d(y, a) - d(x, a) \leq d(y, x) = d(x, y),$$

and the two together give

$$|d(x, a) - d(y, a)| \leq d(x, y). \tag{1}$$

This shows that f is continuous: Given an ϵ, we can take $\delta = \epsilon$.

Now, $\bar{B}(a; r) = f^{-1}(F)$, where F is the closed interval $[0, r]$, and $S(a; r) = f^{-1}(F)$, where F is the single point $\{r\}$.   It is easy to see that both $[0, r]$ and $\{r\}$ are closed in $\mathbf{R}^1$, so Theorem 6.5 shows that the closed ball and the sphere are both closed.

The same kind of argument could be used to show that the open ball is open, except that this fact was used in the proof of Theorem 6.4.

Here is a direct characterization of closed sets.

THEOREM 6.6   *A set $F \subset X$ is closed in X if and only if it has the following property: If $x_n \to x$ and each x_n is in F, then $x \in F$.*

Proof   Suppose that F is closed and that $x_n \to x$, with each $x_n \in F$.   If $x \in X - F$, then since $X - F$ is open, there is a ball $B(x; r) \subset X - F$—which is clearly impossible, since x_n belongs to any such ball if n is large enough.

Now suppose that F is not closed, so that $X - F$ is not open.   Then there is some point $a \in X - F$ such that every ball $B(a; r)$ intersects F. In particular, for each positive integer n, there is a point $x_n \in F \cap B(a; 1/n)$.   Clearly $x_n \to a$, each x_n is in F, but a is not in F; so F does not have the property described in the theorem.

This characterization of closed sets suggests a relation between closed sets and complete metric spaces.

THEOREM 6.7   *Let $X \subset Y$.   If X is complete, then X is a closed subset of Y.   If X is a closed subset of Y and Y is complete, then X is complete.*

Proof   Suppose that X is complete, and let $x_n \to y$ with each $x_n \in X$.   Then the sequence $\{x_n\}$ is Cauchy, and since X is complete, there is a point $x \in X$

with $x_n \to x$. However, a sequence cannot converge to two different points, so $y = x \in X$. This shows that X is closed in Y.

Suppose that X is closed in Y and that Y is complete, and let $\{x_n\}$ be a Cauchy sequence in X. Then $\{x_n\}$ is also a Cauchy sequence in Y, so $x_n \to y$ for some $y \in Y$. Since X is closed in Y, it follows that $y \in X$. Therefore, every Cauchy sequence in X converges in X, and X is complete.

COROLLARY 6.8   *The closed ball and the sphere in $\mathbf{R}^n$ are complete.*

Exercise 3   The union of any number of open sets is open.   The union of a finite number of closed sets is closed.

Exercise 4   The intersection of any number of closed sets is closed.   The intersection of a finite number of open sets is open.

Exercise 5   Show by example (e.g., in $\mathbf{R}^1$) that the assertions of Exercises 3 and 4 cannot be improved.

Exercise 6   The closure $\bar{A}$ of a set $A \subset X$ is the intersection of all closed sets that contain A. Show that $\bar{A}$ is closed and that it consists of all $x \in X$ such that $a_n \to x$ for some sequence $\{a_n\}$ in A. Show that $\bar{\bar{A}} = \bar{A}$.

Exercise 7 $\overline{A \cup B} = \bar{A} \cup \bar{B}$.   Intersection, too?

Exercise 8   A set $A \subset X$ is nowhere dense in X if $\bar{A}$ contains no ball. Show that the union of a finite number of nowhere dense sets is nowhere dense.

Exercise 9   For any $A \subset X$, A not empty, define

$$d(x, A) = \inf\{d(x, y) : y \in A\}.$$

Show that $|d(x, A) - d(y, A)| \leq d(x, y)$, and hence that $d(x, A)$ is continuous. Show that $\bar{A} = \{x : d(x, A) = 0\}$.

Exercise 10   If F_0 and F_1 are two disjoint closed subsets of a metric space X, then there is a continuous real-valued function f on X such that

$$f = 0 \text{ on } F_0, \quad f = 1 \text{ on } F_1, \quad 0 \leq f(x) \leq 1 \qquad \text{for all } x \in X.$$

[*Hint:* Try

$$f(x) = \frac{d(x, F_0)}{d(x, F_0) + d(x, F_1)}.]$$

7 CONNECTED SPACES

It must not be imagined that the subsets of a metric space are divided neatly into the open ones and the closed ones. Most sets are neither open nor closed, and some sets are both.

Exercise 1   A metric space X is always both open and closed in itself. The empty set is always both open and closed in X.

DEFINITION 7.1   *A metric space X is connected if no subset is both open and closed, except X itself and the empty set. If X is not connected, then it is disconnected.*

The definition can be rephrased in two ways by making use of the fact that a set is open if and only if its complement is closed.

DEFINITION 7.2   *A metric space X is disconnected if and only if there exist nonempty open sets G_1 and G_2 such that*

$$G_1 \cap G_2 = \varnothing, \qquad G_1 \cup G_2 = X$$

(where $\varnothing$ is the empty set).

DEFINITION 7.3   *A metric space X is disconnected if and only if there exist nonempty closed sets F_1 and F_2 such that*

$$F_1 \cap F_2 = \varnothing, \qquad F_1 \cup F_2 = X.$$

THEOREM 7.4   *A subset X of $\mathbf{R}^1$ is connected if and only if it is an interval.*

Proof   If X is not an interval, then there exist real numbers a, b, and c such that $a < b < c$, while a and c belong to X and b does not. Define G_1 and G_2 by

$$G_1 = \{x : x \in X \text{ and } x < b\}, \qquad G_2 = \{x : x \in X \text{ and } x > b\}.$$

It is immediately verified that G_1 and G_2 are open, and it is plain from the definition that they are nonempty and satisfy $G_1 \cap G_2 = \varnothing$ and $G_1 \cup G_2 = X$. Therefore, X is disconnected.

Now suppose that X is an interval, but that X is disconnected. Let F_1 and F_2 be as in the third form of the definition, and let $a \in F_1$ and $c \in F_2$ with $a < c$. (If $a > c$, change the notation.) Let

$$b = \sup\{x : x \in F_1 \text{ and } x < c\}.$$

We shall get a contradiction by showing that $b \in F_1 \cap F_2$, which is supposed to be empty.

By the definition of the least upper bound, we can find for each n a point $x_n \in F_1$ satisfying $b - 1/n \le x_n \le b$. Thus, $x_n \to b$ and $x_n \in F_1$, so b must belong to F_1, since F_1 is closed in X.

Exercise 2   The last statement holds water only if $b \in X$—for F_1 is closed in X, not in $\mathbf{R}^1$. How do we know that $b \in X$?

Now, let us show that $b \in F_2$. If $b = c$, we are done. Otherwise $b < c$, and, if n is large enough, then $y_n = b + 1/n$ is also $<c$, hence in X. By the definition of b, y_n cannot be in F_1, so it must be in F_2. We have $y_n \to b$ and $y_n \in F_2$, and since F_2 is closed in X, it follows that $b \in F_2$.

THEOREM 7.5   *If f is a continuous function from X to Y, and X is connected, then $f(X)$ is connected.*

Proof   It is clear that f is a continuous function from X to $f(X)$, so there is no loss in generality in supposing that $Y = f(X)$. If Y is disconnected, then $Y = G_1 \cup G_2$, where G_1 and G_2 are open, etc. Then $X = f^{-1}(G_1) \cup f^{-1}(G_2)$, where $f^{-1}(G_1)$ and $f^{-1}(G_2)$ are open, etc.

Exercise 3   Verify the etc.

Theorems 7.4 and 7.5 give the general version of the basic theorem in Chapter 3 that if f is a continuous real-valued function on an interval I, then $f(I)$ is an interval. Indeed, Theorem 7.4 says that I is connected, Theorem 7.5 says that $f(I)$ is connected, and then Theorem 7.4 says that $f(I)$ is an interval.

The geometric sense of disconnectedness is that the space splits into two parts that are somehow separated from one another. There is another kind of connectedness that is perhaps more intuitive and is easier to handle in many cases.

DEFINITION 7.6   *A path, or curve, or arc in a metric space X is a continuous function φ from a closed bounded interval $[\sigma, \tau]$ into X. The points $\varphi(\sigma)$ and $\varphi(\tau)$ are called the initial and final points of the path, and the path is said to join these two points.*

DEFINITION 7.7   *A metric space X is path connected if any two points can be joined by a path in X.*

THEOREM 7.8   *Every path-connected space is connected.*

Proof   Let G_1 and G_2 disconnect X, choose a and b in G_1 and G_2, and let φ be a path joining a and b. Then $\varphi^{-1}(G_1)$ and $\varphi^{-1}(G_2)$ disconnect the interval

$[\sigma, \tau]$ on which φ is defined, which is impossible, since an interval is connected.

The simplest kind of path in $\mathbf{R}^n$ is of course the straight-line segment.

DEFINITION 7.9   *The line segment joining the points a and b of $\mathbf{R}^n$ is the path φ defined by*

$$\varphi(t) = (1 - t)a + tb \qquad for\ 0 \le t \le 1.$$

Exercise 4   Show that the line segment is a path, that is, is continuous.

THEOREM 7.10   *In $\mathbf{R}^n$ the closed and open balls are path connected.   Indeed, the line segment joining any two points of the ball is a path in the ball.*

Proof   Let a and b be two points of the ball $B(c; r)$ and φ be the line segment joining them.   What has to be shown is that $\varphi(t) \in B(c; r)$ for each t. Writing $c = (1 - t)c + tc$, we have

$$|\varphi(t) - c| = |(1 - t)a + tb - c| \le (1 - t)|a - c| + t|b - c|$$
$$< (1 - t)r + tr = r.$$

The same kind of proof works for the closed ball.

Exercise 5   A rectangle in $\mathbf{R}^n$ is a set R of the form

$$R = \{x : a_i \le x_i \le b_i \text{ for } i = 1, \ldots, n\},$$

where $a = (a_1, \ldots, a_n)$ and $b = (b_1, \ldots, b_n)$ are any two points of $\mathbf{R}^n$ with $a_i < b_i$ for each i.   Show that a rectangle is path connected by showing that the line segment joining any two points of the rectangle is a path in the rectangle.

Exercise 6   The points a and b in the rectangle above are the "lower left" and "upper right" hand vertices of the rectangle.   What are the other vertices?   What is the center?

Exercise 7   Let R be a rectangle in $\mathbf{R}^n$ with center c.   Define an absolute value on $\mathbf{R}^n$ so that $R = \bar{B}(c; 1)$.   Use this to do Exercise 5.

Exercise 8   We have been talking about closed rectangles.   Define open rectangles, and prove that open rectangles are open and that closed rectangles are closed.

THEOREM 7.11   *In $\mathbf{R}^n$, $n > 1$, the sphere $S(a; r)$ is path connected.*

Proof To simplify the notation we shall treat the sphere $S = S(0; 1)$. Let a and b be any two points of S, and suppose first that a and b are not diametrically opposite; that is, that $a \neq -b$. The line segment φ joining a and b is a path in $\mathbf{R}^n$ all right, but it is not a path in S. However, we can "project" it on S in the following way. Define

$$\psi(t) = \frac{\varphi(t)}{|\varphi(t)|} = \frac{(1-t)a + tb}{|(1-t)a + tb|}. \tag{1}$$

It is geometrically clear that the line segment from a to b does not pass through 0 [i.e., that the denominator in (1) is $\neq 0$] if a and b are not diametrically opposite, but let us prove it. If $(1-t)a = -tb$, then, since $|a| = 1$ and $|b| = 1$,

$$1 - t = |(1-t)a| = |-tb| = t.$$

Consequently, $t = \frac{1}{2}$ and $\frac{1}{2}a = -\frac{1}{2}b$, so $a = -b$.

This shows that formula (1) makes perfectly good sense, but it remains to show that ψ is continuous. First of all, the function $|\varphi(t)|$ is continuous, since it is the composite of $|x| = d(x, 0)$ and φ, which are both continuous (see Section 2). Since $|\varphi(t)|$ is continuous and $\neq 0$, it follows that $\alpha(t) = 1/|\varphi(t)|$ is also continuous. The proof is finished by the following exercise, which is entirely similar to the familiar theorem on the product of two continuous real-valued functions.

Exercise 9 Let φ be a continuous function from a metric space X to $\mathbf{R}^n$, and let α be a continuous real-valued function on X. Then the product $\psi(t) = \alpha(t)\varphi(t)$ is continuous from X to $\mathbf{R}^n$.

It remains to treat the case when $a = -b$. But all we have to do is to take any third point c, and join a to c and then c to b. This is the point where the proof breaks down for $n = 1$. When $n = 1$, the sphere $S(0; 1)$ consists of the two points $x = \pm 1$, which are diametrically opposite, and there is no third point to make use of.

In all these examples we have proved that the space is path connected, and we know that every path connected space is connected. The question arises as to whether there are connected spaces that are not path connected. In fact, there are, but examples are not so very easy to produce. Here is one in the plane.

$$X = \left\{(x,y) : 0 < x \leq 1 \text{ and } y = \sin \frac{1}{x}, \text{ or } x = 0 \text{ and } -1 \leq y \leq 1\right\}.$$

Exercise 10 Draw a picture of the set X and show that it is connected but not path connected.

For open subsets of $\mathbf{R}^n$, however, connected and path connected are the same.

THEOREM 7.12   *If G is a connected open subset of $\mathbf{R}^n$, then any two points of G can be joined by a polygonal line in G.*

Proof   A polygonal line is a finite sequence of line segments, each one beginning where the previous one stopped. This is plainly a path. Let a be a fixed point of G, and let G_1 be the set of points in G to which a can be joined by a polygonal line in G, and G_2 be the set of points in G to which a cannot be joined by a polygonal line in G. We shall show that G_1 and G_2 are both open. Let $b \in G_1$, and choose $\delta > 0$ so that $B(b; \delta) \subset G$. If c is any point of $B(b; \delta)$, then the polygonal line from a to b (which exists because $b \in G_1$) followed by the line segment from b to c gives a polygonal line from a to c. Hence $c \in G_1$. The same idea shows that G_2 is open. If $b \in G_2$, let $\delta > 0$ and c be as before. If there were a polygonal line from a to c, then following it by the line segment from c to b we would have one from a to b.

Since G_1 and G_2 are both open, and since plainly their union is G and their intersection is empty, it follows that one of them must be empty. The empty one is not G_1, for $a \in G_1$. Thus G_2 is empty, and so $G_1 = G$, and we are done.

Exercise 11   Why are we done?

Exercise 12   Let $\mathfrak{F}$ be a family of connected subsets of a space X.
(a) If all the sets in $\mathfrak{F}$ have a common point, then the union is connected.
(b) If all the sets in $\mathfrak{F}$ intersect some one of them, then the union is connected.
[*Hint:* Part (b) follows immediately from part (a) by a little trick.]

Exercise 13   For each $x \in X$ let $\mathcal{C}(x)$ be the union of all connected subsets of X that contain x. $\mathcal{C}(x)$ is connected. If x and y are two points, then either $\mathcal{C}(x) = \mathcal{C}(y)$ or $\mathcal{C}(x) \cap \mathcal{C}(y) = \varnothing$.

Exercise 14   If X is an open subset of $\mathbf{R}^n$, then each $\mathcal{C}(x)$ is open in $\mathbf{R}^n$. [Note that $\mathcal{C}(x)$ is formed relative to X, not relative to $\mathbf{R}^n$.]

The sets $\mathcal{C}(x)$ are called the *connected components* of the metric space X.

Exercise 15   What are the connected components of $\mathbf{R}^n - S(a; r)$?

Exercise 16   Every open set in $\mathbf{R}^1$ is the union of a disjoint sequence of open intervals. (*Hint:* The open intervals are the connected components. Why can they be arranged in a sequence?)

8   COMPOSITE FUNCTIONS AND SUBSEQUENCES

If f is a function from a set X to a set Y, and g is a function from Y to Z, then the composite is the function h from X to Z defined by

$$h(x) = g\big(f(x)\big).$$

Usually the composite is designated by $g \circ f$. There is a notation that is convenient in general as an abbreviation, and is particularly convenient in chasing composite functions around. To say that f is a function from X to Y, we write

$$f : X \to Y \qquad \text{or} \qquad X \xrightarrow{f} Y. \tag{1}$$

The setup that produces a composite function is then

$$X \xrightarrow{f} Y \xrightarrow{g} Z. \tag{2}$$

There is one unfortunate aspect of the notation. The setup (2) produces the composite function $g \circ f$, not $f \circ g$. That is, the order is reversed. There are ways to avoid this, but the notation is so well established that they are impractical.

THEOREM 8.1   *Let $X \xrightarrow{f} Y \xrightarrow{g} Z$. If f is continuous at a point a, and g is continuous at $b = f(a)$, then $g \circ f$ is continuous at a.*

It is assumed tacitly, of course, that X, Y, and Z are all metric spaces.

Proof   Let $\epsilon > 0$ be given. First use the fact that g is continuous to choose $\delta_1 > 0$ so that if $d(y, b) < \delta_1$, then $d\big(g(y), g(b)\big) < \epsilon$. Then use the fact that f is continuous to choose $\delta > 0$ so that if $d(x, a) < \delta$, then $d\big(f(x), f(a)\big) < \delta_1$. Now, if $d(x, a) < \delta$, then

$$d\big(g \circ f(x), g \circ f(a)\big) < \epsilon.$$

A special kind of composite function is of particular interest in the next section.

DEFINITION 8.2   *Let x be a sequence in a set X. A subsequence of x is a composite function $x \circ k$, where k is a strictly increasing sequence of positive integers.*

Remember the definitions: A sequence in a set X is a function from the positive integers into X. Therefore, if $\mathbf{N}$ denotes the set of positive integers, then the situation is that

$$\mathbf{N} \xrightarrow{k} \mathbf{N} \xrightarrow{x} X$$

so $x \circ k : \mathbf{N} \to X$ is again a sequence in X.

A subsequence is simply a rule that picks out some of the terms of a sequence (an infinite number, of course) in their proper order. The usual practice is to write x_{k_i} for the point $x(k(i)) = x \circ k(i)$, and to write $\{x_{k_i}\}$ for the subsequence.

THEOREM 8.3   *If $\{x_k\}$ converges to a, then every subsequence converges to a.*

Exercise 1   Prove the theorem, and discuss its relation to Theorem 8.1.

THEOREM 8.4   *Let $\{x_k\}$ be a bounded sequence of real numbers. Then there is a subsequence that converges to the limit superior.*

Proof   The definition of the subsequence is an inductive one based on Lemma 3.2 of Chapter 6. First choose $k_1 > 1$ so that $|x_{k_1} - b| < 1$, where b is the limit superior. This involves using the lemma with $\delta = 1$. Next choose $k_2 > k_1$ with $|x_{k_2} - b| < 1/k_1$. This involves using the lemma with $\delta = 1/k_1$. Suppose that $k_1, k_2, \ldots, k_m$ are already chosen, and choose $k_{m+1} > k_m$ so that $|x_{k_{m+1}} - b| < 1/k_m$—which involves using the lemma with $\delta = 1/k_m$. By the construction, the sequence $\{k_i\}$ is strictly increasing, so the sequence $\{x_{k_i}\}$ is a subsequence. It converges to b, since

$$|x_{k_i} - b| < \frac{1}{k_i} \le \frac{1}{i}.$$

Exercise 2   Show that if $\{k_i\}$ is any strictly increasing sequence of positive integers, then $k_i \ge i$. (This is what is used in the last inequality above and in Exercise 1.)

Exercise 3   Let $\{x_k\}$ be a bounded sequence of real numbers. There is a subsequence that converges to the limit inferior. Furthermore, if a is the limit of any convergent subsequence, then

$$\liminf x_k \le a \le \limsup x_k.$$

THEOREM 8.5   *Every bounded sequence in $\mathbf{R}^n$ has a convergent subsequence.*

Proof   A set $X \subset \mathbf{R}^n$ is bounded if there is a number M such that $|x| \le M$ for every $x \in X$. A sequence $\{x_k\}$ is bounded if $|x_k| \le M$ for every k.

To avoid double subscripts we shall write a point $x \in \mathbf{R}^n$ as $x = (x', t)$, where $x' = (x_1, \ldots, x_{n-1})$ and $t = x_n$, in which case $|x|^2 = |x'|^2 + t^2$. It is immediate that $x_k \to x$ if and only if $x'_k \to x'$ and $t_k \to t$.

Now let $\{x_k\}$ be a bounded sequence in $\mathbf{R}^n$. Then $\{x'_k\}$ is a bounded sequence in $\mathbf{R}^{n-1}$, and by induction there is a subsequence $\{x'_{k_i}\}$ that converges to $x' \in \mathbf{R}^{n-1}$. (Theorem 8.4 starts the induction.) The sequence $\{t_{k_i}\}$ is a bounded sequence of real numbers, so by Theorem 8.4 there is a subsequence $\{t_{k_{i_j}}\}$ that converges to $t \to \mathbf{R}^1$. Now, the sequence $\{x_{k_{i_j}}\}$ is a subsequence of $\{x_k\}$ that converges to (x', t), for $t_{k_{i_j}} \to t$ by definition, and $x'_{k_{i_j}} \to x'$ by Theorem 8.3.

We have used the fact that a subsequence of a subsequence is a subsequence, for which the setup in terms of the notation of this section is simply

$$\mathbf{N} \xrightarrow{i} \mathbf{N} \xrightarrow{k} \mathbf{N} \xrightarrow{x} X.$$

9   COMPACT SPACES

In Chapter 3 we discovered the fundamental theorem that a continuous function on a closed bounded interval is uniformly continuous and has a maximum and minimum, and if it is one to one, then the inverse function is continuous. Now we shall discuss the general analog of this theorem.

DEFINITION 9.1

A metric space X is compact if every sequence in X has a convergent subsequence.

The basic general theorem is as follows:

THEOREM 9.2

Let $f: X \to Y$ be continuous, and let X be compact. Then
  (a) *$f(X)$ is compact.*
  (b) *f is uniformly continuous.*
  (c) *If f is one to one, then the inverse function is continuous.*

It is assumed tacitly that X and Y are metric spaces. Then $f(X)$, as a subset of Y, is also a metric space.

Proof of (a)

Let $\{y_k\}$ be a sequence in $f(X)$. By the definition of $f(X)$, there is a point $x_k \in X$ with $f(x_k) = y_k$. Since X is compact, there is a subsequence $\{x_{k_i}\}$ such that $x_{k_i} \to x$. Then, since f is continuous at x, we have

$$y_{k_i} = f(x_{k_i}) \to f(x).$$

Proof of (b)   If f is not uniformly continuous, then there is some positive ϵ for which there is no δ.   Therefore, for each positive integer k, there are points x_k and y_k in X with

$$d(x_k, y_k) < \frac{1}{k} \qquad \text{and} \qquad d(f(x_k), f(y_k)) \geq \epsilon. \tag{1}$$

(Otherwise $1/k$ would be a δ!)   Use the compactness to choose $\{k_i\}$ so that $x_{k_i} \to x$.   By the first half of formula (1), it follows that $y_{k_i} \to x$. Therefore, since f is continuous at x, it follows that

$$f(x_{k_i}) \to f(x) \qquad \text{and} \qquad f(y_{k_i}) \to f(x),$$

which clearly contradicts the second half of formula (1).

Proof of (c)   If g is the inverse function, then what we have to prove is that if $y_k \to y$, with, of course, y_k and y in $f(X)$, then $g(y_k) \to g(y)$.   Let $x_k = g(y_k)$ and $x = g(y)$.

If it is not true that $x_k \to x$, then, for some $\epsilon > 0$, there are infinitely many x_k that satisfy $d(x_k, x) \geq \epsilon$.   From these we can pick a subsequence that converges to some point z.   Thus, we have a subsequence $\{x_{k_i}\}$ such that

$$x_{k_i} \to z \qquad \text{and} \qquad d(x_{k_i}, x) \geq \epsilon.$$

From the second part of this formula it follows that $z \neq x$, and from the first part it follows that

$$f(z) = \lim f(x_{k_i}) = \lim y_{k_i} = y = f(x),$$

which contradicts the fact that f is one to one.

Now the job is to identify the compact subsets of $\mathbf{R}^n$, and in general to describe compact spaces in some more geometrical way.   It is helpful to divide the property of compactness into two parts.

DEFINITION 9.3   *A metric space X is totally bounded if every sequence in X has a Cauchy subsequence.*

THEOREM 9.4   *A metric space is compact if and only if it is complete and totally bounded.*

Proof   Straight from the definitions it is obvious that if X is complete and totally bounded, then X is compact; and that if X is compact, then it is totally bounded.   What has to be shown is that if X is compact, then it is complete.

If $\{x_k\}$ is a Cauchy sequence, then there is a subsequence $\{x_{k_i}\}$ that converges to some point $x \in X$. And we have the following lemma, valid in any metric space.

LEMMA 9.5

If $\{x_k\}$ is Cauchy and $x_{k_i} \to x$, then $x_k \to x$.

Exercise 1 Prove the lemma.

THEOREM 9.6

A subset of $\mathbf{R}^n$ is compact if and only if it is closed and bounded.

Proof

Theorem 8.5 shows that a bounded set is totally bounded, and Theorem 6.7 shows that a closed subset is complete. This proves half of the theorem.

Theorem 6.7 also shows that a complete subset is closed, so what remains is to show that a totally bounded subset of $\mathbf{R}^n$ is bounded. If the set $X \subset \mathbf{R}^n$ is unbounded, we can produce a sequence with no convergent subsequence as follows. Start with any point $x_1 \in X$. Choose $x_2 \in X$ with $|x_2| > |x_1| + 1$, then $x_3 \in X$ with $|x_3| > |x_2| + 1$, and in general $x_m \in X$ with $|x_m| > |x_{m-1}| + 1$. This sequence has the property that the distance between any two terms is >1, which rules out any chance of a convergent subsequence.

THEOREM 9.7

In $\mathbf{R}^n$ a closed ball, a sphere, and a closed rectangle are all compact.

Proof

From their definition these sets are all bounded, and in Section 6 they are proved closed.

Exercise 2 If Y is a compact subset of $\mathbf{R}^1$, then Y is bounded, and both the least upper bound and the greatest lower bound belong to Y.

Exercise 3 A continuous real-valued function on a compact metric space has a maximum and a minimum.

The "geometric" characterization of a totally bounded space comes next.

DEFINITION 9.8

The diameter of a set A in a metric space X is the number
$$\delta(A) = \sup\{d(x, y) : x \text{ and } y \text{ belong to } A\}.$$

THEOREM 9.9

The metric space X is totally bounded if and only if it has the following property: For every $\epsilon > 0$, X is the union of a finite number of subsets of diameter $\leq \epsilon$.

Proof   Suppose that X is not the union of a finite number of sets of diameter $\leq \epsilon$. We shall produce a sequence with no convergent subsequence. Start with any point x_1. There must be a point x_2 with $d(x_2, x_1) > \epsilon/2$, for otherwise X itself would have diameter $\leq \epsilon$. There must be a point x_3 with

$$d(x_3, x_2) > \epsilon/2 \quad \text{and} \quad d(x_3, x_1) > \epsilon/2,$$

for otherwise X would be the union of the two balls $\bar{B}(x_2, \epsilon/2)$ and $\bar{B}(x_1, \epsilon/2)$, each of which has diameter $\leq \epsilon$. In general, there must be a point x_m with

$$d(x_m, x_j) > \epsilon/2 \quad \text{for } j = 1, \ldots, m-1,$$

for otherwise X would be the union of the balls $\bar{B}(x_j; \epsilon/2)$, each of which has diameter $\leq \epsilon$. The sequence $\{x_m\}$ has no Cauchy subsequence, since the distance between any two terms is $> \epsilon/2$.

Now we shall show that a space with the property described in the theorem must be totally bounded. In the proof we shall use the obvious fact that if X has this property, then so does any subset.

Let $\{x_k\}$ be any sequence in X. Write X as the union of a finite number of sets of diameter ≤ 1. At least one of these contains infinitely many terms of the sequence $\{x_k\}$. Choose any one that does and call it X_1, and choose k_1 so that $x_{k_1} \in X_1$. Write X_1 as the union of a finite number of sets of diameter $\leq \frac{1}{2}$. At least one of these must contain infinitely many terms of the sequence $\{x_k\}$, for X_1 contains infinitely many terms. Choose one of these and call it X_2, and choose $k_2 > k_1$ so that $x_{k_2} \in X_2$. In general, write X_{m-1} as the union of a finite number of sets of diameter $\leq 1/m$. At least one must contain infinitely many terms of the sequence $\{x_k\}$. Choose such a one and call it X_m, and choose $k_m > k_{m-1}$ so that $x_{k_m} \in X_m$. The sequence $\{x_{k_m}\}$ is certainly Cauchy, for if $n > m$, then x_{k_m} and x_{k_n} both belong to X_m, which has diameter $\leq 1/m$; so $d(x_{k_m}, x_{k_n}) \leq 1/m$.

Theorem 9.9 suggests a more geometric way to show that bounded sets in $\mathbf{R}^n$ are totally bounded. It is enough to show that any rectangle

$$R = \{x : a_i \leq x_i \leq b_i \text{ for } i = 1, \ldots, n\},$$

where $a_i < b_i$, is totally bounded. Consider first $\mathbf{R}^2$. The idea is to cut R into four similar rectangles of half the size by the two lines through the center. Then cut each of these into four, and so on. It is clear that this process eventually cuts R into a finite number of rectangles of arbitrarily small diameter. Consider next $\mathbf{R}^3$. This time R is cut into eight smaller rectangles by the three planes through the center, and the argument goes as before. What we have to do is interpret this geometric construction analytically so that it is valid in general.

First, the diameter of the rectangle is $|b - a|$, for if x and y are any two points of R, then $a_i \leq x_i \leq b_i$ and $a_i \leq y_i \leq b_i$, so $|y_i - x_i| \leq b_i - a_i$; hence

$$d(x, y) = \sqrt{\sum_{i=1}^{n} (y_i - x_i)^2} \leq \sqrt{\sum_{i=1}^{n} (b_i - a_i)^2} = d(a, b).$$

The center of the rectangle (and this can be taken as the definition if you like) is the point $c = (a + b)/2$. The smaller rectangles are obtained as follows: Choose any $u = (u_1, \ldots, u_n)$ such that either $u_i = a_i$, or else $u_i = c_i$. (There are 2^n such points u.) For each u, define a corresponding v as follows: If $u_i = a_i$, then $v_i = c_i$; while if $u_i = c_i$, then $v_i = b_i$. Then set

$$R_u = \{x : u_i \leq x_i \leq v_i \text{ for } i = 1, \ldots, n\}.$$

Since there are 2^n points u, there are 2^n rectangles R_u, and it is clear that every point of R belongs to one of them. It is also clear that $v - u = (b - a)/2$, so the diameter of R_u is one half the diameter of R.

Exercise 4 If I is an infinite set, the sphere $S(a; r)$ in $\mathcal{B}(I)$ is not totally bounded. Hence neither is the ball. The case is the same in $\mathcal{C}(I)$.

It often happens in analysis that functions to be minimized are not continuous, but have the following weaker property.

DEFINITION 9.10 *A real-valued function f on a metric space X is lower semicontinuous if*

$$f(x) \leq \liminf f(x_n) \qquad \text{whenever } x_n \to x.$$

Exercise 5 $f : X \to \mathbf{R}^1$ is lower semicontinuous if and only if $f^{-1}(\{x : x \leq \alpha\})$ is closed for every real α.

Exercise 6 If $f : X \to \mathbf{R}^1$ is lower semicontinuous, and X is compact, then f has a minimum.

THEOREM 9.11 *Let X be a compact metric space, and let $\mathfrak{F}$ be a family of open subsets with union X. Then there is a finite subfamily with union X.*

Proof For each $x \in X$, let $f(x)$ be the upper bound of the numbers r such that $B(x; r) \subset G$ for some G in the family $\mathfrak{F}$.

Exercise 7 The function f is lower semicontinuous.

To complete the proof, use Theorem 9.9 to find a finite number of points $x_1, \ldots, x_n$ such that X is the union of the balls $B(x_i; r/2)$, where r is the minimum of the function f which is guaranteed by Exercise 6.   By the definition of r, $B(x_i; r/2)$ is contained in some $G \in \mathfrak{F}$, and the proof is complete.

Exercise 8   The converse of Theorem 9.11 is also true.   If X has the property that whenever it is covered by a family of open sets it must be covered by a finite subfamily, then X is compact.

10   EQUIVALENCE OF ABSOLUTE VALUES ON $\mathbf{R}^n$

Now we can prove easily that any two absolute values on $\mathbf{R}^n$ are equivalent.

THEOREM 10.1   *For any absolute value $\|x\|$ on $\mathbf{R}^n$ there are positive numbers m and M such that*

$$m|x| \leq \|x\| \leq M|x| \quad \text{for all } x \in \mathbf{R}^n.$$

Recall that $\|x\|$ is an absolute value if it has the following three properties:

1. $\|x\| \geq 0$, *and* $\|x\| = 0$ *only if* $x = 0$.
2. $\|\alpha x\| = |\alpha|\,\|x\|$ *if α is a real number.*
3. $\|x + y\| \leq \|x\| + \|y\|$.

Proof   Let e_i be the point in $\mathbf{R}^n$ with ith coordinate 1 and all the rest 0.   Then if $x = (x_1, \ldots, x_n)$, it follows that

$$x = \sum_{i=1}^{n} x_i e_i.$$

Properties 3 and 2 then give

$$\|x\| \leq \sum_{i=1}^{n} |x_i|\,\|e_i\|,$$

and the Cauchy–Schwarz inequality gives

$$\|x\| \leq M|x|, \quad \text{with } M = \sqrt{\sum_{i=1}^{n} \|e_i\|^2}. \tag{1}$$

This proves half of the theorem and also shows that the function $f(x) = \|x\|$ is continuous; for from formula (1) of Section 6 we have

$$\big|\,\|x\| - \|y\|\,\big| \leq \|x - y\| \leq M|x - y|. \tag{2}$$

Since f is continuous and the sphere $S = S(0; 1)$ is compact, f has a minimum m on S. By property 1, $m > 0$. We have $\|y\| \geq m$ for every $y \in S$, that is, for every y with $|y| = 1$. If $x \neq 0$, we can apply this to $y = x/|x|$ to get

$$\left\| \frac{x}{|x|} \right\| \geq m,$$

then multiply both sides by $|x|$ and use property 2 to get

$$\|x\| \geq m|x|.$$

So far this is proved for $x \neq 0$, but both sides are 0 for $x = 0$, so we are done.

11   PRODUCTS

If X_1 and X_2 are sets, then $X_1 \times X_2$ is the set of ordered pairs (x_1, x_2) with $x_1 \in X_1$ and $x_2 \in X_2$. More generally, if $X_1, \ldots, X_n$ are sets, then $X_1 \times \cdots \times X_n$ is the set of n-tuples $(x_1, \ldots, x_n)$ with $x_j \in X_j$. If X is a single set, then X^n is the set of n-tuples $(x_1, \ldots, x_n)$ with $x_j \in X$. In the particular case where $X = \mathbf{R}$ is the set of real numbers, this terminology agrees with the terminology we have been using, for $\mathbf{R}^n$ was defined to be the set of all n-tuples of real numbers.

If X_j is a metric space with metric d_j, there are various natural ways to define a metric on $X = X_1 \times \cdots \times X_n$. The one suggested by $\mathbf{R}^n$ is

$$d(x, y) = \sqrt{\Sigma d_j(x_j, y_j)^2}. \tag{1}$$

It is plain that $d(x, x) = 0$, $d(x, y) > 0$ if $x \neq y$, and $d(y, x) = d(x, y)$; so what has to be established is the triangle inequality $d(x, z) \leq d(x, y) + d(y, z)$. To see this, let a, b, and c be the points of $\mathbf{R}^n$ with coordinates $a_j = d_j(x_j, y_j)$, $b_j = d_j(y_j, z_j)$, and $c_j = d_j(x_j, z_j)$. From the triangle inequality in X_j we have $0 \leq c_j \leq a_j + b_j$, and, therefore, $|c| \leq |a + b| \leq |a| + |b|$, which is just what is required.

This argument suggests a way to define all kinds of metrics on X. Choose any absolute value $\| \quad \|$ on $\mathbf{R}^n$. If x and y are two points of X, let $a(x, y)$ be the point in $\mathbf{R}^n$ with coordinates $a_j = d_j(x_j, y_j)$ and define

$$d(x, y) = \|a(x, y)\|. \tag{2}$$

The argument above shows that each of these is a metric on X, and the theorem of the last section (equivalence of absolute values on $\mathbf{R}^n$) shows that all of them are equivalent. The ones most commonly seen are the initial one in formula (1) and

$$d'(x, y) = \Sigma d_j(x_j, y_j),$$
$$d''(x, y) = \max_j d_j(x_j, y_j).$$

Whenever we speak of the product of metric spaces it is to be understood that the metric is the initial one of formula (1) unless otherwise stated, although in most questions this one can be replaced by any of the equivalent metrics of formula (2).

Exercise 1 The product of complete spaces is complete.

Exercise 2 The product of totally bounded spaces is totally bounded.

Exercise 3 The product of compact spaces is compact.

Exercise 4 The product of connected spaces is connected.

Exercise 5 The product of path-connected spaces is path connected.

12   STONE–WEIERSTRASS APPROXIMATION THEOREM

In this section we shall prove a far-reaching generalization of the Weierstrass approximation theorem (Section 8 of the last chapter).

THEOREM 12.1   *(Stone–Weierstrass Approximation Theorem).   Let X be a compact metric space, and let $\mathcal{C}$ be any class of continuous real-valued functions on X with the following properties:*

(a) Each constant function is in $\mathcal{C}$.

(b) If f and g are in $\mathcal{C}$, then so are $f + g$ and fg.

(c) For any two distinct points x and y of X, there is at least one function f in $\mathcal{C}$ with $f(x) \neq f(y)$.

Then every continuous real-valued function on X can be approximated uniformly by functions in $\mathcal{C}$.

The initial Weierstrass theorem is essentially the case where X is a closed bounded interval in $\mathbf{R}^1$ and $\mathcal{C}$ is the set of polynomials.   Two other fundamental examples are as follows:

Example 1 X is any compact set in $\mathbf{R}^n$ and $\mathcal{C}$ is the set of polynomials in n variables.   In this case the theorem says that if $f: X \to \mathbf{R}^1$ is continuous, then there is a sequence $\{f_k\}$ of polynomials such that $f_k \to f$ uniformly on X.

Exercise 1   In this example conditions (a) and (b) are obvious. Establish condition (c). (It is obvious, too!)

Example 2   Let $\mathcal{Q}$ be the set of "trigonometric polynomials" on the interval $[-\pi, \pi]$, that is, the set of functions of the form

$$f(x) = \sum_{k=0}^{n} a_k \cos kx + \sum_{k=1}^{n} b_k \sin kx, \qquad -\pi \le x \le \pi.$$

The number n is not fixed. The sums are arbitrary finite sums. The numbers a_k and b_k are arbitrary real numbers.

Exercise 2   Show that conditions (a) and (b) hold when $\mathcal{Q}$ is the set of trigonometric polynomials and $X = [-\pi, \pi]$.

In this example we have to be a little careful because condition (c) in the theorem is not satisfied when $X = [-\pi, \pi]$. Every trigonometric polynomial takes the same value at $-\pi$ as at π. Therefore, we must "identify" $-\pi$ and π; that is, wrap the interval $[-\pi, \pi]$ around the unit circle in the plane. In this case the theorem says that every continuous function f on $[-\pi, \pi]$ such that $f(-\pi) = f(\pi)$ can be approximated uniformly by trigonometric polynomials.

Exercise 3   Give a rigorous proof of the contention in the last paragraph.

Exercise 4   What kind of functions on $[0, \pi]$ can be approximated by trigonometric polynomials that involve just the sines? Or just the cosines? (*Hint:* Apply the result of Example 2 and note that the sines are odd functions and the cosines are even functions.)

Proof of the Theorem   The space $\mathcal{C}(X)$ of all continuous real-valued function on X is a metric space with the absolute value

$$\|f\| = \sup\{|f(x)| : x \in X\}$$

and the distance $d(f, g) = \|f - g\|$. Convergence in the space $\mathcal{C}(X)$ is just uniform convergence of functions. Therefore, the conclusion of the theorem is just that

$$\bar{\mathcal{Q}} = \mathcal{C}(X).$$

LEMMA 12.2 $\bar{\mathcal{Q}}$ *satisfies conditions* (a), (b), *and* (c) *in Theorem 12.1.*

Proof   Conditions (a) and (c) are self-evident because $\bar{\mathcal{Q}} \supset \mathcal{Q}$. As for (b), suppose that f and g in $\bar{\mathcal{Q}}$, and choose sequences $\{f_n\}$ and $\{g_n\}$ in $\mathcal{Q}$ so that $f_n \to f$ and $g_n \to g$.

Exercise 5   Use the usual familiar proofs to show that $f_n + g_n \to f + g$ and $f_n g_n \to fg$ and conclude that $f + g$ and fg are in $\bar{\mathcal{Q}}$.

LEMMA 12.3   *If* $f \in \bar{\mathcal{Q}}$, *then* $|f| \in \bar{\mathcal{Q}}$.

Proof   Let $\epsilon > 0$ be given and let M be the maximum of $|f|$. Use the original Weierstrass approximation theorem to find a polynomial

$$p(t) = \sum_{k=1}^{m} a_k t^k$$

which approximates the continuous function $|t|$ to within ϵ on the interval $[-M, M]$. Then for any point $x \in X$ we have

$$\big| |f(x)| - p(f(x)) \big| < \epsilon,$$

or, in other words,

$$\big| |f| - p \circ f \big| < \epsilon.$$

By virtue of conditions (a) and (b) (for $\bar{\mathcal{Q}}$) and the fact that $f \in \bar{\mathcal{Q}}$, it follows that the function

$$p \circ f = \sum_{k=1}^{m} a_k f^k$$

is in $\bar{\mathcal{Q}}$. Thus, $|f| \in \bar{\bar{\mathcal{Q}}} = \bar{\mathcal{Q}}$.

LEMMA 12.4   *If* $f_1, \ldots, f_m$ *are in* $\bar{\mathcal{Q}}$, *then the functions*
$$\max(f_1, \ldots, f_m) \qquad and \qquad \min(f_1, \ldots, f_m)$$
are in $\bar{\mathcal{Q}}$.

Proof   The maximum can be obtained by taking first the maximum of f_1 and f_2, and then the maximum of this with f_3, and so on, so it suffices to deal with two functions, in which case we have the formulas

$$\max(f, g) = \frac{f + g + |f - g|}{2}, \qquad \min(f, g) = \frac{f + g - |f - g|}{2}.$$

The formulas are verified by simply noting that $|f - g| = f - g$ if $f \geq g$, while $|f - g| = g - f$ if $g \geq f$ (all at an arbitrary point x, of course). This lemma results from these formulas and the previous lemma.

LEMMA 12.5   *If* F_0 *and* F_1 *are disjoint closed subsets of* X, *then there is a function* $f \in \bar{\mathcal{Q}}$ *such that*

$$f = 0 \text{ on } F_0, \quad f = 1 \text{ on } F_1, \quad 0 \leq f(x) \leq 1 \qquad \text{for all } x \in X.$$

Proof

Let x and y be fixed points of F_0 and F_1 and use condition (c) to find $f \in \bar{\mathfrak{a}}$ with $f(x) \neq f(y)$. Let $\alpha = f(x)$ and $\beta = f(y)$, and consider the function

$$g = \frac{f - \alpha}{\beta - \alpha}.$$

Clearly, we have $g(x) = 0$ and $g(y) = 1$. Now let $h = \min(g, 1)$, and then $k = \max(h, 0)$. It is immediate that $k(x) = 0$, $k(y) = 1$, and $0 \leq k(z) \leq 1$ for all $z \in X$; and from Lemma 12.4 [and condition (a)] it follows that $k \in \bar{\mathfrak{a}}$. (Thus, we have proved the lemma when F_0 and F_1 consist of the single points x and y!)

Let x be a fixed point of F_0. For each $y \in F_1$, use what has just been proved to find a function $f_y \in \bar{\mathfrak{a}}$ such that

$$f_y(x) = 0, \quad f_y(y) = 1, \quad 0 \leq f_y(z) \leq 1 \qquad \text{for all } z \in X,$$

and let

$$G_y = \{z : f_y(z) > \tfrac{1}{2}\}.$$

Now, G_y is open by Theorem 6.4, and the family of the G_y covers the set F_1 (since $y \in G_y$). Also F_1 is compact, for it is plain that any closed subset of a compact space is compact. Hence, it follows from Theorem 9.11 that a finite number of the G_y cover F_1. Call them $G_{y_1}, \ldots, G_{y_m}$. This means that at each point of F_1 at least one of the functions $f_{y_1}, \ldots, f_{y_m}$ is $> \tfrac{1}{2}$, and hence that the function

$$g = \max(f_{y_1}, \ldots, f_{y_m})$$

is $> \tfrac{1}{2}$ at each point of F_1. On the other hand, each of the functions f_y is 0 at x, so $g(x) = 0$. For the function $h = \min(2g, 1)$, we have

$$h(x) = 0, \quad h = 1 \text{ on } F_1, \quad 0 \leq h(z) \leq 1 \qquad \text{for all } z \in X.$$

(Thus, we have proved the lemma when F_0 consists of the single point x and F_1 is arbitrary!)

For each point $x \in F_0$, use what has just been proved to find a function $f_x \in \bar{\mathfrak{a}}$ such that

$$f_x(x) = 0, \quad f_x = 1 \text{ on } F_1, \quad 0 \leq f_x(z) \leq 1 \qquad \text{for all } z \in X,$$

and set

$$G_x = \{z : f_x(z) < \tfrac{1}{2}\}.$$

By the same reasoning as before, a finite number, $G_{x_1}, \ldots, G_{x_n}$, of the G_x covers the set F_0. Consequently, the function

$$g = \min(f_{x_1}, \ldots, f_{x_n})$$

is $< \frac{1}{2}$ at each point of F_0, is equal to 1 at each point of F_1, and is between 0 and 1 everywhere. Therefore, the function $f = \max(2g - 1, 0)$ does the job required by the lemma.

One final lemma will almost complete the proof.

LEMMA 12.6   *If $g \in \mathcal{C}(X)$, $g \geq 0$, then there is an $f \in \bar{\alpha}$ with*

$$0 \leq f \leq g \qquad and \qquad \|g - f\| \leq \tfrac{2}{3}\|g\|.$$

Proof   Define the closed sets F_0 and F_1 by

$$F_0 = \{x : g(x) \leq \tfrac{1}{3}\|g\|\} \qquad and \qquad F_1 = \{x : g(x) \geq \tfrac{2}{3}\|g\|\},$$

and use Lemma 12.5 (multiplied by $\tfrac{1}{3}\|g\|$) to find a function $f \in \bar{\alpha}$ such that

$$f = 0 \text{ on } F_0, \quad f = \tfrac{1}{3}\|g\| \text{ on } F_1, \quad 0 \leq f(x) \leq \tfrac{1}{3}\|g\| \qquad \text{for all } x \in X.$$

In the three cases, $x \in F_0$, $x \in F_1$, and $x \in$ neither, it is easily checked that

$$0 \leq g(x) - f(x) \leq \tfrac{2}{3}\|g\|,$$

and this proves the lemma.

To complete the proof we must show that every $g \in \mathcal{C}(X)$ must lie in $\bar{\alpha}$. It is enough to show that every $g \geq 0$ in $\mathcal{C}(X)$ must lie in $\bar{\alpha}$, for if we know this then we shall know that both

$$g^+ = \max(g, 0) \qquad and \qquad g^- = \min(g, 0)$$

lie in $\bar{\alpha}$, and therefore that $g = g^+ + g^-$ lies in $\bar{\alpha}$. [The fact that g^+ and g^- belong to $\mathcal{C}(X)$ follows from Lemma 12.4 with $\bar{\alpha}$ replaced by $\mathcal{C}(X)$.]

Suppose, therefore, that $g \in \mathcal{C}(X)$ and that $g \geq 0$. Let f_1 be the function $f \in \bar{\alpha}$ given by Lemma 12.6, and put $g_1 = g - f_1$. Now use the lemma again with g_1 in place of g, let f_2 be the corresponding function in $\bar{\alpha}$, and let $g_2 = g_1 - f_2$. Once we have found $f_1, \ldots, f_n$ and $g_1, \ldots, g_n$, we use the lemma with g_n in place of g, find the corresponding function f_{n+1} in $\bar{\alpha}$, and put $g_{n+1} = g_n - f_{n+1}$. With this construction we get that $g_{n+1} \geq 0$ and that

$$\|g_{n+1}\| \leq \tfrac{2}{3}\|g_n\| \leq (\tfrac{2}{3})^2\|g_{n-1}\| \leq (\tfrac{2}{3})^k\|g_{n-(k-1)}\| \leq (\tfrac{2}{3})^{n+1}\|g\|.$$

Furthermore, we get that

$$g = g_1 + f_1 = g_2 + f_1 + f_2 = g_n + \sum_{k=1}^{n} f_k.$$

Since the sum on the right lies in $\bar{\alpha}$, the two formulas together show that $g \in \bar{\bar{\alpha}} = \bar{\alpha}$.

Exercise 6 The original Weierstrass theorem applied to an arbitrary interval, not just a compact one, and involved uniform approximation on all compact subsets. Invent a theorem that applies to any metric space that is the union of a sequence of open sets, each of which has compact closure. Show that any open set in $\mathbf{R}^n$ is a metric space with this property.

It was shown in Example 1, as an application of the Stone–Weierstrass theorem, that every continuous function on $[-\pi, \pi]$ which takes the same value at the end points can be approximated uniformly by trigonometric polynomials. Of course, a function that does not take the same value at the end points cannot be approximated uniformly (or even pointwise) by trigonometric polynomials, because all trigonometric polynomials do take the same value at the end points. In formulas (3) and (4) of Section 4 another metric was defined on the space $C([-\pi, \pi])$ of continuous functions on $[-\pi, \pi]$ by means of the inner product and absolute value

$$\langle f, g \rangle = \int_{-\pi}^{\pi} f(x) g(x) \, dx, \qquad |f|^2 = \sqrt{\langle f, f \rangle}. \tag{1}$$

THEOREM 12.7

Let f be continuous on $[-\pi, \pi]$. For each $\epsilon > 0$ there is a trigonometric polynomial p such that

$$|f - p| < \epsilon.$$

Proof

For each positive integer k, let φ_k be a continuous function on the line that is equal to 1 for $|x| \leq \pi - 1/k$, equal to 0 for $|x| \geq \pi - 1/2k$, and between 0 and 1 everywhere. On the one hand, we have

$$|f - \varphi_k f|^2 = \int_{-\pi}^{\pi} |f|^2 (1 - \varphi_k)^2 \, dx \leq \frac{2}{k} \|f\|^2.$$

On the other hand, $\varphi_k f$ does take the same value (which is 0) at the two end points, so Example 1 shows that there is a trigonometric polynomial p_k such that

$$|\varphi_k f - p_k|^2 \leq 2\pi \|\varphi_k f - p_k\|^2 \leq \frac{1}{k}.$$

The two inequalities together prove the theorem as a result of the triangle inequality for the absolute value $|\ \ |$.

8 { Functions from $\mathbf{R}^1$ to $\mathbf{R}^n$

1 LINES, HALF-LINES, AND DIRECTIONS

**DEFINITION
1.1**

The line passing through the two distinct points a and b of $\mathbf{R}^n$ is the set of all points x of the form

$$x = (1 - t)a + tb = a + t(b - a), \qquad t \text{ real.}$$

Let us check that this agrees with the usual notion of line in $\mathbf{R}^2$. The slope of the line in $\mathbf{R}^2$ passing through the points a and x is

$$\frac{x_2 - a_2}{x_1 - a_1}.$$

Therefore, x is on the line passing through a and b if and only if

$$\frac{x_2 - a_2}{x_1 - a_1} = \frac{b_2 - a_2}{b_1 - a_1}. \tag{1}$$

Suppose that

$$x = (1 - t)a + tb = a + t(b - a).$$

Then $x - a = t(b - a)$, so

$$x_1 - a_1 = t(b_1 - a_1) \qquad \text{and} \qquad x_2 - a_2 = t(b_2 - a_2),$$

from which it is plain that formula (1) holds. On the other hand, if formula (1) holds, then $x - a = t(b - a)$ with

$$t = \frac{x_1 - a_1}{b_1 - a_1}.$$

Exercise 1 The above argument requires that $b_1 \neq a_1$. Provide an argument for the case $b_1 = a_1$.

Exercise 2   Discuss the situation in $\mathbf{R}^3$.

DEFINITION 1.2   *The half-line starting at the point a and passing through the point $b \neq a$ is the set of all points x of the form*

$$x = (1 - t)a + tb = a + t(b - a), \qquad t \geq 0.$$

Exercise 3   Show that the definition agrees with the geometric notion in $\mathbf{R}^2$.

DEFINITION 1.3   *A direction in $\mathbf{R}^n$ is a half-line starting at 0. The direction of a point $b \in \mathbf{R}^n$ is the direction that contains it; that is, it is the half-line that starts at 0 and passes through b. (Of course, $b \neq 0$.)*

Each point $b \neq 0$ determines a direction, but of course many points determine the same direction. Exactly one of these, $b/|b|$, has absolute value 1. Therefore, a direction can be defined alternatively as follows.

DEFINITION 1.4   *A direction in $\mathbf{R}^n$ is a point of absolute value 1.*

The first definition is a little more intuitive, but the second is usually more convenient in practice. Both definitions are used, and the context determines which one is relevant.

Let h be the half-line

$$h = \{x : x = a + t(b - a), t \geq 0\},$$

and let

$$\theta = \frac{b - a}{|b - a|}.$$

Then clearly

$$h = \{x : x = a + t\theta, t \geq 0\}. \tag{2}$$

The points a and θ are called the *initial point* and the *direction* of the half-line h.

THEOREM 1.5   *A half-line determines its initial point and direction uniquely.*

Proof   Suppose that h is given by (2) and that also

$$h = \{x : x = a' + s\theta', s \geq 0\}.$$

Then

$$a' = a + t_0\theta \qquad \text{and} \qquad a = a' + s_0\theta',$$

so $s_0\theta' = -t_0\theta$. Since both θ and θ' have absolute value 1, it follows that $|s_0| = |t_0|$; then since both s_0 and t_0 are ≥ 0, it follows that $s_0 = t_0$, and hence that $\theta' = -\theta$—or else that $s_0 = t_0 = 0$; therefore, $a' = a$.

Exercise 4   Show that if $a' = a$, then it follows immediately that $\theta' = \theta$.

To finish the proof, we have to rule out the possibility that $\theta' = -\theta$.   To do this, consider the point

$$a'' = a + t\theta, \qquad \text{with } t > t_0.$$

This point is on h, so we must have

$$a + t\theta = a' + s\theta' = a + t_0\theta - s\theta = a + (t_0 - s)\theta.$$

Therefore, $t = t_0 - s$, which is impossible, since $t > t_0$ and $s \geq 0$.

Exercise 5   Let h be the half-line with initial point a and direction θ.   Let a' be a point of h, and let h' be the half-line with initial point a' and direction $-\theta$.   Show that

$$h \cap h' = \{x : x = (1 - t)a + ta', \ 0 \leq t \leq 1\}.$$

Thus, $h \cap h'$ is the line segment from a to a', which, of course, is what it should be.

Exercise 6   Let a, b, and c be three points in $\mathbf{R}^n$.   Define what it means to say that b is between a and c.   Show that b is between a and c if and only if

$$d(a, c) = d(a, b) + d(b, c).$$

A point a and a direction θ also determine a line—the set

$$l = \{x : x = a + t\theta, \ t \text{ real}\}.$$

The line is the union of the two half-lines with initial point a and directions $\pm\theta$.

Exercise 7   A line does not determine a direction, but a pair of opposite directions.   If a and b are any two points of the line, then the two opposite directions are $\pm(b - a/|b - a|)$.

DEFINITION   *An oriented line is a line on which one of the two directions is designated as the*
1.6   *"positive" direction.*

Remark   Geometrical statements often appear obvious simply because of the language. No one could seriously doubt that a half-line determines its initial point and direction.   However, the language by itself does not constitute a proof; rather, the theorems that can be proved justify the language.

2   DERIVATIVES AND INTEGRALS

The general topic of the rest of the book is the study of functions from one space $\mathbf{R}^m$ to another $\mathbf{R}^n$. The easiest case to begin with is the case $m = 1$, which has many similarities, as well as some sharp differences, with the case where both m and n are 1, which has been discussed already. The study of functions from $\mathbf{R}^1$ to $\mathbf{R}^n$ is, of course, the study of paths, curves, or arcs in $\mathbf{R}^n$, but this is not always the fruitful point of view to take.

DEFINITION 2.1
A function $f: I \to \mathbf{R}^n$, where I is an interval in $\mathbf{R}^1$, is differentiable at an interior point a of I if the limit

$$\lim_{\substack{x \to a \\ x \neq a}} \frac{f(x) - f(a)}{x - a}$$

exists. The value of the limit is called the derivative of f at a and is denoted by $f'(a)$.

The definition is formally identical with the initial definition for the case $n = 1$. However, $f(a)$, $f(x)$, and $[f(x) - f(a)]/x - a$ are now all points of $\mathbf{R}^n$, and so is the limit $f'(a)$. Differentiability is brought back to the one-dimensional case by the following theorem.

THEOREM 2.2
Let $f: I \to \mathbf{R}^n$, and let

$$f(x) = (f_1(x), \ldots, f_n(x)).$$

Then f is differentiable at the point $a \in I$ if and only if each f_k is differentiable at a. If f is differentiable at a, then

$$f'(a) = (f_1'(a), \ldots, f_n'(a)).$$

Proof
This should be a routine, if somewhat burdensome, calculation by now. To make sure that the calculation is routine, we shall carry out the part which says that if f is differentiable at a, then each f_k is differentiable at a, and $f'(a) = (f_1'(a), \ldots, f_n'(a))$. For any point $y \in \mathbf{R}^n$ we have $|y_k| \leq |y|$. Consequently, if $f'(a) = (b_1, \ldots, b_n)$, then

$$\left| \frac{f_k(x) - f_k(a)}{x - a} - b_k \right| \leq \left| \frac{f(x) - f(a)}{x - a} - b \right|.$$

Given $\epsilon > 0$ we can find $\delta > 0$ such that if $|x - a| < \delta$ and $x \neq a$, then the right-hand side is $< \epsilon$. This shows that f_k is differentiable and its derivative is b_k.

Exercise 1   Carry out the other half of the proof.

Exercise 2   If f is differentiable at a, then f is continuous at a.

Exercise 3   Use the example $f(x) = (\cos x, \sin x)$, $0 \leq x \leq 2\pi$, to show that the mean-value theorem does not hold for functions with values in $\mathbf{R}^n$.

An entirely similar situation prevails with regard to integrals.

DEFINITION 2.3   *Let $p = (x_0, \ldots, x_n)$ and $(\xi_1, \ldots, \xi_n)$ be a partition of the closed bounded interval I.   If $f: I \to \mathbf{R}^n$, then*

$$S(p; f) = \sum_{i=1}^{n} f(\xi_i)(x_i - x_{i-1}).$$

The function f is integrable on I, and the integral is the point $L \in \mathbf{R}^n$ if for every positive number ϵ there is a positive number δ such that if $|p| < \delta$, then

$$|S(p; f) - L| < \epsilon.$$

THEOREM 2.4   *Let $f: I \to \mathbf{R}^n$, where $I = [a, b]$, and let $f(x) = (f_1(x), \ldots, f_n(x))$. Then f is integrable on I if and only if each f_k is integrable on I.   If f is integrable on I, then the kth coordinate of $\int_a^b f(x) \, dx$ is $\int_a^b f_k(x) \, dx$.*

Proof   This time we shall prove the other half of the theorem.   Suppose that each f_k is integrable on I, let

$$\int_a^b f_k(x) \, dx = L_k,$$

and let $L = (L_1, \ldots, L_n)$.   For any given $\epsilon > 0$, Theorem 6.3 of Chapter 4 provides $\delta > 0$ such that if p is any partition with $|p| < \delta$, then $|S(p; f_k) - L_k| < \epsilon$, in which case

$$|S(p; f) - L| < \sqrt{n} \, \epsilon.$$

Exercise 4   Prove the other half of the theorem.

In this context the upper and lower sums that were prevalent in Chapter 4 are meaningless.   It makes no sense to say that one point of $\mathbf{R}^n$ is "larger" than another.

THEOREM 2.5   *Let $f: I \to \mathbf{R}^n$ be bounded and continuous at all but a finite number of points.   Then*

(a) *$F(x) = \int_a^x f(t) \, dt$ is a primitive of f.*
(b) *If G is any primitive of f, then $G(b) - G(a) = \int_a^b f(t) \, dt$.*

Exercise 5 Prove the theorem on the basis of Theorems 2.2 and 2.4 and the results of Chapter 4.

THEOREM 2.6 *Let $f: I \to \mathbf{R}^n$ be bounded and continuous at all but a finite number of points. Then*

$$\left| \int_a^b f(x) \; dx \right| \leq \int_a^b |f(x)| \; dx.$$

Proof The triangle inequality in $\mathbf{R}^n$ shows that $|S(p; f)| \leq S(p; |f|)$ for every partition p.

Exercise 6 Show that if $f: I \to \mathbf{R}^n$ is Riemann integrable, then so is $|f|$, and deduce that Theorem 2.6 holds for any Riemann integrable f.

Exercise 7 Show that if $f: I \to \mathbf{R}^1$ is Riemann integrable, then so is f^2.

Exercise 8 Show that if f, $g: I \to \mathbf{R}^n$ are Riemann integrable, then so is $\langle f, g \rangle$. (*Hint:* $4\langle f, g \rangle = |f + g|^2 - |f - g|^2$.)

DEFINITION 2.7 *If $f: X \to \mathbf{R}^n$ is a function from any set X into $\mathbf{R}^n$, and $f(x) = (f_1(x), \; \ldots \;, f_n(x))$, then the function $f_k: X \to \mathbf{R}^1$ is called the kth coordinate function of f.*

3 TANGENT LINES, VELOCITY, AND ACCELERATION

A path in $\mathbf{R}^n$ was defined to be a continuous function from a closed bounded interval into $\mathbf{R}^n$. This is not in perfect accord with geometric intuition—for a function contains too much information. Think of it this way. Suppose that an object is moving around in $\mathbf{R}^n$ during a time interval I. For each $t \in I$, let $f(t)$ denote the location of the object at the time t. The function f describes the motion completely, so it must contain not only the information about where the point goes, but with what speed, with what acceleration, and so on. Intuition is not very explicit as to whether two paths are the same if they go to the same places, but differ in speed, acceleration, and so on. For instance, are the paths to school the same if on the one hand you go there directly, and on the other you sit down for a while at the sidewalk cafe on the way? Is the path to the dentist's office the same as the path home? In both cases most people would say no—that the one is somewhat more pleasant than the other.

It is not at all a simple matter to give a rule that defines when two functions represent the same path. In fact, it all depends on the problem. In some problems two given functions should be considered different representations of the same path, while in others the same two functions should be considered different paths. Therefore, we shall stick with the original definition that the path *is* the function. It is not really harmful that the function contains too much information (as it would be if it contained too little). Because of the extra information, however, the function may be "bad" at some points where the intuitive path is not bad at all. An example is given below.

DEFINITION 3.1

The tangent line to the path $f : I \to \mathbf{R}^n$ at the point $f(a)$ is the oriented line that passes through $f(a)$ and has the direction of $f'(a)$—provided f is differentiable at a and $f'(a) \neq 0$.

The fact that the tangent line is oriented reflects the fact that there is a "positive direction" on the path itself. (The path to the dentist's office is not the same as the path home.) The unoriented tangent line is the set

$$l = \{x : x = f(a) + tf'(a)\},$$

and the orientation is the direction of $f'(a)$, which is

$$\frac{f'(a)}{|f'(a)|}.$$

Exercise 1   If $f : [0, 1] \to \mathbf{R}^n$ is the path to the dentist's office, what is the path home? Show that the tangent line to the path home at any point is the same as the tangent line to f, but with the opposite orientation.

Example 1   If $f(t) = (t, t^3)$ in $\mathbf{R}^2$, then $f(0) = (0, 0)$ and $f'(0) = (1, 0)$, so the tangent line to f at the origin is the set of points x of the form $x = (0, 0) + t(1, 0) = (t, 0)$, which is just the x_1 axis. From the intuitive point of view, the function $g(t) = (t^3, t^9)$ represents the same path, but $g'(0) = (0, 0)$; so the tangent line to g at the origin is undefined by Definition 3.1. Again, from the intuitive point of view, the function $h(t) = (\sqrt[3]{t}, t)$ represents the same path, but h is not even differentiable at 0. This is an illustration of how g and h may be "bad" at the origin even though the intuitive path is not bad at all.

One situation in which it is very tempting to say that $f : I \to \mathbf{R}^n$ and $g : J \to \mathbf{R}^n$ represent the same intuitive path is when there is a strictly increasing function $\varphi : J \to I$ such that $g = f \circ \varphi$. (From the point of view of motion, φ simply gives a different way of measuring time.) It is to be hoped that f and g have the same tangent line, but Example 1 shows the need for caution, for it is of just this nature with $\varphi(t) = t^3$.

THEOREM 3.2

Let $J \xrightarrow{\varphi} I \xrightarrow{f} \mathbf{R}^n$, and let $g = f \circ \varphi$. If $\varphi'(a) > 0$ and $f'(\varphi(a)) \neq 0$, then f and g have the same tangent line at the point $g(a) = f(\varphi(a))$.

Proof   Since the two lines pass through the same point, all that is necessary is to show that they have the same direction, that is, that $g'(a)$ is a positive multiple of $f'(\varphi(a))$. This results from the following theorem on the derivative of a composite function.

THEOREM 3.3   *Let* $J \xrightarrow{\varphi} I \xrightarrow{f} \mathbf{R}^n$. *If* φ *is differentiable at* a, *and* f *is differentiable at* $\varphi(a)$, *then* $f \circ \varphi$ *is differentiable at* a, *and*

$$(f \circ \varphi)'(a) = \varphi'(a)f'(\varphi(a)).$$

Exercise 2   Prove Theorem 3.3 by considering the coordinate functions of f separately (Theorem 2.2), and then by using the chain rule for composite functions in Section 6 of Chapter 2.

From the geometrical point of view the justification of Definition 3.1 is the following.

THEOREM 3.4   *Let* $f: I \to \mathbf{R}^n$ *be differentiable at* a, *with* $f'(a) \neq 0$. *The direction of the tangent line at* $f(a)$ *is the limit of the directions of the chords from* $f(a)$ *to* $f(b)$ *as* $b \to a$ *from the right.*

Proof   The chord from $f(a)$ to $f(b)$ is the line joining these two points with the direction

$$\frac{f(b) - f(a)}{|f(b) - f(a)|} = \frac{f(b) - f(a)}{b - a} \frac{b - a}{|f(b) - f(a)|}. \tag{1}$$

As $b \to a$ from the right, the first factor has the limit $f'(a)$ and the second factor has the limit $1/|f'(a)|$.

Exercise 3   What happens when $b \to a$ from the left?

Exercise 4   If $f'(a) \neq 0$, then $f(b) \neq f(a)$ for all b sufficiently close to a, so formula (1) makes sense for all b close to a.

Exercise 5   In Definition 3.1 we speak of the tangent line to the path f at the point $f(a)$. Give several examples of various kinds to show that we should really speak of the tangent line at a, not the tangent line at $f(a)$—even though this is not so pleasing to the intuition.

Notice that the phrase "limit of directions" needs no definition. A direction is a point of absolute value 1, so nothing more is involved than a limit of points.

**DEFINITION
3.5**

Let an object move in $\mathbf{R}^n$ *during a time interval I. For each* $t \in I$, *let* $f(t)$ *be the location of the object at time t. The velocity at time* $t = a$ *is* $f'(a)$, *the speed is* $|f'(a)|$, *and the acceleration is* $f''(a)$.

The definition is formally the same as the one in Chapter 1 for motion on a line. Furthermore, the velocity at time $t = a$ is the limit of the average velocity over the time interval from $t = a$ to $t = x$ as the time interval goes to 0. Indeed, this average velocity is (by definition) the difference between the final and initial positions divided by the time interval, which is nothing but

$$\frac{f(x) - f(a)}{x - a}.$$

Notice that the direction of the velocity is the direction of the tangent line.

4   GEOMETRIC MODELS OF $\mathbf{R}^n$

Although $\mathbf{R}^n$ is not by its definition a geometric object (it is a set of n-tuples), it is convenient to have a geometric model in mind to suggest terminology, results, and sometimes proofs. So far the model has been the everyday three-dimensional space, which is a model of $\mathbf{R}^3$, and by analogy of $\mathbf{R}^n$. The everyday space becomes a model of $\mathbf{R}^3$ by assigning to the triple $(x_1, x_2, x_3) \in \mathbf{R}^3$ the point with these coordinates relative to given coordinate axes.

A second model is also useful. It is obtained by assigning to the triple (x_1, x_2, x_3) not the point with these coordinates, but rather the line segment from the origin to that point.

We have seen in Section 3 that the velocity of a moving object at a given time is a point in $\mathbf{R}^n$. The information that is expected from a velocity consists of a direction (the direction of the motion) and a positive number (the magnitude of the velocity, or speed). Now, a point $x \neq 0$ in $\mathbf{R}^n$ does carry just this information. It determines the direction $x/|x|$ and the positive number $|x|$, and, in turn, it is determined by these two quantities. However, from the intuitive geometric point of view, a point does not carry the flavor of a magnitude and a direction. It is rather a line segment starting at the origin that carries this flavor.

The geometric language for $\mathbf{R}^n$ derives about equally from the two models, so both must be kept in mind, and the good model to use in a given situation must be determined by the context. This is a bother at first, but becomes fairly easy with practice. Of course, it is not necessary to keep any model in mind. Everything is on a perfectly sound, if somewhat mysterious, footing right in $\mathbf{R}^n$ itself.   Often a point in $\mathbf{R}^n$ is called a vector.   This is a rather neutral term that does not suggest either model, as does the term point, although perhaps it does have a slight bias toward the line-segment model.

Back in Section 1 of Chapter 7 (Exercise 2), the formula

$$\langle x, y \rangle = |x| \, |y| \cos \theta \tag{1}$$

was established for dimension 2, where θ is the angle between the half-lines $0x$ and $0y$.   With the line-segment interpretation, we can say simply that θ is the angle between x and y.   The formula can serve as the definition of the angle between two vectors in $\mathbf{R}^n$.

DEFINITION 4.1

The angle between the nonzero vectors x and y of $\mathbf{R}^n$ is the angle θ defined by

$$\cos \theta = \frac{\langle x, y \rangle}{|x| \, |y|}, \qquad 0 \le \theta \le \pi. \tag{2}$$

The definition makes sense because on the one hand the cosine is one to one on the interval $0 \le \theta \le \pi$ and takes every value between -1 and 1; and on the other the number $\langle x, y \rangle / |x| \, |y|$ does lie between -1 and 1 by virtue of the Cauchy–Schwarz inequality.

If one of the two vectors is 0, the angle is undefined.   This is as it should be, because then the line segment has length 0, and the angle between some line segment and a segment of length 0 does not make geometrical sense.   Notice that the angle from x to y cannot be distinguished from the angle from y to x.

DEFINITION 4.2

The vectors x and y of $\mathbf{R}^n$ are perpendicular, or orthogonal, if $\langle x, y \rangle = 0$.

If both x and y are $\ne 0$, then the definition is consistent with Definition 4.1. The two are perpendicular if the angle between them is $90°$.   However, the vector 0 is perpendicular to every vector.   There is no reason why this should seem geometrically "right."   It is a technical convenience which is made part of the definition.   Notice that 0 is the only vector that is perpendicular to every vector—indeed, it is the only one that is perpendicular to itself, for, if $x \ne 0$, then $\langle x, x \rangle = |x|^2 \ne 0$.

As an illustration of these concepts we shall consider a couple of problems about motion.   To do so, we need Theorem 4.3.

THEOREM 4.3

Let $f : I \to \mathbf{R}^n$ and $g : I \to \mathbf{R}^n$.   Then

$$\langle f, g \rangle' = \langle f', g \rangle + \langle f, g' \rangle.$$

The meaning of the theorem is this.   It is assumed that both f and g are differentiable at some point a, and that $h(t) = \langle f(t), g(t) \rangle$, so that $h : I \to \mathbf{R}^1$.

The assertion is that h is differentiable at a and that

$$h'(a) = \langle f'(a), g(a) \rangle + \langle f(a), g'(a) \rangle.$$

Proof   If $f(t) = (f_1(t), \ldots, f_n(t))$ and $g(t) = (g_1(t), \ldots, g_n(t))$, then

$$h(t) = \sum_{k=1}^{n} f_k(t) g_k(t);$$

so

$$h'(t) = \sum_{k=1}^{n} f_k'(t) g_k(t) + \sum_{k=1}^{n} f_k(t) g_k'(t)$$
$$= \langle f'(t), g(t) \rangle + \langle f(t), g'(t) \rangle.$$

Example 1   If a point moves on a sphere, then the velocity at any point is perpendicular to the radius drawn to that point.

Proof   If $f(t)$ lies on the sphere $S(a; r)$ for each t, then

$$r^2 = |f(t) - a|^2 = \langle f(t) - a, f(t) - a \rangle.$$

Differentiation by the formula of the last theorem gives

$$0 = 2\langle f'(t), f(t) - a \rangle,$$

and $f'(t)$ is the velocity, while $f(t) - a$ has the direction of the radius.

Another way of saying the same thing is that if a path lies on a surface (in this case the sphere), then the tangent line to the path is tangent to the surface.   But tangents to surfaces have not yet been discussed.

Example 2   If a point moves at constant speed, then the velocity and acceleration vectors are perpendicular.

Proof   If the speed is v, then we have

$$v^2 = |f'(t)|^2 = \langle f'(t), f'(t) \rangle,$$

and differentiation gives

$$0 = 2\langle f''(t), f'(t) \rangle.$$

Example 3   Let a point move at the constant speed v around the circle $S(0; 1)$ in the plane. By Example 1 we have $0 = \langle f', f \rangle$, and differentiating again we get $0 = \langle f'', f \rangle + \langle f', f' \rangle = \langle f'', f \rangle + v^2$. Now, if f' is perpendicular to f and f'' is perpendicular to f', it follows that f'' is a multiple of f, say $f'' = \alpha f$. (This is where the fact that the dimension is 2 comes in.)   We get

$$-v^2 = \langle f'', f \rangle = \langle \alpha f, f \rangle = \alpha,$$

so finally

$$f'' = -v^2 f. \tag{3}$$

This is the formula for centrifugal force.   It says that when a point moves around a circle of radius 1 at constant speed v, there is an acceleration that is directed toward the center of the circle and has magnitude v^2.

Exercise 1   Explain why the acceleration is directed toward the center of the circle.

Exercise 2   A point moves in the plane at speed 1 along the curve $y = x^2$.   Find the acceleration at the point (x, y) of the curve.

5   MISSILES, MOONS, AND SO ON

Let us shoot off a missile and try to find out what becomes of it.

The basic law governing motion in the three-dimensional space is again the second law of Newton (see Section 4 of Chapter 1)—but in vector form: *The acceleration of a moving object is proportional to the force acting on it.*   In other words, there is a constant (depending on the object, and called its *mass*) such that if $x(t)$ is the position of the object at time t, and $f(t)$ is the force acting on it at time t, then

$$mx''(t) = f(t) \qquad \text{for all } t. \tag{1}$$

Notice that for the formula to make sense the force must be a vector, which is consistent with the usual concept of a force as a quantity with a magnitude and a direction.

The second physical law that has to be used is the law of gravity, which will determine the force in our present problem.   It says that two objects of masses m and M at a distance r apart attract one another with a force of magnitude

$$f = \frac{mM}{r^2}, \tag{2}$$

provided the units for measuring force and acceleration are chosen suitably. [It is remarkable that the mass that appears in formula (2) is the same as the mass that appears in formula (1).   The verification of this fact was the object of the famous experiment of Galileo's in which he dropped stones from the leaning tower of Pisa—or so the story goes.]

In applying these two laws to discover where the missile goes, we shall make the assumption that the rocket engines have been shut off and that the missile and the earth are coasting along under the influence of their mutual gravitational attraction alone. This is absurd, of course. The main force acting on the earth is not the attraction of the missile but the attraction of the sun. However, the calculations are interesting, and the results are in fact realistic in the similar problem of the orbit of the earth (or other planets or meteors) around the sun.

Let m and M be the masses of the missile and the earth. Choose coordinates in the three-dimensional space, and let $y(t)$ and $z(t)$ be the positions of the missile and the earth at time t. The distance between the two at time t is then $|y(t) - z(t)|$, and the direction from the missile to the earth is $z(t) - y(t)/|z(t) - y(t)|$. According to formula (2), the force on the missile at time t is

$$f(t) = mM \frac{z(t) - y(t)}{|z(t) - y(t)|^3},$$

and the force on the earth is the opposite (i.e., negative) of this. Thus, formula (1) gives the two equations

$$my'' = mM \frac{z - y}{|z - y|^3}, \qquad Mz'' = mM \frac{y - z}{|z - y|^3}, \tag{3}$$

which hold at each point t.

It simplifies matters a little to calculate the path of the missile relative to the earth rather than to calculate the two paths separately. This means to calculate the function $x = y - z$ rather than to calculate y and z separately. The equations (3) give the following equation for x:

$$x'' = -\mu \frac{x}{|x|^3}, \qquad \text{where } \mu = m + M. \tag{4}$$

Before going on we should understand exactly what is at issue here. We are assuming that the function x has two continuous derivatives and satisfies equation (4) at each point of some open interval I, and we are trying to discover what we can about x under these conditions. In particular, equation (4) implies that $x(t) \neq 0$ for each $t \in I$. [From the physical point of view this condition is obvious, for if $x(t_0) = 0$, then the missile has crashed at time t_0, and there is nothing more to be said.]

Exercise 1   Consideration of x rather than y and z amounts to putting the origin of the coordinate system at the center of the earth so that the coordinate system moves right along with the earth. Why can't this be done at the beginning?

The first step is to show that the point $x(t)$ remains in a fixed plane through the origin for all t. To see this, write equation (4) for the ith and jth coordinates.

$$x_i'' = -\mu \frac{x_i}{|x|^3}, \qquad x_j'' = -\mu \frac{x_j}{|x|^3}.$$

Multiply the first by x_j and the second by x_i and subtract, to get

$$0 = x_i'' x_j - x_j'' x_i = (x_i' x_j - x_j' x_i)',$$

and hence that $x_i' x_j - x_j' x_i$ is constant. If h is the constant vector with coordinates

$$h_1 = x_2' x_3 - x_3' x_2, \qquad h_2 = x_3' x_1 x_1' x_3, \qquad h_3 = x_1' x_2 - x_2' x_1, \tag{5}$$

then

$$\langle x, h \rangle = x_1 h_1 + x_2 h_2 + x_3 h_3 = 0.$$

(Everything cancels out.) This means that for every t, $x(t)$ lies in the plane that passes through the origin and is perpendicular to the fixed vector h, provided of course that $h \neq 0$.

Exercise 2   If $h = 0$, then $x(t)$ lies on a line through the origin. [*Hint:* On any interval where $x_2 \neq 0$, the fact that $h_3 = 0$ gives $(x_1|x_2)' = 0$; hence $x_1 = cx_2$. Similarly, the fact that $h_1 = 0$ gives $x_3 = dx_2$. Now use connectedness and the fact that $x(t) \neq 0$ for each $t \in I$ to show that $x(t)$ lies on the line with equations $x_1 = cx_2$ and $x_3 = dx_2$ for all $t \in I$.]

Now we shall choose the coordinates in $\mathbf{R}^3$ so that $x(t)$ lies in the plane $x_3 = 0$ for each $t \in I$. Then we can simply forget about the three-dimensional space and treat the problem in $\mathbf{R}^2$ instead. Equation (4) remains the same. The complex multiplication on the plane (especially multiplication by i) is what will let us form the right expressions. Remember that the inner product is expressed in terms of the complex product by the formula

$$\langle z, w \rangle = \mathrm{Re}\ z\bar{w},$$

so in particular we have

$$\langle iz, z \rangle = 0 \qquad \text{for every } z \in \mathbf{R}^2.$$

Take the inner product with ix on both sides of (4). The result is that $\langle ix, x'' \rangle = 0$, and hence that $\langle ix, x' \rangle' = \langle ix, x'' \rangle + \langle ix', x' \rangle = 0$. Therefore. $\langle ix, x' \rangle$ is a constant, which we shall call k:

$$\langle ix, x' \rangle = k\ (= \text{constant}). \tag{6}$$

Now let $r = |x|$ and $\theta = x/|x|$, so

$$x = r\theta \qquad \text{and} \qquad x' = r'\theta + r\theta'$$

[This is all right because $x(t) \neq 0$ for every $t \in I$.]   According to Example 1, Section 4, θ' and θ are perpendicular, so θ' is a (real) multiple of $i\theta$, say $\theta' = \alpha i\theta$, in which case $x' = r'\theta + \alpha r i\theta$; then $k = \langle ix, x' \rangle = r'\langle ix, \theta \rangle + \alpha r\langle ix, i\theta \rangle = \alpha r^2$. Thus,

$$\theta' = \frac{k}{r^2} i\theta.$$

This formula and the basic equation (4) show that $kix'' = -\mu\theta'$, and integration gives

$$kix' = -\mu(\theta + e), \tag{7}$$

where e is a constant vector.   Now, $k = \langle ix, x' \rangle = -\langle x, ix' \rangle$.   Therefore,

$$k^2 = -\langle x, kix' \rangle = \mu\langle x, \theta + e \rangle.$$

If we let $\epsilon = |e|$ and let φ be the angle between θ and the fixed vector e, we get the final equation

$$r(1 + \epsilon \cos \varphi) = \frac{k^2}{\mu} \tag{8}$$

for the equation of the curve on which the missile moves relative to the earth (in polar coordinates).   The constant μ is the sum of the masses, and the constants k and ϵ are determined by any initial position $x_0 = x(t_0)$ and velocity $v_0 = x'(t_0)$ by

$$k = \langle ix_0, v_0 \rangle, \qquad e = -\frac{k}{\mu} iv_0 - \frac{x_0}{|x_0|}, \qquad \epsilon = |e|. \tag{9}$$

If the coordinates in the plane of the motion are chosen so that e lies on the x_1 axis, then in rectangular coordinates [(x, y) instead of (x_1, x_2) to avoid subscripts] the equation becomes

$$(1 - \epsilon^2)x^2 + \frac{2k^2\epsilon}{\mu} x + y^2 = \frac{k}{\mu^2}. \tag{10}$$

It is shown in analytic geometry that this is an ellipse if $\epsilon < 1$, a hyperbola if $\epsilon > 1$, and a parabola if $\epsilon = 1$.   (To study the curve, simply complete the square in the x's.)   Thus, an orbit must be one of these three curves—which one depends on the initial position and velocity.

Exercise 3   Discuss the curves with equation (10).   What is the story if $k = 0$?

Exercise 4   The center of mass of the earth and the missile is defined to be the point

$$w = \frac{my + Mz}{m + M}.$$

Show that the center of mass moves with constant velocity (i.e., constant speed along a straight line) and find the formula for w.

Exercise 5   Use the foregoing results to describe the paths $y - w$ and $z - w$ (i.e., the motion of the missile and the earth relative to the center of mass). With Exercises 4 and 5 you will have the individual paths y and z.

Exercise 6   If x is any path in the plane and $r(t)$ and $\varphi(t)$ are polar coordinates of $x(t)$, then

$$\langle ix, x' \rangle = r^2 \varphi'. \tag{11}$$

Therefore, equation (6) and the formula for area in polar coordinates show that the area swept out by the radius drawn from the earth to the missile during the time interval $[t_1, t_2]$ is

$$\text{area} = \frac{k}{2}(t_2 - t_1). \tag{12}$$

The ideas in this section had a profound effect on the early development of physics—but they emerged in the reverse order. By detailed analysis of actual observations, Kepler deduced that the motion of a planet about the sun is governed by the following celebrated laws:

1. *The orbit of the planet is planar and is an ellipse with focus at the sun.*
2. *The radius from the sun to the planet sweeps out equal areas in equal times.*
3. *The square of the period (period = length of time for one orbit) is proportional to the cube of the mean distance from the sun.*

From these laws Newton was able to deduce the law of gravitation (2).

As for law 1, we have shown that the orbit is planar and that in fact it is an ellipse. (Hyperbola and parabola are ruled out because they require that the planet disappear in the distance.) We have not discussed the focus, but if you know what the focus of an ellipse is, you can show easily that the focus is the sun.

As for law 2, this is just formula (12). And as for law 3, we shall omit it, but the interested student may be able to work it out for himself.

Kepler's laws were the monumental result of several years of extraordinary labor, and he was highly pleased with them.

Johann Kepler

"The die is cast, the book is written, to be read either now or by posterity,
I care not which; it may well wait a century for a reader, as God has waited six
thousand years for an observer."

6   ARC LENGTH

The ideas needed to discuss arc length in $\mathbf{R}^n$ are already present in the two-
dimensional case, which we have discussed briefly in Section 7 of Chapter 4,
and in the theory of the Riemann integral.

**DEFINITION
6.1**

*Let $f : [a, b] \to \mathbf{R}^n$ be a path in $\mathbf{R}^n$. If $p = (t_0, \ldots, t_m)$ is a partition
of $[a, b]$, let*

$$l(p) = \sum_{i=1}^{m} |f(t_i) - f(t_{i-1})|.$$

*The path has length L if $\lim_{|p| \to 0} l(p) = L$ in the sense that for each positive
number ϵ there is a positive number δ such that if $|p| < \delta$, then $|l(p) -
L| < \epsilon$.*

The points $f(t_i)$ are the vertices of an inscribed polygon, and the number $l(p)$ is
the sum of the distances between successive vertices, that is, the length of the
polygon. Thus, the length of the path is the limit of the lengths of approximating inscribed polygons.

Exercise 1   Define the length of a path in any metric space.

**LEMMA
6.2**

If r is a refinement of p, then $l(r) \geq l(p)$.

Proof   If r is obtained from p by adding just one point t which is between t_{i-1} and t_i, then the term $|f(t_i) - f(t_{i-1})|$ which occurs in $l(p)$ is replaced by the sum $|f(t_i) - f(t)| + |f(t) - f(t_{i-1})|$. The triangle inequality gives

$$|f(t_i) - f(t_{i-1})| \leq |f(t_i) - f(t)| + |f(t) - f(t_{i-1})|,$$

and this shows that $l(r) \geq l(p)$.

**THEOREM
6.3**

The length of the path f exists if and only if the numbers $l(p)$ are bounded, and, if this is the case, then

$$L = \sup\{l(p) : p \text{ is a partition}\}.$$

Proof   First we shall show that if the length L exists, then for every partition q we must have $l(q) \leq L$. Let ϵ be a given positive number and choose δ in accordance with the definition. Fix a partition p with $|p| < \delta$. If r is the common refinement of p and q, then $|r| < \delta$; so by the choice of δ and by Lemma 6.2 we have

$$l(q) \leq l(r) \leq L + \epsilon.$$

Since this holds for every $\epsilon > 0$, it follows that $l(q) \leq L$.

   The proof of the other half is very much like the proof of Theorem 6.1 of Chapter 4. Suppose that the upper bound S of the $l(p)$ is finite and let $\epsilon > 0$ be given. Fix a partition q such that

$$l(q) \geq S - \epsilon,$$

and let N be the number of points in q and δ_0 be the smallest interval in q. Choose $\delta < \delta_0$ so that if $|t - s| < \delta$, then $|f(t) - f(s)| < \epsilon$. We shall show that if $|p| < \delta$, then

$$l(p) \geq S - (1 + 2N)\epsilon, \tag{1}$$

which will prove the theorem because we have automatically that $l(p) \leq S$.

   If r is the common refinement of p and q, then on the one hand we have $l(r) \geq l(q) \geq S - \epsilon$, and on the other hand we can compare $l(r)$ and $l(p)$. Suppose that t is a point of r that is not in p. Then the two adjacent points of r, call them t' and t'', must belong to p by virtue of the fact that $|p| < \delta < \delta_0$. If we replace the term $|f(t'') - f(t')|$, which occurs in p, by the sum $|f(t'') - f(t)| + |f(t) - f(t')|$, which occurs in r, we increase the sum by at most 2ϵ. If we do this for each point of r that is not in p, we increase the sum by at most $2N\epsilon$.   Thus, we have

$$S - \epsilon \leq l(r) \leq l(p) + 2N\epsilon.$$

This proves inequality (1) and hence the theorem.

THEOREM 6.4   *Let $f:[a,\ b] \to \mathbf{R}^n$ be a path, and let c be a point between a and b.   The arc length on $[a,\ b]$ exists if and only if the arc lengths on $[a,\ c]$ and $[c,\ b]$ both exist—in which case it is of course the sum.*

Exercise 2   This theorem is obvious from the preceding one.   Prove it.

THEOREM 6.5   *Let $f:[a,\ b] \to \mathbf{R}^n$ be a path.   If f' is uniformly continuous on $(a,\ b)$, then the arc length exists and is given by*

$$L = \int_a^b |f'(t)|\ dt.$$

Exercise 3   The proof is just like the proof of Theorem 7.4 of Chapter 4.   Carry it out.

Remark   By combining Theorems 6.4 and 6.5, we can claim that the arc length of the path f exists and is given by the integral any time the interval $[a,\ b]$ can be partitioned into a finite number of subintervals in such a way that f' is uniformly continuous on each open subinterval.   Such paths are usually called piecewise smooth.   Every polygon is, of course, piecewise smooth.

It is sometimes advantageous to choose the parametric equations for a given geometric path in such a way that the arc length along the path is the parameter.   We shall have to use this idea later, so let us see what is involved.

Let $f:[a,\ b] \to \mathbf{R}^n$ be a path of finite length L.   For $a \leq t \leq b$, let $L(t)$ be the length of the path over the interval $[a,\ t]$.

Exercise 4   Show that $L(t)$ is continuous and nondecreasing on $a \leq t \leq b$, and that if f is not constant on any subinterval, then $L(t)$ is strictly increasing.   In this latter case, the inverse function $A = L^{-1}$ is continuous and strictly increasing on the interval $0 \leq s \leq L$.

DEFINITION 6.6   *Let $f:[a,\ b] \to \mathbf{R}^n$ be a path of finite length L that is not constant on any subinterval.   The path $g = f \circ A$ is called the parametric representation of f by arc length.*

Exercise 5   Let g be the parametric representation of f by arc length.   As s goes from 0 to L, the point $g(s)$ traces out the same geometric path as does $f(t)$ when t goes from a to b.   For each s, the arc length of g on $[0,\ s]$ is equal to s.

We shall need the results of Exercises 4 and 5 only when the initial path f satisfies the conditions:

1. *f' is uniformly continuous on (a, b).*
2. *$f'(t) \neq 0$ on (a, b).*

In this case the results are quite obvious from the integral formula for arc length (Theorem 6.5).

Exercise 6   Establish the results of Exercises 4 and 5 in this special case by using the integral formula for arc length.   Show that $|g'(s)| = 1$ for each s.

Exercise 7   Parametric representation by arc length is more of theoretical importance than of practical importance.   To get an idea of what is involved, try to find the parametric representation by arc length in the case of the circle $x^2 + y^2 = r^2$, the parabola $y = x^2$, and the ellipse $x^2/a^2 + y^2/b^2 = 1$.

Algebra and Geometry in $\mathbf{R}^n$

1 SUBSPACES

The subject of Chapter 8 was paths in $\mathbf{R}^n$ and functions from $\mathbf{R}^1$ to $\mathbf{R}^n$. The next step is surfaces in $\mathbf{R}^n$ and functions from $\mathbf{R}^m$ to $\mathbf{R}^n$. But first it is necessary to take a close look at the simplest case in which the surfaces are planes and the functions are linear.

DEFINITION 1.1

A subspace of $\mathbf{R}^n$ is a subset V with the property that if x and y are any two points of V, and α is any real number, then

$$x + y \in V \qquad and \qquad \alpha x \in V.$$

The simplest subspace is the set $V = \{0\}$ (consisting of the vector 0 alone). The next is the set of all multiples of some nonzero vector v—in other words, the line through v and the origin.

Exercise 1 Show that the line through v and 0 is a subspace and that it is the smallest subspace containing v.

Next consider two vectors v and w that are not on a line through 0, and let

$$V = \{\alpha v + \beta w : \alpha \text{ and } \beta \text{ are real}\}. \tag{1}$$

Exercise 2 Show that V is a subspace and that it is the smallest subspace containing both v and w.

The subspace V in formula (1) is called the two-dimensional plane passing through v, w, and 0.

Exercise 3   Discuss this last statement geometrically in $\mathbf{R}^3$.

DEFINITION 1.2   *If S is any subset of $\mathbf{R}^n$, then $[S]$ is the set of all vectors of the form*

$$x = \alpha_1 v_1 + \alpha_2 v_2 + \cdots + \alpha_m v_m, \tag{2}$$

where $v_1, \ldots, v_m$ are vectors in S, and $\alpha_1, \ldots, \alpha_m$ are real. $[S]$ is called the span of S, or the subspace generated by S.

THEOREM 1.3   *For any set S, $[S]$ is a subspace, and it is the smallest subspace containing S.*

Proof   If x and y belong to $[S]$, then

$$x = \alpha_1 v_1 + \cdots + \alpha_m v_m \quad \text{and} \quad y = \beta_1 w_1 + \cdots + \beta_k w_k,$$

where each v_i and each w_j belong to S. Then

$$x + y = \alpha_1 v_1 + \cdots + \alpha_m v_m + \beta_1 w_1 + \cdots + \beta_k w_k$$

and

$$\alpha x = \alpha \alpha_1 v_1 + \cdots + \alpha \alpha_m v_m.$$

This shows that $x + y$ and αx belong to $[S]$, and hence that $[S]$ is a subspace. If V is any subspace containing S, then certainly the definition requires that V contain every vector x of the form (2), and hence that V contain $[S]$.

A given subspace $V \neq \{0\}$ is always spanned by many different sets of vectors. For example, a line through the origin is spanned by any nonzero vector on it; a two-dimensional plane through the origin is spanned by any two vectors in it that do not lie on the same line through the origin.

Exercise 4   Suppose that the two vectors do lie on a line through the origin. What do they span?

DEFINITION 1.4   *The dimension of a subspace $V \neq \{0\}$ is the smallest number of vectors that span it. In other words, the dimension of V is m if there exists some set of m vectors that spans V, but no set of $m-1$ vectors spans V. The dimension of $\{0\}$ is 0.*

Another way to state the definition is that the dimension of V is $\leq m$ if there exists a set of m vectors that spans V.

Example   Let $V = \mathbf{R}^n$ and let

$$e_1 = (1, 0, 0, \ldots, 0), \qquad e_2 = (0, 1, 0, \ldots, 0), \qquad e_3 = (0, 0, 1, \ldots, 0),$$

and in general let e_j be the vector with jth coordinate 1 and all others 0. It is immediately seen that if $x = (x_1, \ldots, x_n)$, then

$$x = \sum_{j=1}^{n} x_j e_j. \tag{3}$$

This shows that the vectors $e_1, \ldots, e_n$ span $\mathbf{R}^n$, and hence that the dimension is $\leq n$. It is true that the dimension is equal to n, but this is not so easy to show. It follows from the results of Section 2, which provide a simple condition that m vectors span a space of dimension m and not one of some smaller dimension.

2   BASES

DEFINITION 2.1   *Vectors $v_1, \ldots, v_m$ are linearly dependent if there exist real numbers $\alpha_1, \ldots, \alpha_m$ not all 0 such that*

$$\sum_{j=1}^{m} \alpha_j v_j = 0. \tag{1}$$

The vectors are linearly independent if they are not linearly dependent.

Example 1   Formula (3) in the last section shows that the vectors $e_1, \ldots, e_n$ described there are linearly independent.

Exercise 1   A single vector is linearly dependent if and only if it is 0. Two vectors are linearly dependent if and only if they lie on a line through 0.

The example and the exercise suggest that m vectors are linearly independent if and only if they span a subspace of dimension m. To prove it we shall make use of the following definition and lemma.

DEFINITION 2.2   *A vector x is a linear combination of the vectors $v_1, \ldots, v_m$ if there exist real numbers $\alpha_1, \ldots, \alpha_m$ such that*

$$x = \sum_{j=1}^{m} \alpha_j v_j.$$

In terms of this definition, the span of a set of vectors consists of all linear combinations of these vectors.

LEMMA 2.3

If $v_1, \ldots, v_m$ are linearly dependent, then one of them is a linear combination of preceding ones.

Proof

Let equation (1) hold, and let α_k be the last nonzero α_j. Then

$$v_k = -\sum_{j=1}^{k-1} \frac{\alpha_j}{\alpha_k} v_j.$$

THEOREM 2.4

If $v_1, \ldots, v_k$ are linearly independent and lie in the span of $w_1, \ldots, w_m$, then $k \leq m$.

Proof

Let V be the span of $w_1, \ldots, w_m$, and suppose, to get a contradiction' that $m < k$. Let $S^0 = \{w_1, \ldots, w_m\}$. We shall prove by induction that for each $j = 0, 1, \ldots, m$ there is a set S^j composed of $v_1, \ldots, v_j$ and of $m - j$ of the w's which still spans V. This will indeed be a contradiction, for it will imply that $S^m = \{v_1, \ldots, v_m\}$ spans V, so that

$$v_{m+1} = \sum_{j=1}^{m} \alpha_j v_j,$$

which is impossible if the v's are linearly independent.

Suppose that we have already $S^j = \{v_1, \ldots, v_j, u_{j+1}, \ldots, u_m\}$, where $u_{j+1}, \ldots, u_m$ are certain of the w's. Since S^j spans V, we have

$$v_{j+1} = \sum_{i=1}^{j} \alpha_i v_i + \sum_{i=j+1}^{m} \alpha_i u_i,$$

which implies that the vectors $v_1, \ldots, v_{j+1}, u_{j+1}, \ldots, u_m$ are linearly dependent. According to the lemma, one of them is a linear combination of preceding ones. It cannot be a v that is a linear combination of preceding ones, since the v's are linearly independent. Therefore, some u, say u_l, is a linear combination of $v_1, \ldots, v_{j+1}$ and of the other u's; then $S^{j+1} = S^j \cup \{v_{j+1}\} - \{u_l\}$ still spans V.

Theorem 2.4 makes it easy to show that the dimension of $\mathbf{R}^n$ is n. It has been seen that the vectors $e_1, \ldots, e_n$ are linearly independent and that they span $\mathbf{R}^n$. The fact that they span means (by the definition) that the dimension is $\leq n$, and the fact that they are linearly independent means (by Theorem 2.4) that no fewer number can span. Similar arguments can be used to discuss subspaces in general.

DEFINITION 2.5

A basis of a subspace V is a set of linearly independent vectors that span V.

THEOREM 2.6

Every set of linearly independent vectors in V is part of a basis of V.

Proof

Let $v_1, \ldots, v_k$ in V be linearly independent. If they do not already span V, then some $x \in V$ is not a linear combination of $v_1, \ldots, v_k$, in which case Lemma 2.3 shows that $v_1, \ldots, v_k, x$ are linearly independent. Write $x = v_{k+1}$ and start again. $v_1, \ldots, v_{k+1}$ are linearly independent. If they do not span V, then there exists $v_{k+2} \in V$ so that $v_1, \ldots, v_{k+2}$ are linearly independent. In due course the process must stop, because Theorem 2.4 shows that more than n vectors in $\mathbf{R}^n$ cannot be linearly independent.

COROLLARY 2.7

Every subspace $V \neq \{0\}$ has a basis.

Proof

Use the theorem, starting with any nonzero vector in V.

In order to avoid treating the case $V = \{0\}$ as an exceptional one in every theorem, it is convenient to consider the empty set as a basis of $\{0\}$. This requires occasional special proofs, which will always be left to the reader.

THEOREM 2.8

Let the dimension of V be m. Then every basis of V contains exactly m vectors, and every set of m linearly independent vectors in V is a basis of V.

Remark

Up to the point of Corollary 2.7 it was not clear that the definition of the dimension really made sense. That is, it was not clear that every subspace of $\mathbf{R}^n$ is spanned by a finite number of vectors. The corollary does make this clear.

Exercise 2

Prove Theorem 2.8—on the basis, of course, of Theorem 2.4.

Exercise 3

Show that if $V \subset W$ and $V \neq W$, then dim $V <$ dim W, where dim V is the dimension of V.

THEOREM 2.9

The vectors $v_1, \ldots, v_m$ in V are a basis of V if and only if every vector $x \in V$ can be written in one and only one way in the form

$$x = \sum_{j=1}^{m} \alpha_j v_j, \qquad (2)$$

where $\alpha_1, \ldots, \alpha_m$ are real numbers. The number α_j is called the jth coordinate of x relative to the basis $v_1, \ldots, v_m$.

Proof

The fact that each $x \in V$ can be written in the form (2) is part of the definition—the part that says that $v_1, \ldots, v_m$ span V. The uniqueness

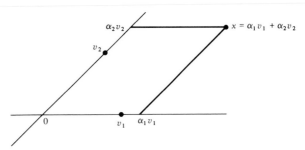

Figure 1

is proved as follows.   If

$$x = \sum_{j=1}^{m} \beta_j v_j,$$

then

$$0 = \sum_{j=1}^{m} (\alpha_j - \beta_j)v_j;$$

then $\alpha_j - \beta_j$ must be 0 because $v_1, \ldots, v_m$ are linearly independent.

Geometrically, we can think of a basis of V as determining a set of coordinate axes in V.   The jth coordinate axis is the line through v_j and the origin.   In two dimensions it looks as shown in Figure 1.   To get the first coordinate of x relative to the basis v_1, v_2, draw the line through x parallel to v_2.   The point where it meets the line through v_1 is a multiple of v_1, say $\alpha_1 v_1$.   The number α_1 is the first coordinate of x.

Exercise 4   In a similar geometric way, describe how to get the coordinates of a vector $x \in \mathbf{R}^3$ relative to a basis v_1, v_2, v_3.

A plane in $\mathbf{R}^n$ is a set that is "parallel" to a subspace.   To describe this properly we shall need a couple of definitions.

DEFINITION
2.10
If X is a subset of $\mathbf{R}^n$ and a is a point, then $X + a$ is the set of all points $x + a$ with $x \in X$.   It is called the translate of X by a. $X - a = X + (-a)$ is the set of all points $x - a$ with $x \in X$.

DEFINITION
2.11
A subset Π of $\mathbf{R}^n$ is a plane of dimension m if for some point a, $\Pi - a$ is a subspace of dimension m.

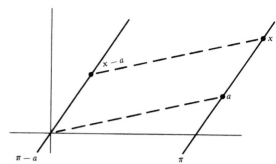

Figure 2

Exercise 5   If Π is a plane, then $\Pi - a$ is a subspace if and only if $a \in \Pi$.   If a and b are any two points of a plane Π, then $\Pi - a = \Pi - b$.

DEFINITION   *If Π is a plane and $a \in \Pi$, the subspace $\Pi - a$ is called the subspace parallel*
2.12   *to Π.*

By Exercise 5, the subspace parallel to a plane Π does not depend on the point a used to define it (Figure 2).

Exercise 6   A line (in the sense of Chapter 8) is a plane of dimension 1, and conversely.

One typical way to determine a plane of dimension m is to use $m + 1$ points on it.   A line is determined by two points, an ordinary plane is determined by three points, and so on.

Example 2   Find the plane determined by the points $a_0, \ldots, a_m$.

Let Π be the plane and let V be the parallel subspace.   Then V is spanned by the points $a_1 - a_0, \ldots, a_m - a_0$, so

$$\Pi - a_0 = V = \left\{ x : x = \sum_{j=1}^{m} t_j(a_j - a_0) \right\}.$$

Hence

$$\Pi = \left\{ x : x = a_0 + \sum_{j=1}^{m} t_j(a_j - a_0) \right\}, \tag{3}$$

or

$$\Pi = \left\{ x : x = \sum_{j=0}^{m} t_j a_j \quad \text{with} \quad \sum_{j=0}^{m} t_j = 1 \right\}. \tag{4}$$

To pass back and forth between (3) and (4) simply note that the coefficient of a_0 in (3) is

$$t_0 = 1 - \sum_{j=1}^{m} t_j.$$

Formula (4) is a little the better of the two because it does not give preference to a_0 over the other points, but formula (3) is sometimes easier to use.

Exercise 7   Find the plane in $\mathbf{R}^3$ determined by the three points $(1, 0, 0)$, $(0, 1, 0)$, and $(0, 0, 1)$.

Of course, $m + 1$ points determine a plane of dimension $< m$ unless the points are independent in a suitable sense. Three colinear points, for instance, determine a line, not a two-dimensional plane.

DEFINITION 2.13   *The points $a_0, \ldots, a_m$ are affinely dependent if there exist numbers $t_0, \ldots, t_m$ not all 0 such that*

$$\sum_{j=0}^{m} t_j a_j = 0 \qquad and \qquad \sum_{j=0}^{m} t_j = 0.$$

THEOREM 2.14   *The following are equivalent:*
     (a) *$a_0, \ldots, a_m$ are affinely independent.*
     (b) *$a_1 - a_0, \ldots, a_m - a_0$ are linearly independent.*
     (c) *$a_0, \ldots, a_m$ determine a plane of dimension m.*
     (d) *The points $(a_0, 1), \ldots, (a_m, 1)$ in $\mathbf{R}^{n+1}$ are linearly independent, where $(a, 1)$ is the point whose first n coordinates are the coordinates of a and whose last coordinate is 1.*

Exercise 8   Prove the theorem.

A second typical way to determine a plane of dimension m in $\mathbf{R}^n$ is to use one point on the plane and $n - m$ directions that are perpendicular to the plane.

Example 3   Find the plane through a given point a and perpendicular to given directions $\theta_1, \ldots, \theta_k$, $k = n - m$.

This means, of course, that the parallel subspace is perpendicular to the given directions. Thus, if Π is the plane and x is a point of Π, then $x - a$ is perpendicular to each θ_j. Therefore, the equations of Π are simply

$$\langle x - a, \theta_j \rangle = 0, \qquad j = 1, \ldots, k. \tag{5}$$

Exercise 9   Find the plane in **R**^{3} that passes through the point $(1, -2, 3)$ and is perpendicular to the direction of $(-1, -8, 6)$.

Exercise 10   Use Theorem 3.7 below to show that if $\theta_1, \ldots, \theta_k$ are linearly independent, then the equations (5) define a plane of dimension $n - k$.

3   ORTHONORMAL BASES

In general the coordinate axes determined by an arbitrary basis are not perpendicular to one another. In problems that involve only addition and scalar multiplication, this is usually immaterial. One basis is as good as another. But in problems involving the distance or the inner product, the calculations are usually much simpler if the coordinate axes are perpendicular.

DEFINITION 3.1   *A set of vectors is orthonormal if each vector has length* 1 *and any two are perpendicular to one another.   In other words,* $e_1, \ldots, e_m$ *is orthonormal if*

$$\langle e_i, e_j \rangle = \begin{cases} 1 & \text{if } i = j, \\ 0 & \text{if } i \neq j. \end{cases}$$

Exercise 1   Show that the usual basis of **R**^{n}, that is, $e_1 = (1, 0, \ldots, 0)$, $e_2 = (0, 1, 0, \ldots, 0)$, $\ldots$, is orthonormal.

Remark   A vector of length 1 is sometimes said to be "normalized."   The word orthonormal is a combination of orthogonal and normalized.

THEOREM 3.2   *Let* $e_1, \ldots, e_m$ *be orthonormal, and let*

$$x = \sum_{j=1}^{m} \alpha_j e_j. \tag{1}$$

Then

$$\alpha_k = \langle x, e_k \rangle. \tag{2}$$

Proof   Take the inner product on both sides of (1) with e_k.   Then $\langle x, e_k \rangle$ appears on the left and α_k appears on the right (since $\langle e_j, e_k \rangle = 0$ for all $j \neq k$, while $\langle e_k, e_k \rangle = 1$).

COROLLARY 3.3   *An orthonormal set is linearly independent.*

Proof   If $x = 0$ in formula (1), then formula (2) shows that each α_k is 0.

Example   Rotate the coordinate axes in $\mathbf{R}^2$ through an angle θ.   Find the formula relating the new coordinates of a point x to the old ones.

If e_1, e_2 is the original orthonormal basis and e_1', e_2' is the new one, then

$$e_1 = (1, 0), \quad e_2 = (0, 1), \quad e_1' = (\cos\theta, \sin\theta), \quad e_2' = (-\sin\theta, \cos\theta).$$

If (x_1, x_2) are the original coordinates of x and (x_1', x_2') are the new ones, then

$$x = x_1 e_1 + x_2 e_2 \quad \text{and} \quad x = x_1' e_1' + x_2' e_2'.$$

According to Theorem 3.2, we have

$$\begin{aligned} x_1' &= \langle x, e_1' \rangle = x_1 \langle e_1, e_1' \rangle + x_2 \langle e_2, e_1' \rangle = x_1 \cos\theta + x_2 \sin\theta, \\ x_2' &= \langle x, e_2' \rangle = x_1 \langle e_1, e_2' \rangle + x_2 \langle e_2, e_2' \rangle = -x_1 \sin\theta + x_2 \cos\theta. \end{aligned} \tag{3}$$

The calculations are the same for rotations of the three-dimensional space and, in general, for passing from one orthonormal basis of $\mathbf{R}^n$ to another.   If

$$x = \sum_{j=1}^{n} x_j e_j \quad \text{and} \quad x = \sum_{j=1}^{n} x_j' e_j',$$

then

$$x_k' = \langle x, e_k' \rangle = \sum_{j=1}^{n} x_j \langle e_j, e_k' \rangle,$$

so what has to be calculated is each of the inner products $\langle e_j, e_k' \rangle$.

Exercise 2   The number $\langle e_j, e_k' \rangle$ is the cosine of the angle between the x_j and x_k' axes.

THEOREM 3.4   *Let V be a subspace with orthonormal basis $e_1, \ldots, e_m$.   Each vector $x \in \mathbf{R}^n$ can be written uniquely as*

$$x = x' + x'', \tag{4}$$

where x' is in V and x'' is perpendicular to V.   In fact,

$$x' = \sum_{j=1}^{m} \langle x, e_j \rangle e_j. \tag{5}$$

Exercise 3   Prove the uniqueness part of the theorem.   Show that x' is closer to x than any other point of V, and give a geometric statement of the theorem.

Proof   If we define x' by formula (5), then certainly $x' \in V$, and what we have to show is that $x'' = x - x'$ is perpendicular to V (i.e., is perpendicular to every vector in V).   Theorem 3.2 shows that $\langle x', e_k \rangle = \langle x, e_k \rangle$ for each k, and hence that $\langle x'', e_k \rangle = 0$ for each k.   If y is any vector in V, then y is a linear combination of the e_k's, say

$$y = \sum_{j=1}^{m} \alpha_k e_k,$$

and then

$$\langle x'', y \rangle = \sum_{j=1}^{m} \alpha_k \langle x'', e_k \rangle = 0.$$

THEOREM 3.5   *Every subspace has an orthonormal basis.   Indeed, any orthonormal subset in V is part of an orthonormal basis of V.*

Proof   Let $e_1, \ldots, e_k$ be orthonormal and lie in V, and let W be the span of $e_1, \ldots, e_k$.   It is enough to show that if $W \neq V$, then there exists $e_{k+1} \in V$ such that $e_1, \ldots, e_{k+1}$ is still orthonormal.

Exercise 4   Prove that this is enough.

If $W \neq V$, take any vector x that is in V but not in W, and apply Theorem 3.4 to write $x = x' + x''$, where x' is in W and x'' is perpendicular to W. Now, x'' is in V, because both x and x' are in V; and $x'' \neq 0$, because x is not in W.   Hence, we can take $e_{k+1} = x''/|x''|$.

Remark   What we have shown is that every orthonormal subset of V is part of an orthonormal basis.   To finish the first part of the theorem, choose $x \neq 0$ in V and start with the orthonormal set $e_1 = x/|x|$.   As usual, the case $V = \{0\}$ is exceptional.   To include it in the theorem we must do what?

DEFINITION 3.6   *If A is any subset of $\mathbf{R}^n$, its orthogonal complement, written $A^\perp$, is the set of all vectors $y \in \mathbf{R}^n$ that are perpendicular to every vector in A.*

Exercise 5   For any subsets A and B we have
(a) If $A \subset B$, then $A^\perp \supset B^\perp$.
(b) $A^\perp$ is a subspace (whether A is or not).
(c) $[A]^\perp = A^\perp$.

Notice that part (c) was already used in the proof of Theorem 3.4 in the case where $A = \{e_1, \ldots, e_m\}$, and $V = [A]$.

THEOREM 3.7   *For any subspace V of $\mathbf{R}^n$, $\dim V + \dim V^\perp = n$.*

Proof   The abbreviation "dim" stands for the dimension.   Use Theorem 3.5 to find an orthonormal basis $e_1, \ldots, e_m$ of V, and then use the theorem again to find $e_{m+1}, \ldots, e_n$ so that $e_1, \ldots, e_n$ is an orthonormal basis of $\mathbf{R}^n$.   What we shall show is that $e_{m+1}, \ldots, e_n$ is then an orthonormal basis of $V^\perp$, which will imply that dim $V = n - m$, as required.

First of all, $e_{m+1}, \ldots, e_n$ do lie in $V^\perp$ by virtue of part (c) of Exercise 5.   On the other hand, for any $x \in \mathbf{R}^n$ we have

$$x = \sum_{j=1}^{n} \langle x, e_j \rangle e_j.$$

If x lies in $V^\perp$, then the first m terms are 0.   Therefore, $e_{m+1}, \ldots, e_n$ span $V^\perp$, and since they are linearly independent (being orthonormal), they do form a basis.

THEOREM 3.8   *For any subspace V, $(V^\perp)^\perp = V$.*

Proof   It is plain from the definition that $V \subset (V^\perp)^\perp$.   On the other hand, the dimensions of the two are the same.   So by Exercise 3 of Section 2 the two subspaces must be the same.

Exercise 6   For any subset A, $(A^\perp)^\perp = [A]$.

Theorem 3.8 gives another way to look at subspaces.   If V has dimension m, let $e_{m+1}, \ldots, e_n$ be a basis of $V^\perp$ (orthonormal if you like, but it does not make any difference).   Let

$$V_j = \{e_j\}^\perp = f_j^{-1}(0), \qquad \text{where } f_j(x) = \langle x, e_j \rangle. \tag{6}$$

The theorem says that

$$V = V_{m+1} \cap \cdots \cap V_n. \tag{7}$$

There are two interesting consequences.

Exercise 7   If e is a nonzero vector in $\mathbf{R}^n$, then $\{e\}^\perp$ has dimension $n - 1$.

Combining the exercise with formula (7), we get the first interesting consequence.

THEOREM 3.9   *Every subspace of dimension m in $\mathbf{R}^n$ is the intersection of $n - m$ subspaces of dimension $n - 1$.*

In the three-dimensional space, for instance, every line is the intersection of two planes. Of course, the two planes can be chosen in infinitely many different ways.

The second interesting consequence is the following one.

THEOREM 3.10

Every subspace of $\mathbf{R}^n$ *is a closed set.*

Proof

Because of formula (7) it is sufficient to prove that each V_j is a closed set. (Why?) For this, it is sufficient to prove that the function f_j defined in formula (6) is continuous. (Why?) Now, if

$$f(x) = \langle e, x \rangle,$$

then by Cauchy–Schwarz we have

$$|f(x) - f(y)| = |\langle e, x - y \rangle| \le |e|\,|x - y|,$$

which certainly does show that f is continuous. (Why?)

When we come to define the tangent space to a surface at a point, it will be natural sometimes to define it as the span of a certain set of vectors (as in the initial discussion of subspaces), and other times to define it as the orthogonal complement of a certain set of vectors (as in Theorems 3.8 and 3.9).

We shall finish up this section by describing an orthonormal basis in an infinite-dimensional space. The space is the space $\mathcal{C}([-\pi, \pi])$ of continuous real-valued functions on the interval $[-\pi, \pi]$ with the inner product

$$\langle f, g \rangle = \int_{-\pi}^{\pi} f(x)g(x)\,dx, \tag{8}$$

and the corresponding absolute value and distance

$$|f| = \sqrt{\langle f, f \rangle}, \qquad d(f, g) = |f - g|. \tag{9}$$

Since all questions of convergence refer to this absolute value rather than the usual one ($\|f\| = \sup\{|f(x)| : x \in [-\pi, \pi]\}$) on a space of continuous functions, we shall call the resulting metric space $\mathcal{H}$, rather than $\mathcal{C}([-\pi, \pi])$.

Exercise 8   Define $e_0 = 1/\sqrt{2\pi}$ and

$$e_n = \frac{1}{\sqrt{\pi}} \cos \frac{n}{2} x \text{ for } n \text{ even}, \qquad e_n = \frac{1}{\sqrt{\pi}} \sin \frac{n+1}{2} x \text{ for } n \text{ odd},$$

$$n = 1, 2, \ldots.$$

Show that the e_n are orthonormal in $\mathcal{H}$.

What we want to do is to show that the e_n form an orthonormal basis of $\mathcal{H}$ in a suitable sense.

THEOREM 3.11

For each $f \in \mathcal{H}$ we have

$$f = \sum_{n=0}^{\infty} \alpha_n e_n, \qquad \text{where } \alpha_n = \langle f, e_n \rangle, \tag{10}$$

and the series converges in the space $\mathcal{H}$.

The series (10) is called the *Fourier series of the function f.* The coefficients are called the *Fourier coefficients.* Note that they are just the coefficients that are given by Theorem 3.2. It is not claimed that the series converges uniformly, or even pointwise, but rather that the partial sums converge in the metric space $\mathcal{H}$.

Exercise 9 Let V_m be the space spanned by $e_0, \ldots, e_m$, and let s_m be the mth partial sum of the series (10). Use Theorem 3.4 and Exercise 3 to show that $f - s_m$ is orthogonal to V_m and that s_m is the closest point of V_m to f.

Proof of the Theorem

With the aid of Exercise 8 and Theorem 12.7 of Chapter 7, the proof of the theorem is immediate. The latter says that every $f \in \mathcal{H}$ can be approximated in $\mathcal{H}$ by trigonometric polynomials, and this means precisely that

$$d(f, V_m) \to 0.$$

On the other hand, the fact that s_m is the closest point of V_m to f means that

$$d(f, V_m) = |f - s_m|.$$

Consequently, $s_m \to f$, and so the series converges to f.

The inner product and absolute value are expressed in terms of coordinates relative to this orthonormal basis in just the usual way.

THEOREM 3.12

If f and g are in $\mathcal{H}$ with Fourier coefficients $\{a_n\}$ and $\{b_n\}$, then

$$\langle f, g \rangle = \sum_{n=0}^{\infty} a_n b_n, \qquad |f|^2 = \sum_{n=0}^{\infty} a_n^2, \tag{11}$$

and the series converge absolutely.

Exercise 10   If s_m and t_m are the partial sums, then

$$\langle s_m, t_m \rangle = \sum_{n=0}^{m} a_n b_n, \qquad |s_m|^2 = \sum_{n=0}^{m} a_n^2.$$

Remark   Things like this, and also Exercise 9, can be proved either by repeating the proofs we have already had or else by noticing that V_m is a finite-dimensional space; so it is the same as R^{m+1} and there is after all nothing really to prove.

Exercise 11   If $s_m \to f$ and $t_m \to g$, then $\langle s_m, t_m \rangle \to \langle f, g \rangle$.   Deduce from this and Exercise 10 that formula (11) holds.

Exercise 12   What remains of the proof of the theorem is to show that the series (11) converge absolutely.   For the second one there is no problem, since each term is nonnegative.   Show that the first converges absolutely by using the Cauchy–Schwarz inequality to prove that

$$\sum_{n=0}^{m} |a_n b_n| \le |f|\,|g|.$$

Questions about pointwise convergence of Fourier series are extremely difficult.   For a great many years, in fact, one of the celebrated unsolved problems of analysis was to decide whether the Fourier series of a continuous function must converge at at least one point.   A couple of years ago the Swedish mathematician, L. Carleson, succeeded in showing that it must, and, in fact, that it must converge at "almost all" points, the "almost all" being a technical term that is explained in Chapter 13.

Exercise 13   Calculate the Fourier series for the functions x, $|x|$, and e^x.   Discuss the pointwise convergence as far as you can.

4   LINEAR TRANSFORMATIONS

The simplest functions from $\mathbf{R}^m$ to $\mathbf{R}^n$ are the linear functions (or linear transformations, or linear operators—the three terms mean the same thing).

DEFINITION 4.1   *A linear transformation from* $\mathbf{R}^m$ *to* $\mathbf{R}^n$ *is a function* $T: \mathbf{R}^m \to \mathbf{R}^n$ *such that*

$$T(x + y) = T(x) + T(y), \qquad T(\alpha x) = \alpha T(x), \tag{1}$$

for all x and y in $\mathbf{R}^m$ *and all real* α.

THEOREM 4.2

Let $e_1, \ldots, e_m$ be a basis of $\mathbf{R}^m$, and let $v_1, \ldots, v_m$ be arbitrary vectors in $\mathbf{R}^n$.   There is one and only one linear transformation $T:\mathbf{R}^m \to \mathbf{R}^n$ such that

$$T(e_j) = v_j \quad \text{for } j = 1, \ldots, m.$$

Proof

Formula (1) implies that

$$\text{If } x = \sum_{j=1}^{m} x_j e_j, \quad \text{then } T(x) = \sum_{j=1}^{m} x_j T(e_j), \tag{2}$$

from which it is plain how to define T:

$$\text{If } x = \sum_{j=1}^{m} x_j e_j, \quad \text{then } T(x) = \sum_{j=1}^{m} x_j v_j. \tag{3}$$

The definition makes sense because every vector x in $\mathbf{R}^m$ can be written in one and only one way in the form indicated (Theorem 2.9), and (2) shows that (3) is the only possible definition.   What remains is to show that T is linear, that is, that condition (1) holds.

Exercise 1

Show that the function T defined by (3) is linear.

Once bases are fixed in both $\mathbf{R}^m$ and $\mathbf{R}^n$ the linear transformations can be represented in a concrete way.   Let $e_1, \ldots, e_m$ and $f_1, \ldots, f_n$ be bases in $\mathbf{R}^m$ and $\mathbf{R}^n$, and let T be a linear transformation from $\mathbf{R}^m$ to $\mathbf{R}^n$.   If α_{ij} is the ith coordinate of $T(e_j)$ relative to the basis $f_1, \ldots, f_n$, then (by the definition of coordinates relative to a basis)

$$T(e_j) = \sum_{i=1}^{n} \alpha_{ij} f_i \quad \text{for } j = 1, \ldots, m. \tag{4}$$

On the other hand, if the numbers α_{ij} are given, then formula (4) determines a unique linear transformation T.   Indeed, call the right-hand side v_j and apply Theorem 4.2.

DEFINITION 4.3

The double sequence $\{\alpha_{ij}\}$ is called the matrix of the linear transformation T relative to the bases $e_1, \ldots, e_m$ and $f_1, \ldots, f_n$.

What has been shown is that there is a one-to-one correspondence between linear transformations and matrices *once the bases are fixed.*

Ordinarily, the numbers in a matrix are displayed in a rectangular array with α_{ij} in the ith row, jth column.

$$\begin{pmatrix} \alpha_{11} & \alpha_{12} & \alpha_{13} & \cdot \ \cdot \ \cdot & \alpha_{1m} \\ \alpha_{21} & \alpha_{22} & \alpha_{23} & \cdot \ \cdot \ \cdot & \alpha_{2m} \\ \cdot & \cdot & \cdot & & \cdot \\ \cdot & \cdot & \cdot & & \cdot \\ \cdot & \cdot & \cdot & & \cdot \\ \alpha_{n1} & \alpha_{n2} & \alpha_{n3} & \cdot \ \cdot \ \cdot & \alpha_{nm} \end{pmatrix}$$

The way to remember it is that the coordinates of $T(e_j)$ go down the jth column.

The relation between a linear transformation and its matrix can be expressed in another way.

THEOREM
4.4

Let $\{\alpha_{ij}\}$ be the matrix of a linear transformation T relative to bases e_1, $\ldots$, e_m and $f_1, \ldots, f_n$. Let

$$x = \sum_{j=1}^{m} x_j e_j \quad and \quad y = \sum_{i=1}^{n} y_i f_i.$$

Then $y = T(x)$ if and only if

$$y_i = \sum_{j=1}^{m} \alpha_{ij} x_j \quad for \ i = 1, \ldots, n. \tag{5}$$

Proof

Formulas (2) and (4) give

$$T(x) = \sum_{j=1}^{m} x_j T(e_j) = \sum_{j=1}^{m} \sum_{i=1}^{n} x_j \alpha_{ij} f_i.$$

The coordinate that multiplies f_i is the number y_i in (5).

Note that formula (4) expresses the relation between a linear transformation and its matrix by showing how the basis vectors transform, while formula (5) expresses the relation by showing how coordinates transform. The equations (5) are called a system of n *linear equations in m unknowns*. The theorem shows that the study of linear transformations is equivalent to the study of systems of linear equations. Usually, a given problem can be studied from either point of view. Sometimes the one is convenient, sometimes the other.

There is still another way to express the relation between a linear transformation and its matrix that works when the bases are orthonormal. Take the inner product with f_k on both sides of formula (4). If $f_1, \ldots, f_n$ is orthonormal, then all terms but one drop out:

$$\langle T(e_j), f_k \rangle = \sum_{i=1}^{n} \alpha_{ij} \langle f_i, f_k \rangle = \alpha_{kj}.$$

When T is a linear transformation, it is customary to write Tx instead of $T(x)$ as long as no confusion is likely.   In this notation the result above is as follows:

**THEOREM
4.5**

Let $\{\alpha_{ij}\}$ be the matrix of T relative to bases $e_1, \ldots, e_m$ and $f_1, \ldots, f_n$. If $f_1, \ldots, f_n$ is orthonormal, then

$$\alpha_{ij} = \langle T e_j, f_i \rangle. \tag{6}$$

Exercise 2   If $a \in \mathbf{R}^m$, then the function

$$Tx = \langle x, a \rangle \tag{7}$$

is linear from $\mathbf{R}^m$ to $\mathbf{R}^1$.   Conversely, every linear transformation from $\mathbf{R}^m$ to $\mathbf{R}^1$ has this form and a is unique.

Exercise 3   Let V be a subspace of $\mathbf{R}^n$.   For each $x \in \mathbf{R}^n$, let Px be the closest point in V to x.   (See Theorem 3.4 and Exercise 3 which follows it.)   Show that P is a linear transformation.   Show how to choose a basis of $\mathbf{R}^n$ so that the matrix of P looks as shown in Figure 3, where everything is 0 in the areas marked by big zeros.   How many 1's are there?   (Note that since $P : \mathbf{R}^n \to \mathbf{R}^n$, there is only one basis to choose, which plays the role of both the e's and the f's.) P is called the projection on the subspace V.

Exercise 4   Define the graph of a function $f : \mathbf{R}^m \to \mathbf{R}^n$.   (It is a subset of $\mathbf{R}^{n+n}$.)

Exercise 5   Show that a function $f : \mathbf{R}^m \to \mathbf{R}^n$ is linear if and only if its graph is a subspace of $\mathbf{R}^{m+n}$.

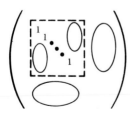

Figure 3

5   SUMS AND PRODUCTS

If f and g are functions from a set X into $\mathbf{R}^n$, there is a natural way to define the sum $f + g$ and scalar product αf:

$$(f + g)(x) = f(x) + g(x), \qquad (\alpha f)(x) = \alpha f(x). \tag{1}$$

THEOREM
5.1

If S and T are linear transformations from $\mathbf{R}^m$ to $\mathbf{R}^n$, then $S + T$ and αS are linear. If S and T have matrices $\{s_{ij}\}$ and $\{t_{ij}\}$ relative to given bases, then $S + T$ and αS have the matrices $\{u_{ij}\}$ and $\{v_{ij}\}$ defined by

$$u_{ij} = s_{ij} + t_{ij}, \qquad v_{ij} = \alpha s_{ij}. \tag{2}$$

Exercise 1   Prove the theorem.

Formula (2) is used to define the sum of two matrices without any reference to linear transformations. If M and N are the matrices $\{s_{ij}\}$ and $\{t_{ij}\}$, then $M + N$ is the matrix $\{u_{ij}\}$ defined in (2), and αM is the matrix $\{v_{ij}\}$. This definition makes no reference to linear transformations, and it is rather natural; but the reason for making it is Theorem 5.1, which says that the matrix of a sum $S + T$ is then the sum of the matrices.

The composite of two linear transformations S and T (which makes sense if $\mathbf{R}^l \overset{S}{\to} \mathbf{R}^m \overset{T}{\to} \mathbf{R}^n$) is usually called the *product* and is written TS rather than $T \circ S$.

Exercise 2   If S and T are linear, then TS is linear (when the spaces are right so that it makes sense).

The reason for calling the composite a product is simply that it acts like a product in many respects.

Exercise 3 $R(S + T) = RS + RT$, $(S + T)R = SR + TR$, and $(RS)T = R(ST)$, whenever the formulas make sense. For example, the first formula makes sense when $S, T:\mathbf{R}^l \to \mathbf{R}^m$ and $R:\mathbf{R}^m \to \mathbf{R}^n$. Discuss when the others make sense.

Exercise 4   The linear transformation $I:\mathbf{R}^n \to \mathbf{R}^n$ defined by $Ix = x$ for each x is called the *identity*. It acts like the number 1 in the sense that $SI = S$ and $IT = T$ if the products are defined.

Exercise 5   Suppose that $T:\mathbf{R}^m \to \mathbf{R}^n$ is one to one and onto so that as a function it has an inverse $T^{-1}:\mathbf{R}^n \to \mathbf{R}^m$. Show that if T is linear, then T^{-1} is linear and acts like the reciprocal of T in the sense that $TT^{-1} = I$ and $T^{-1}T = I$. Are the two I's the same?   (This is an unfair question.   In the next section we shall show that m must equal n, so in fact they are.   However, all these considerations apply equally well to linear transformations from one subspace V to another one W, in which case the I's do not have to be the same.)

In some respects the product of linear transformations is quite unlike the product of numbers.

Exercise 6   Give an example of S, $T:\mathbf{R}^2 \rightarrow \mathbf{R}^2$ such that $ST = 0$, while $TS \neq 0$.   This shows that on the one hand the product of two nonzero transformations can be zero, and on the other that the product is not the same when the order is reversed.

Let us calculate the matrix of a product.   Choose bases $d_1, \ldots, d_l$ of $\mathbf{R}^l$, $e_1, \ldots, e_m$ of $\mathbf{R}^m$, and $f_1, \ldots, f_n$ of $\mathbf{R}^n$, and let S and T have matrices $\{s_{ij}\}$ and $\{t_{ij}\}$ relative to these bases.   Then

$$Sd_j = \sum_k s_{kj}e_k \qquad \text{and} \qquad Te_k = \sum_i t_{ik}f_i,$$

so

$$TSd_j = \sum_{i,k} s_{kj}t_{ik}f_i = \sum_i \left\{ \sum_k t_{ik}s_{kj} \right\} f_i,$$

which gives Theorem 5.2:

THEOREM 5.2   *If S and T have matrices $\{s_{ij}\}$ and $\{t_{ij}\}$ relative to given bases in the three spaces, then TS has the matrix $\{u_{ij}\}$ defined by*

$$u_{ij} = \sum_k t_{ik}s_{kj}. \tag{3}$$

As in the case of sums, formula (3) is used to define the product of two matrices.   If $M = \{s_{ij}\}$ and $N = \{t_{ij}\}$, then $NM = \{u_{ij}\}$, where u_{ij} is given by (3).   The reason for the definition is to make the matrix of a product equal to the product of the matrices.

Exercise 7   What are the conditions on two matrices M and N in order that the product NM make sense?

In assigning a matrix to a linear transformation $T:\mathbf{R}^m \rightarrow \mathbf{R}^n$, we have to choose two bases, one in $\mathbf{R}^m$ and the other in $\mathbf{R}^n$.   When $m = n$, that is, when T is a linear transformation from $\mathbf{R}^n$ to $\mathbf{R}^n$, we often (though not always) choose just one basis and let it play the role of both.   Thus, when we say that $M = \{m_{ij}\}$

is the matrix of T relative to the basis $e_1, \ldots, e_n$, we mean that

$$Te_i = \sum_{j=1}^{n} m_{ji}e_j.$$

Exercise 8   The matrix of the identity $I:\mathbf{R}^n \to \mathbf{R}^n$ (relative to any one basis) is the matrix with 1's along the diagonal and 0's everywhere else. It also is called I. For any matrices M and N, we have $MI = M$ and $IN = N$ if the products make sense.

A square matrix M is invertible if there is a matrix M^{-1} such that $MM^{-1} = I$ and $M^{-1}M = I$.

Exercise 9   A square matrix M is invertible if and only if the corresponding linear transformation (relative to any basis) is invertible.   Why does M have to be square?

Exercise 10   Two square matrices M and N are the matrices of the same linear transformation $T:\mathbf{R}^n \to \mathbf{R}^n$ (M relative to one basis, N relative to some other basis) if and only if there is an invertible matrix U such that $M = UNU^{-1}$.

6   NULL SPACE AND RANGE

DEFINITION 6.1   *The null space of a linear transformation* $T:\mathbf{R}^m \to \mathbf{R}^n$ *is the set*
$$N_T = T^{-1}(0) = \{x \in \mathbf{R}^m : Tx = 0\}.$$

The range is the set
$$R_T = T(\mathbf{R}^m) = \{y \in \mathbf{R}^n : y = Tx \text{ for some } x \in \mathbf{R}^m\}.$$

Exercise 1   Prove the following theorem.

THEOREM 6.2   *The null space of a linear* $T:\mathbf{R}^m \to \mathbf{R}^n$ *is a subspace of* $\mathbf{R}^m$; *the range is a subspace of* $\mathbf{R}^n$.   *T is one to one if and only if* $N_T = \{0\}$.

Here is a basic dimension formula.

THEOREM 6.3   *Let T be a linear transformation from $\mathbf{R}^m$ to $\mathbf{R}^n$.   Then*
$$\dim N_T + \dim R_T = m. \qquad (1)$$

Proof   Choose a basis $e_1, \ldots, e_k$ of N_T, and then choose $e_{k+1}, \ldots, e_m$ so that $e_1, \ldots, e_m$ is a basis of $\mathbf{R}^m$ (Theorem 2.6).   Then the dimension of N_T is

k, so what must be proved is that the dimension of R_T is $m - k$. For this it is sufficient to prove that $Te_{k+1}, \ldots, Te_m$ is a basis of R_T.

If $y \in R_T$, then $y = Tx$ for some

$$x = \sum_{j=1}^{m} x_j\, e_j,$$

and then

$$y = Tx = \sum_{j=k+1}^{m} x_j\, Te_j.$$

(The first k terms drop out because the first k e's are in the null space.) This shows that every $y \in R_T$ is a linear combination of $Te_{k+1}, \ldots, Te_m$— hence that they span R_T.

It remains to show that they are linearly independent. If

$$\sum_{j=k+1}^{m} \alpha_j Te_j = 0,$$

then $\Sigma^{m}_{j=k+1}\, \alpha_j e_j$ is in the null space; therefore,

$$\sum_{j=k+1}^{m} \alpha_j e_j = \sum_{j=1}^{k} \alpha_j e_j. \tag{2}$$

But this is impossible unless all α_j are 0, because the e's are linearly independent.

Exercise 2   What can you say about the matrix of T relative to the basis used above for $\mathbf{R}^m$ and any basis for $\mathbf{R}^n$?

COROLLARY 6.4   *If T is one to one, then $m \leq n$.   If T is onto, then $n \leq m$.   Hence, if T is both one to one and onto, then $m = n$.*

Proof   If T is one to one, then dim $N_T = 0$, so $m = \dim R_T \leq n$. If T is onto, then $n = \dim R_T \leq m$.

THEOREM 6.5   *Let T be a linear transformation from $\mathbf{R}^n$ to $\mathbf{R}^n$ (same dimension).   Then T is one to one if and only if it is onto.*

Proof   Now the m and n are equal, so if T is one to one, then formula (1) shows that dim $R_T = m = n$. But a proper subspace of $\mathbf{R}^n$ cannot have the same dimension (Exercise 3, Section 2). If T is onto, then formula (1) shows that dim $N_T = 0$, and then Theorem 6.2 shows that T is one to one.

With each linear transformation $T: \mathbf{R}^m \to \mathbf{R}^n$ is associated a linear transformation $T^*: \mathbf{R}^n \to \mathbf{R}^m$ called the *adjoint*, which provides important information about T itself.

DEFINITION 6.6

The adjoint of a linear $T: \mathbf{R}^m \to \mathbf{R}^n$ is the linear transformation $T^*: \mathbf{R}^n \to \mathbf{R}^m$ defined by the equation

$$\langle Tx, y \rangle = \langle x, T^*y \rangle \quad \text{for all } x \in \mathbf{R}^m \text{ and } y \in \mathbf{R}^m. \tag{3}$$

It is not at all obvious that the definition makes sense, so some remarks are called for. Let y be a fixed vector in $\mathbf{R}^n$ and consider the function $L(x) = \langle Tx, y \rangle$. It is plain that this is a linear function from $\mathbf{R}^m$ to $\mathbf{R}^1$, so by Exercise 3 of Section 5 there is a unique point $a \in \mathbf{R}^m$ such that $L(x) = \langle x, a \rangle$. We define T^*y to be this point a. Then equation (3) is satisfied, and what remains is to show that the function T^* is linear. We have

$$\langle x, T^*(y + z) \rangle = \langle Tx, y + z \rangle = \langle Tx, y \rangle + \langle Tx, z \rangle = \langle x, T^*y \rangle + \langle x, T^*z \rangle$$
$$= \langle x, T^*y + T^*z \rangle.$$

It follows that $T^*(y + z) = T^*y + T^*z$, for two distinct vectors cannot have the same inner product with every vector x. (Their difference would be perpendicular to itself!) The fact that $T^*(\alpha x) = \alpha T^*x$ is proved similarly.

Exercise 3   Do it.

In terms of matrices, the matrix of T^* is obtained from the matrix of T by interchanging rows and columns, provided the bases are orthonormal.

THEOREM 6.7

Let T be a linear transformation from $\mathbf{R}^m$ to $\mathbf{R}^n$. Let T and T^* have matrices $\{\alpha_{ij}\}$ and $\{\alpha_{ij}^*\}$ relative to orthonormal bases $e_1, \ldots, e_m$ and $f_1, \ldots, f_n$. Then

$$\alpha_{ji}^* = \alpha_{ij}. \tag{4}$$

Proof

According to Theorem 4.5, we have

$$\alpha_{ji}^* = \langle T^*f_i, e_j \rangle = \langle e_j, T^*f_i \rangle = \langle Te_j, f_i \rangle = \alpha_{ij}.$$

Exercise 4   The theorem provides another way to define the adjoint. If T has matrix $\{\alpha_{ij}\}$ relative to orthonormal bases $e_1, \ldots, e_m$ and $f_1, \ldots, f_n$, let T^* be the linear transformation from $\mathbf{R}^n$ to $\mathbf{R}^m$ with matrix $\{\alpha_{ij}^*\}$ relative to the same bases, where $\alpha_{ji}^* = \alpha_{ij}$. Show that if the definition is made this way, then T^* satisfies equation (3).

Exercise 5

$$(T^*)^* = T.$$

THEOREM 6.8

$$N_{T^*} = (R_T)^\perp \quad and \quad R_{T^*} = (N_T)^\perp.$$

Proof

Suppose that $y \in N_{T^*}$. Then

$$0 = \langle x, T^*y \rangle = \langle Tx, y \rangle \quad \text{for every } x \in \mathbf{R}^m,$$

which says exactly that y is perpendicular to the range of T. On the other hand, if y is perpendicular to the range of T, then

$$0 = \langle Tx, y \rangle = \langle x, T^*y \rangle \quad \text{for every } x \in \mathbf{R}^m.$$

This says that T^*y is perpendicular to every $x \in \mathbf{R}^m$, and hence that $T^*y = 0$.

This establishes the first formula in the theorem. If we take orthogonal complements and use Theorem 3.8, we get

$$R_T = (N_{T^*})^\perp. \tag{5}$$

Replacement of T by T^* and the use of Exercise 5 give the second formula in the theorem. Note that taking orthogonal complements in the second formula in the theorem gives

$$N_T = (R_{T^*})^\perp. \tag{6}$$

THEOREM 6.9

$$\dim R_T = \dim R_{T^*}.$$

Proof

According to Theorems 6.8 and 3.7, we have

$$\dim N_T + \dim R_{T^*} = m.$$

Comparing this with Theorem 6.3, we get the theorem.

THEOREM 6.10

If T is a linear transformation from $\mathbf{R}^n$ to $\mathbf{R}^n$ (same dimension), then the following are equivalent:
    (a) *T is one to one.*
    (b) *T is onto.*
    (c) *T^* is one to one.*
    (d) *T^* is onto.*

Proof

Theorem 6.5 shows that (a) and (b) are equivalent, and also that (c) and (d) are equivalent. Theorem 6.9 shows that (b) and (d) are equivalent.   [Or, if you like, Theorem 6.8 shows that (a) and (d) are equivalent and that (b) and (c) are equivalent.]

DEFINITION
6.11

The rank of a linear transformation is the dimension of its range.

7   MATRICES AND LINEAR EQUATIONS

The theorems of the last section have some intriguing applications to matrices and to the theory of systems of linear equations.   Let M be the matrix $\{\alpha_{ij}\}$; that is,

$$M = \begin{pmatrix} \alpha_{11} & \alpha_{12} & \cdots & \alpha_{1m} \\ \alpha_{21} & \alpha_{22} & \cdots & \alpha_{2m} \\ \cdot & \cdot & & \cdot \\ \cdot & \cdot & & \cdot \\ \cdot & \cdot & & \cdot \\ \alpha_{n1} & \alpha_{n2} & \cdots & \alpha_{nm} \end{pmatrix}.$$

Each of the columns can be considered as a vector in $\mathbf{R}^n$.   The column rank of the matrix is the maximum number of linearly independent columns, or, what is the same, the dimension of the subspace that they span.   Similarly, the rows can be considered as vectors in $\mathbf{R}^m$, and the row rank is the dimension of the subspace that they span.   The following theorem is quite surprising.

THEOREM
7.1

The row rank of any matrix is equal to its column rank.   This number is called the rank of the matrix.

Proof

Let $e_1, \ldots, e_m$ and $f_1, \ldots, f_n$ be the natural bases of $\mathbf{R}^m$ and $\mathbf{R}^n$ [that is, $e_1 = (1, 0, \ldots, 0)$, etc.], and let T be the linear transformation with matrix M relative to these bases.   Then the jth column of M is precisely Te_j, so the column rank is the dimension of the range of T. Since the matrix of T^* is obtained by interchanging rows and columns, the row rank of M is the dimension of the range of T^*—and Theorem 6.9 says that the two are equal.

Exercise 1   If M is the matrix of a linear transformation T (relative to any bases), then rank M = rank T.

Exercise 2   If $M = C^{-1}NC$, where M and N are matrices and C is an invertible matrix, then rank M = rank N.

Let E denote the system of equations

$$E: \sum_{j=1}^{m} \alpha_{ij}x_j = y_i \qquad \text{for } i = 1, \ldots, n,$$

and let E_0 denote the special system when $y = 0$. (E_0 is called the *homogeneous system* corresponding to the system E.) Similarly, let E^* denote the system

$$E^*: \sum_{i=1}^{n} \alpha_{ij}z_i = w_j \qquad \text{for } j = 1, \ldots, m.$$

The system E is simply another form of the equation $Tx = y$, and the system E^* is another form of the equation $T^*z = w$. Therefore, we have the following theorems [formula (5) of Section 6 and Theorem 6.10].

THEOREM 7.2

The equations E have a solution if and only if y is orthogonal to every solution of the equations E_0^*.

THEOREM 7.3

If $m = n$, the following are equivalent:
  (a) *The equations E_0 have only the solution $x = 0$.*
  (b) *The equations E have one and only one solution for every y.*
  (c) *The equations E_0^* have only the solution $z = 0$.*
  (d) *The equations E^* have one and only one solution for every w.*

Consider the case of two equations in two unknowns:

$$\begin{aligned} a_{11}x_1 + a_{12}x_2 &= y_1, \\ a_{21}x_1 + a_{22}x_2 &= y_2. \end{aligned} \qquad (1)$$

Multiply the first equation by a_{22} and the second by a_{12} and subtract. This causes the terms with x_2 to drop out and gives

$$(a_{11}a_{22} - a_{12}a_{21})x_1 = a_{22}y_1 - a_{12}y_2. \qquad (2)$$

Multiply the first equation by a_{21} and the second by a_{11} and subtract. This causes the terms with x_1 to drop out and gives

$$-(a_{11}a_{22} - a_{12}a_{21})x_2 = a_{21}y_1 - a_{11}y_2. \qquad (3)$$

DEFINITION 7.4

The determinant of the matrix

$$M = \begin{pmatrix} a_{11} & a_{12} \\ a_{21} & a_{22} \end{pmatrix},$$

written det M, *is the number* $a_{11}a_{22} - a_{12}a_{21}$.

**THEOREM
7.5**

The matrix M is invertible if and only if its determinant is $\neq 0$.   If this is the case, then the solution to equations (1) *is given by*

$$x_1 = \frac{\det \begin{pmatrix} y_1 & a_{12} \\ y_2 & a_{22} \end{pmatrix}}{\det M}, \qquad x_2 = \frac{\det \begin{pmatrix} a_{11} & y_1 \\ a_{21} & y_2 \end{pmatrix}}{\det M}. \tag{4}$$

Notice that the matrices in the numerators in (4) are obtained as follows: For x_1, replace the first column of M by y_1 and y_2; for x_2, replace the second column by y_1 and y_2.

Proof

First suppose that M is invertible.   Then the equations (1) have a solution for every y, and the solution must satisfy (2) and (3).   This is plainly impossible if the determinant is 0, because in that case the left sides of (2) and (3) are 0 no matter what y is.

Exercise 3   Fill the small gap.

Now suppose that the determinant is $\neq 0$.   Then (2) and (3) show that if $y = 0$, then $x = 0$, and Theorem 7.3 shows that M is invertible.

Exercise 4   If $v = (a_{11}, a_{21})$ and $w = (a_{12}, a_{22})$ are the two columns of M, then $|\det M|$ is the area of the parallelogram with vertices $0, v, w, v + w$.   Use this to deduce all of Theorem 7.5 except for formula (4).

The determinant can be defined for square matrices of any size.   The definition is more complicated than Definition 7.4, but Theorem 7.5 and the analog of Exercise 2 remain true.   (See Section 11.)

8   CONTINUITY OF LINEAR TRANSFORMATIONS

**THEOREM
8.1**

Every linear transformation $T: \mathbf{R}^m \to \mathbf{R}^n$ is continuous.   Indeed, there is a number M such that

$$|Tx| \leq M|x| \qquad \text{for all } x \in \mathbf{R}^m. \tag{1}$$

Proof

The inequality (1) does imply that T is continuous, for

$$|Tx - Ty| = |T(x - y)| \leq M|x - y|. \tag{2}$$

To prove the inequality (1), notice that $\|x\| = |Tx| + |x|$ is an absolute value on $\mathbf{R}^m$, and use the fact that any two absolute values are equivalent to get $\|x\| \leq N|x|$; hence $|Tx| \leq (N - 1)|x|$.

DEFINITION 8.2

If $T:\mathbf{R}^m \to \mathbf{R}^n$ *is a linear transformation, then*

$$\|T\| = \sup\{|Tx|:|x| = 1\}. \tag{3}$$

THEOREM 8.3

The number $\|T\|$ *is the least number* M *for which formula* (1) *holds. Moreover,*

$$\|T\| = \sup\{|Tx|:|x| \leq 1\}. \tag{4}$$

Proof

Taking $|x| = 1$ in formula (1), we get $|Tx| \leq M$; hence $\|T\| \leq M$. On the other hand, if $x \neq 0$, then $x/|x|$ has absolute value 1, so

$$\left| T\frac{x}{|x|} \right| \leq \|T\|;$$

hence

$$|Tx| \leq \|T\|\,|x| \qquad \text{for all } x. \tag{5}$$

This shows that formula (1) holds with $M = \|T\|$, and it was shown in the first sentence of the proof that (1) cannot hold with any smaller number—which proves the first part of the theorem.

To prove the second part, notice that by the definition of things the sup on the right side of (4) is $\geq \|T\|$, while by formula (5) it is $\leq \|T\|$.

Exercise 1 If $\{\alpha_{ij}\}$ is the matrix of T relative to any orthonormal bases, then

$$\|T\|^2 \leq \sum_{i,j} \alpha_{ij}^2. \tag{6}$$

(*Hint:* Write $y_i = \sum_{j=1}^m \alpha_{ij}x_j$ and use Cauchy–Schwarz.) This gives a different proof of Theorem 8.1 that does not use the fact that any two absolute values on $\mathbf{R}^m$ are equivalent.

We shall write $\mathcal{L}_{mn}$ for the space of linear transformations from $\mathbf{R}^m$ to $\mathbf{R}^n$.

Exercise 2 $\|T\|$ is an absolute value on $\mathcal{L}_{mn}$; that is,

$$\|S + T\| \leq \|S\| + \|T\| \qquad \text{and} \qquad \|\alpha T\| = |\alpha|\,\|T\|.$$

The exercise shows that $\mathcal{L}_{mn}$ is a metric space with the metric

$$d(S, T) = \|S - T\|. \tag{7}$$

THEOREM 8.4

The space $\mathcal{L}_{mn}$ *is complete.*

Proof

If $\{T_k\}$ is a Cauchy sequence and $x \in \mathbf{R}^m$, then formula (5) shows that

$$|T_kx - T_lx| \leq \|T_k - T_l\|\,|x|,$$

which implies that $\{T_k x\}$ is a Cauchy sequence in $\mathbf{R}^n$. Since $\mathbf{R}^n$ is complete, we can define

$$Tx = \lim_{k \to \infty} T_k x. \tag{8}$$

This defines T as a function from $\mathbf{R}^m$ to $\mathbf{R}^n$, and the first problem is to show that T is linear. For this we have

$$T(x + y) = \lim_{k \to \infty} (T_k x + T_k y) = \lim_{k \to \infty} T_k x + \lim_{k \to \infty} T_k y = Tx + Ty,$$

and, similarly, $T(\alpha x) = \alpha Tx$.

The next problem is to show that $T_k \to T$. Let $\epsilon > 0$ be given and choose k_0 so that if $k > k_0$ and $l > k_0$, then $\| T_k - T_l \| \leq \epsilon$. We shall show that if $k > k_0$, then $\| T_k - T \| \leq 2\epsilon$. Indeed, for any x with $|x| \leq 1$, we can choose l in accordance with definition (8) so that

$$|Tx - T_l x| < \epsilon \qquad \text{and} \qquad l > k_0.$$

Now, if $k > k_0$, then we have

$$|T_k x - Tx| \leq |T_k x - T_l x| + |T_l x - Tx| \leq \| T_k - T_l \| \, |x| + |T_l x - Tx$$
$$\leq \epsilon |x| + \epsilon \leq 2\epsilon \qquad \text{if } |x| \leq 1,$$

from which it follows that $\| T_k - T \| \leq 2\epsilon$, as claimed.

Exercise 3   The theorem can also be proved by choosing bases in $\mathbf{R}^m$ and $\mathbf{R}^n$, identifying $\mathcal{L}_{mn}$ with the space of m by n matrices, and then the latter with $\mathbf{R}^{mn}$—and, finally, using the fact that any two absolute values on $\mathbf{R}^{mn}$ are equivalent. Carry out this program.

Exercise 4   If $S, T \in \mathcal{L}_{mn}$, then

$$(S + T)^* = S^* + T^*, \qquad (\alpha T)^* = \alpha T^*, \qquad \| T^* \| = \| T \|. \tag{9}$$

THEOREM 8.5   *A linear transformation T is one to one if and only if there is a number $m > 0$ such that*

$$|Tx| \geq m|x| \qquad \text{for all } x. \tag{10}$$

Proof   If (10) holds, then the null space of T is $\{0\}$, and T is one to one. On the other hand, if T is one to one, then $\| x \| = |Tx|$ is an absolute value on $\mathbf{R}^m$, and the equivalence of any two absolute values implies that (10) holds.

THEOREM 8.6   *Let T be one to one and satisfy*

$$|Tx| \geq m|x|.$$

If $\|S - T\| \le \epsilon < m$, then S is also one to one and satisfies

$$|Sx| \ge (m - \epsilon)|x|. \tag{11}$$

Proof

$$|Sx| \ge |Tx| - |Tx -. Sx| \ge m|x| - \epsilon|x|.$$

THEOREM 8.7

The one-to-one linear transformations from $\mathbf{R}^m$ to $\mathbf{R}^n$ form an open subset of $\mathfrak{L}_{mn}$. The linear transformations from $\mathbf{R}^m$ onto $\mathbf{R}^n$ form an open subset of $\mathfrak{L}_{mn}$.

Proof

The first part comes from Theorem 8.6. As for the second part, if T maps $\mathbf{R}^m$ onto $\mathbf{R}^n$, then T^* is one to one, so there exists m such that

$$|T^*y| \ge m|y| \qquad \text{for all } y \in \mathbf{R}^n.$$

Now, if $\|S - T\| < m$, then by Exercise 4, $\|S^* - T^*\| < m$, so by Theorem 8.6, S^* is one to one, and then S is onto.

Exercise 5   If $\mathbf{R}^l \xrightarrow{S} \mathbf{R}^m \xrightarrow{T} \mathbf{R}^n$, then $\|TS\| \le \|T\|\,\|S\|$.

THEOREM 8.8

The invertible linear transformations from $\mathbf{R}^n$ to $\mathbf{R}^n$ form an open subset $\mathcal{I}$ of $\mathfrak{L}_{nn}$, and the function $\mathfrak{R}:\mathcal{I} \to \mathcal{I}$ (reciprocal) defined by

$$\mathfrak{R}(T) = T^{-1}$$

is continuous.

Proof

It is already shown in Theorem 8.7 that $\mathcal{I}$ is open. To show that $\mathfrak{R}$ is continuous, let $T \in \mathcal{I}$ satisfy $|Tx| \ge m|x|$. If $\|S - T\| \le m/2$, then by Theorem 8.6, $|Sx| \ge (m/2)|x|$, from which it follows that $\|S^{-1}\| \le 2/m$. Now we have $S^{-1} - T^{-1} = T^{-1}(T - S)S^{-1}$; therefore,

$$\|S^{-1} - T^{-1}\| \le \|T^{-1}\|\,\|T - S\|\,\|S^{-1}\| \le \frac{2}{m^2}\|T - S\|. \tag{12}$$

Exercise 6   Let $T:\mathbf{R}^m \to \mathbf{R}^n$. There exists $S:\mathbf{R}^n \to \mathbf{R}^m$
(a) such that $ST = I$ if and only if T is one to one.
(b) such that $TS = I$ if and only if T is onto.
[Note that the I in (a) is the identity on $\mathbf{R}^m$, while the I in (b) is the identity on $\mathbf{R}^n$.]

Exercise 7   The function $r(T) = \text{rank } T$ is lower semicontinuous on $\mathfrak{L}_{mn}$.

9   SELF-ADJOINT TRANSFORMATIONS

There are two important special kinds of linear transformations that we shall look at briefly—the self-adjoint transformations, which correspond geometrically to stretchings in various directions, and the orthogonal transformations, which correspond to rotations and reflections. These are intrinsically interesting. In addition they provide a good hold on general linear transformations, for every nonsingular linear transformation is a product of one that is self-adjoint and one that is orthogonal.

DEFINITION 9.1   *A linear transformation $H: \mathbf{R}^n \to \mathbf{R}^n$ is self-adjoint if $H = H^*$, or equivalently if*

$$\langle Hx, y \rangle = \langle x, Hy \rangle \qquad \text{for all } x \text{ and } y \text{ in } \mathbf{R}^n. \tag{1}$$

If a linear transformation T effects a stretching in a certain direction e, then T should simply carry e into a multiple of itself. In this case e is called an *eigenvector* of T.

DEFINITION 9.2   *An eigenvector of a linear transformation $T: \mathbf{R}^n \to \mathbf{R}^n$ is a nonzero vector e such that $Te = \lambda e$ for some real number λ. The number λ is called the corresponding eigenvalue.*

The basic theorem is as follows:

THEOREM 9.3   *If $H: \mathbf{R}^n \to \mathbf{R}^n$ is self-adjoint, then $\mathbf{R}^n$ has an orthonormal basis composed of eigenvectors of H.*

If $e_1, \ldots, e_n$ is an orthonormal basis of eigenvectors of H, and $\lambda_1, \ldots, \lambda_n$ are the corresponding eigenvalues, then geometrically H just effects a stretching by an amount λ_i in the direction e_i. (Of course, λ_i may be negative.) In terms of matrices, relative to the basis $e_1, \ldots, e_n$ the matrix of H has $\lambda_1, \ldots, \lambda_n$ along the diagonal and 0's everywhere else. In terms of linear equations, with the coordinates relative to the basis $e_1, \ldots, e_n$ the equation $y = Hx$ is equivalent to the system

$$y_i = \lambda_i x_i, \qquad i = 1, \ldots, n,$$

which is, of course, trivial to solve.

Proof of the Theorem   The main job is to show that H does have an eigenvector!

Exercise 1   Give an example of a linear transformation on the plane that has no eigenvector.

What we shall show is that if M is the maximum of $\langle Hx, x \rangle$ on the unit sphere, and e is the point where the maximum is assumed, then e is an eigenvector with eigenvalue M. The maximum makes sense because $\langle Hx, x \rangle$ is continuous and the unit sphere is compact. The proof is very quick, but it is a trick. Define

$$\langle x, y \rangle_0 = M \langle x, y \rangle - \langle Hx, y \rangle.$$

The definition of M gives $\langle x, x \rangle_0 \geq 0$ and the fact that H is self-adjoint gives $\langle x, y \rangle_0 = \langle y, x \rangle_0$, so this is an inner product in the sense of Theorems 2.3 and 1.2 of Chapter 7. Therefore, the Cauchy–Schwarz inequality gives

$$\langle x, y \rangle_0^2 \leq |\langle x, x \rangle_0|\, |\langle y, y \rangle_0|.$$

If $x = e$, then $\langle x, x \rangle_0 = 0$, so we have

$$\langle Me - He, y \rangle = \langle e, y \rangle_0 = 0 \qquad \text{for every } y,$$

which implies that $Me - He = 0$, or that $He = Me$.

Now induction carries the rest of the proof. Set $V = \{e\}^\perp$. Clearly, $H(V) \subset V$, for if $x \in V$, then $\langle Hx, e \rangle = \langle x, He \rangle = \langle x, Me \rangle = 0$. (Here again we use the fact that H is self-adjoint.) Therefore, the restriction of H to V is a self-adjoint linear transformation on V—and V, of course, is the same as $\mathbf{R}^{n-1}$. By induction on the dimension, V has an orthonormal basis of eigenvectors, which, combined with e, gives an orthonormal basis of $\mathbf{R}^n$.

Exercise 2 To be perfectly proper about this inductive proof, we should start by defining a self-adjoint transformation $H: V \to V$, where V is any subspace of $\mathbf{R}^n$, and then state the theorem in the form that if $H: V \to V$ is self-adjoint, then V has an orthonormal basis composed of eigenvectors. Give the definition by using formula (1) and the proper proof.

The eigenvalues of a linear transformation are uniquely determined. They are simply the numbers λ such that $Te = \lambda e$ for some nonzero vector e. The eigenvectors are determined, too, but the orthonormal basis in Theorem 9.3 is not determined. For instance, if $H = I$, then $\lambda = 1$ is the only eigenvalue, but every nonzero vector e is an eigenvector. The next two exercises show that the eigenvalues that appear in Theorem 9.3 are in fact all the eigenvalues and show how much uniqueness there is in the orthonormal basis.

Exercise 3 Let $T: \mathbf{R}^n \to \mathbf{R}^n$. Let $e_1, \ldots, e_n$ be linearly independent eigenvectors with eigenvalues $\lambda_1, \ldots, \lambda_n$. Then T has no other eigenvalues.

Exercise 4 Let $H: \mathbf{R}^n \to \mathbf{R}^n$ be self-adjoint. If e and f are eigenvectors with distinct eigenvalues, then $e \perp f$. As far as the uniqueness in Theorem 9.3 is concerned, the conclusion of the two exercises is as follows.   Let $H: \mathbf{R}^n \to \mathbf{R}^n$ be self-adjoint, and let $\lambda_1, \ldots, \lambda_k$ be the distinct eigenvalues.   Let V_j be the null space of $H - \lambda_j I$, that is, the set of eigenvectors with eigenvalue λ_j together with 0. Choose any orthonormal basis whatever of V_j.   Then the union of these is an orthonormal basis of $\mathbf{R}^n$ composed of eigenvectors, and this is the only way to get such a basis.

Exercise 5   Prove the assertions above.

A closer examination of the proof of Theorem 9.3 gives some interesting information about the eigenvalues.   Note first that if λ is any eigenvalue of the self-adjoint H and e is a corresponding eigenvector of length 1, then $\langle He, e \rangle = \langle \lambda e, e \rangle = \lambda$.   Therefore, the eigenvalue M produced in the proof is the largest eigenvalue.   By carrying out each step in the induction, we shall pick out the eigenvalues in decreasing order.   Indeed, let

$$\lambda_1 = \sup_{|x| = 1} \langle Hx, x \rangle,$$

and let e_1 be a point where the maximum is assumed, so that λ_1 is the largest eigenvalue and e_1 is a corresponding eigenvector.   Now let $V_1 = \{e_1\}^\perp$, let

$$\lambda_2 = \sup\{\langle Hx, x \rangle : |x| = 1 \text{ and } x \in V_1\},$$

and let e_2 be a point where the maximum is assumed.   By the same proof, λ_2 is an eigenvalue and e_2 is a corresponding eigenvector.   In general, when λ_1, $\ldots, \lambda_k$ and $e_1, \ldots, e_k$ have been picked out, let $V_k = \{e_1, \ldots, e_k\}^\perp$, let

$$\lambda_{k+1} = \sup\{\langle Hx, x \rangle : |x| = 1 \text{ and } x \in V_k\},$$

and let e_{k+1} be a point where the maximum is assumed.   In this way we obtain eigenvalues $\lambda_1 \geq \lambda_2 \geq \cdots \geq \lambda_n$ with corresponding orthonormal eigenvectors $e_1, \ldots, e_n$.   Exercise 3 shows that H has no other eigenvalues.

This method picks out the eigenvalues successively, starting with the largest. There is an important formula that picks out the kth largest directly, without making use of the previous ones and without making use of the eigenvectors.

THEOREM 9.4   *If H is self-adjoint, then the kth largest eigenvalue of H is given by*

$$\lambda_k = \inf_W (\sup\{\langle Hx, x \rangle : |x| = 1 \text{ and } x \in W\}), \qquad (2)$$

where the inf *is taken over all subspaces W of dimension $n - k + 1$.*

Proof   Let $\mu_1 \geq \mu_2 \geq \cdots \geq \mu_n$ be the eigenvalues with corresponding orthonormal eigenvectors $e_1, \ldots, e_n$.   Since V_k has dimension $n - k$, it is plain that $\lambda_{k+1} \leq \mu_{k+1}$. Now, if W is any subspace of dimension $n - k$, then $W \cap [e_1, \ldots , e_{k+1}] \neq \{0\}$ (why?); so there exists a point $x \in W$ with $|x| = 1$ and

$$x = \sum_{j=1}^{k+1} x_j e_j,$$

and we have

$$\langle Hx, x \rangle = \sum_{j=1}^{k+1} \mu_j x_j^2 \geq \mu_{k+1}.$$

This shows that for each W the sup in formula (2) is $\geq \mu_{k+1}$, so $\lambda_{k+1} \geq \mu_{k+1}$.

DEFINITION 9.5

The statement $H \geq K$ means that both H and K are self-adjoint and that $\langle Hx, x \rangle \geq \langle Kx, x \rangle$ for every x. If $H \geq 0$, then H is called positive definite.

Exercise 6 Let H and K be self-adjoint with eigenvalues $\lambda_1 \geq \cdots \geq \lambda_n$ and $\mu_1 \geq \cdots \geq \mu_n$. If $H \geq K$, then $\lambda_j \geq \mu_j$ for each j. (*Hint:* Use Theorem 9.4.)

This result is quite important in numerical work with eigenvalues. These cannot be computed explicitly in general, but Exercise 6 can be used to obtain good bounds. Given H, you look for K and L such that $L \geq H \geq K$ and such that the eigenvalues of K and L can be computed explicitly.

Exercise 7 If H is self-adjoint, then

$$\|H\| = \sup_{|x|=1} |\langle Hx, x \rangle|.$$

Theorem 9.3 can be used to construct interesting functions of self-adjoint transformations. For example,

THEOREM 9.6

Every positive definite transformation has a unique positive definite square root.

Proof Let H be self-adjoint with eigenvalues $\lambda_1, \ldots , \lambda_n$ and corresponding eigenvectors $e_1, \ldots , e_n$. If $H \geq 0$, then by Exercise 6 each $\lambda_j \geq 0$; so we can define μ_j to be the nonnegative square root of λ_j, and then define K by $Ke_j = \mu_j e_j$. Then $K^2 e_j = K(Ke_j) = \mu_j Ke_j = \mu_j^2 e_j = \lambda_j e_j = He_j$ for each j, which implies that $K^2 = H$. Exercise 6 shows that K is positive definite.

Exercise 8 Prove the uniqueness.

Remark The term positive definite refers to the fact that the sign of $\langle Hx, x \rangle$ is definite. It is not sometimes positive and sometimes negative. The transformation H is *negative definite* if $\langle Hx, x \rangle \leq 0$ for all x. It is *indefinite* if $\langle Hx, x \rangle$ is sometimes positive and sometimes negative. The transformation H is strictly positive definite, written $H > 0$, if $\langle Hx, x \rangle > 0$ for all $x \neq 0$, and is strictly negative definite if $\langle Hx, x \rangle < 0$ for all $x \neq 0$.

Exercise 9 A self-adjoint H is strictly positive definite if and only if all eigenvalues are > 0.

Exercise 10 Are there any linear transformations that are both positive definite and negative definite?

Exercise 11 The converse of Theorem 9.3 is also true. If $\mathbf{R}^n$ has an orthonormal basis composed of eigenvectors of H, then H is self-adjoint. What can you say about T if $\mathbf{R}^n$ has a basis (but not orthonormal) composed of eigenvectors of T?

Exercise 12 The projection P on a subspace V of $\mathbf{R}^n$ is defined as follows: Each $x \in \mathbf{R}^n$ can be written uniquely in the form $x = x' + x''$, and then $Px = x'$. Show that P is self-adjoint and that $P^2 = P$. What are the eigenvalues? Show that if Q is self-adjoint and $Q^2 = Q$, then Q is the projection on some subspace.

Exercise 13 Let $H: \mathbf{R}^n \to \mathbf{R}^n$ be self-adjoint with eigenvalues $\lambda_1 \geq \cdots \geq \lambda_n$. Let P be the projection on a subspace V of dimension m, and let $\mu_1 \geq \cdots \geq \mu_n$ be the eigenvalues of $K = PHP$. Show that

$$\lambda_k \geq \mu_{k+n-m} \qquad \text{for } k \leq m.$$

[*Hint:* Show first that in Theorem 9.4 the inf is the same if it is taken over all subspaces W of dimension $\geq n - k + 1$. Show next that if W has dimension $n - k + 1$, then $W' = V \cap W$ has dimension $\geq m - k + 1 = n - (k + n - m) + 1$. Now use Theorem 9.4.]

10 ORTHOGONAL TRANSFORMATIONS

The orthogonal transformations correspond geometrically to rotations and reflections, at least in $\mathbf{R}^n$.

DEFINITION 10.1 *A linear transformation* $U: \mathbf{R}^m \to \mathbf{R}^n$ *is orthogonal if* $\langle Ux, Uy \rangle = \langle x, y \rangle$ *for every x and y in* $\mathbf{R}^m$.

THEOREM 10.2

The following conditions are equivalent on a linear $U:\mathbf{R}^m \to \mathbf{R}^n$.
  (a) *U is orthogonal.*
  (b) *$U^*U = I$.*
  (c) *For every orthonormal basis $e_1, \ldots, e_m$ of $\mathbf{R}^m$, the set $Ue_1, \ldots, Ue_m$ is orthonormal in $\mathbf{R}^n$.*
  (d) *For some orthonormal basis $e_1, \ldots, e_m$ of $\mathbf{R}^m$, the set $Ue_1, \ldots, Ue_m$ is orthonormal in $\mathbf{R}^n$.*

Proof

We shall show that (d) implies (a) and leave the other implications as an exercise. If $e_1, \ldots, e_m$ is orthonormal and

$$x = \sum_{i=1}^{m} x_j e_j \quad \text{and} \quad y = \sum_{i=1}^{m} y_j e_j,$$

then

$$\langle x, y \rangle = \sum_{i,j} x_i y_j \langle e_i, e_j \rangle = \sum_i x_i y_i,$$

because $\langle e_i, e_j \rangle$ is 0 if $i \neq j$ and is 1 if $i = j$. The same argument applied to $Ue_1, \ldots, Ue_m$ in place of $e_1, \ldots, e_m$ shows that

$$\langle Ux, Uy \rangle = \Sigma x_i y_i = \langle x, y \rangle.$$

Exercise 1 Show that (a) $\Rightarrow$ (b) $\Rightarrow$ (c) $\Rightarrow$ (d) in order to complete the proof of the theorem.

Exercise 2 A linear $U:\mathbf{R}^m \to \mathbf{R}^n$ is orthogonal if and only if it preserves distance in the sense that

$$|Ux| = |x| \quad \text{for all } x \in \mathbf{R}^m.$$

(*Hint:* Prove the identity $4\langle z, w \rangle = |z + w|^2 - |z - w|^2$ and apply it to x and y and to Ux and Uy.)

Exercise 3 What can you say about UU^* when U is orthogonal? What can you say in the case where $m = n$?

THEOREM 10.3

If $T:\mathbf{R}^m \to \mathbf{R}^n$ is one to one, then $T = UH$, where $U:\mathbf{R}^m \to \mathbf{R}^n$ is orthogonal and $H:\mathbf{R}^m \to \mathbf{R}^m$ is positive definite.

Proof

To see what U and H must be, suppose that the theorem is true. Then $T^* = H^*U^* = HU^*$, so $T^*T = HU^*UH = H^2$. Now T^*T is strictly positive definite from $\mathbf{R}^m$ to $\mathbf{R}^m$, for

$$\langle T^*Tx, x \rangle = \langle Tx, Tx \rangle = |Tx|^2 > 0 \quad \text{for } x \neq 0.$$

Therefore, we can start back at the beginning and use Theorem 9.6 to define H to be the positive-definite square root of T^*T and then define U to be TH^{-1}. What has to be shown is that U is orthogonal, and this follows from

$$U^*U = H^{-1}T^*TH^{-1} = H^{-1}H^2H^{-1} = I.$$

Exercise 4   Why is the H above invertible?

Exercise 5   In the proof we used the fact that $(H^{-1})^* = (H^*)^{-1}$. Prove this for any $H: \mathbf{R}^m \to \mathbf{R}^m$.

Next we give a theorem to show that an orthogonal transformation $U: \mathbf{R}^n \to \mathbf{R}^n$ can be interpreted geometrically as either a rigid motion that leaves the origin fixed (i.e., a rotation) or else as a reflection followed by such a rigid motion. The following notation is customary for the set of orthogonal transformations.

DEFINITION 10.4 $\mathbf{O}_n$ *is the set of orthogonal transformations from* $\mathbf{R}^n$ *to* $\mathbf{R}^n$.

THEOREM 10.5   *Every* $U \in \mathbf{O}_n$ *can be joined by a path in* $\mathbf{O}_n$ *either to the identity or to the reflection* J *defined by*

$$Jx = (x_1, \ldots, x_{n-1}, -x_n),$$

which is the reflection across the subspace $x_n = 0$.

Proof   We show first that U can be joined by a path in $\mathbf{O}_n$ to a V with the property that $Ve_n = \pm e_n$. If e_n and Ue_n are linearly dependent, then $Ue_n = \pm e_n$, since both e_n and Ue_n have length 1. Otherwise, the two span a two-dimensional space V_2 in which we shall make a suitable rotation. Let $f = Ue_n$ and let g be a unit vector in V_2 that is perpendicular to f. (There are two of these, and we just choose one.) Since e_n is in V_2 and has length 1, there is a number θ, $0 \leq \theta < 2\pi$, such that

$$e_n = f \cos \theta - g \sin \theta. \tag{1}$$

Now define R_t, $0 \leq t \leq \theta$, as follows: $R_t x = x$ if $x \in V_2^{\perp}$ and

$$R_t f = f \cos t - g \sin t,$$
$$R_t g = f \sin t + g \cos t.$$

Exercise 6   Verify that R_t is orthogonal.

$R_0 = I$ and $R_\theta f = e_n$ [formula (1)], so $U_t = R_t U$ is a path from $U_0 = U$ to $U_\theta = V$, where V does have the required property that $Ve_n = \pm e_n$. (In fact, $Ve_n = e_n$, but we have to take account of the initial possibility that $Ue_n = -e_n$.)

The proof is finished by induction on the dimension. With the usual identification, $\mathbf{R}^{n-1}$ is the space spanned by $e_1, \ldots, e_{n-1}$. If V' is the restriction of V to $\mathbf{R}^{n-1}$, then V' is an orthogonal transformation from $\mathbf{R}^{n-1}$ to $\mathbf{R}^{n-1}$. (Why?) Therefore, induction gives a path V'_t in $\mathbf{O}_{n-1}$ which joins V' to either I' or J'. Let V_t be defined by $V_t x' = V'_t x'$ if $x' \in \mathbf{R}^{n-1}$ and $V_t e_n = e_n$ (or $V_t e_n = -e_n$ if it happens that $Ue_n = -e_n$). Then V_t is a path in $\mathbf{O}_n$ that joins V to one of the following:

$$Wx = Ix, \quad Wx = (J'x', -x_n), \quad Wx = (J'x', x_n), \quad Wx = (x', -x_n).$$

Exercise 7   Show how to join the first two to I and the second two to J.

In the next section we shall show that the two possibilities in Theorem 10.5 are mutually exclusive, or, in other words, that I cannot be joined to J by a path in $\mathbf{O}_n$. Another way to interpret this is that $\mathbf{O}_n$ has exactly two connected components, one containing I and the other containing J. Note that if we carry out the inductive steps in this proof, they show that the initial U can be joined to either I or J by a path that consists of a finite number of rotations, each rotation taking place in a two-dimensional subspace and leaving the orthogonal complement of this two-dimensional subspace fixed.

There is a corresponding theorem for the set $\mathscr{I}$ of invertible transformations on $\mathbf{R}^n$.

THEOREM 10.6   *Each $T \in \mathscr{I}$ can be joined by a path in $\mathscr{I}$ to either the identity I or to the reflection J.*

Proof   Use Theorem 10.3 to write $T = UH$ and Theorem 10.5 to find a path U_t in $\mathbf{O}_n$ that joins U to either I or to J. Set $H_t = (1 - t)H + tI$ and then $T_t = U_t H_t$.

Exercise 8   Show that each H_t is strictly positive definite, and therefore that T_t is a path in $\mathscr{I}$ which joins T to either I or J.

Again the same comment holds. I cannot be joined to J by a path in $\mathscr{I}$, so $\mathscr{I}$ has exactly two connected components—one containing I and the other containing J. This is shown in the next section.

11   DETERMINANTS

In this section we shall give the definition of the determinant of an n by n matrix and shall establish the properties that will be needed later.

DEFINITION
11.1

A permutation of a set I is a one-to-one function from I onto itself. If μ and ν are permutations of I, the product $\mu\nu$ is the composite $\mu \circ \nu$. A permutation μ is a transposition if there exist two distinct points i and j in I such that $\mu(i) = j$, $\mu(j) = i$ and $\mu(k) = k$ for all other points of I.

We shall be interested only in the case where I is the set $\{1, \ldots, n\}$. It is clear that every permutation is a product of transpositions—first interchange 1 and $\mu(1)$, then interchange 2 and $\mu(2)$, and so on. This can be done in many ways. For instance, if μ is a transposition, then $\mu^2 = 1$ (the identity), so in any product of transpositions, factors μ^2 can be inserted at random without changing the product.

THEOREM
11.2

In the various expressions of a permutation μ as a product of transpositions, the number of factors is always even, or else the number of factors is always odd.

Proof

If f is a real-valued function on $\mathbf{R}^n$ and μ is a permutation of $I = \{1, \ldots, n\}$, define the function f_μ by

$$f_\mu(x_1, \ldots, x_n) = f(x_{\mu(1)}, \ldots, x_{\mu(n)}).$$

Exercise 1   Show that $(f_\mu)_\nu = f_{\nu\mu}$.

Exercise 2   Let f be the particular function

$$f(x) = \prod_{i<j} (x_i - x_j), \tag{1}$$

where $\prod$ denotes the product over the indices in question. If $n = 3$, for example, then $f(x) = (x_1 - x_2)(x_1 - x_3)(x_2 - x_3)$. Show that if μ is a transposition, then $f_\mu(x) = -f(x)$.

The two exercises together show that if μ is the product of an even number of transpositions, then $f_\mu(x) = f(x)$, while if μ is a product of an odd number of transpositions, then $f_\mu(x) = -f(x)$. This characterizes the evenness or oddness of the number of transpositions directly in terms of μ.

DEFINITION
11.3

The sign of the permutation μ, written $\epsilon(\mu)$, is the number 1 if μ is the product of an even number of transpositions and is the number -1 if μ is the product of an odd number of transpositions.

Exercise 3

$$\epsilon(\mu\nu) = \epsilon(\mu)\epsilon(\nu).$$

DEFINITION
11.4

The determinant of the n by n matrix $A = \{a_{ij}\}$ is the number

$$\det A = \sum_{\mu} \epsilon(\mu) a_{1\mu(1)} a_{2\mu(2)} \cdots a_{n\mu(n)}, \tag{2}$$

where the sum is taken over all permutations μ of $I = \{1, \ldots, n\}$.

Exercise 4   Write out the sum in full in the 2 by 2 case and in the 3 by 3 case.   [There are $n!$ permutations of the set $\{1, \ldots, n\}$, so there are $n!$ terms in the sum (2). Already this becomes a little burdensome at $n = 4$, for there are 24 terms.]

THEOREM
11.5

If B is obtained from A by making a permutation π of the columns, then $\det B = \epsilon(\pi) \det A$.

Proof   If $b_{ij} = a_{i\pi(j)}$, then

$$\det B = \sum_{\mu} \epsilon(\mu) b_{1\mu(1)} \cdots b_{n\mu(n)} = \sum_{\mu} \epsilon(\mu) a_{1\pi\mu(1)} \cdots a_{n\pi\mu(n)}.$$

If we write $\pi\mu = \nu$ and use the fact that $\epsilon(\mu) = \epsilon(\pi)\epsilon(\nu)$ and the fact that as μ runs through all the permutations of $\{1, \ldots, n\}$, so does ν, we get

$$\det B = \epsilon(\pi) \sum_{\nu} \epsilon(\nu) a_{1\nu(1)} \cdots a_{n\nu(n)} = \epsilon(\pi) \det A.$$

If A is a matrix, then A^* is, of course, the matrix obtained by interchanging rows and columns.   In other words, if A is the matrix of a linear transformation T relative to an orthonormal basis, then A^* is the matrix of T^*.

THEOREM
11.6

$$\det A^* = \det A.$$

Proof   We have

$$\det A^* = \sum_{\mu} \epsilon(\mu) a^*_{1\mu(1)} \cdots a^*_{n\mu(n)} = \sum_{\mu} \epsilon(\mu) a_{\mu(1)1} \cdots a_{\mu(n)n}.$$

If we write $\mu^{-1} = \nu$ and use the fact that $\epsilon(\mu^{-1}) = \epsilon(\mu)$ and the fact that as μ runs through all permutations, so does μ^{-1}, we get

$$\det A^* = \sum_\nu \epsilon(\nu) a_{1\nu(1)} \cdots a_{n\nu(n)} = \det A.$$

COROLLARY 11.7

If B is obtained from A by making a permutation π of the rows, then $\det B = \epsilon(\pi) \det A$.

Proof

The permutation of the rows can be effected by first interchanging rows and columns, then making the permutation on the columns, and finally interchanging rows and columns again. The first and last operations leave the determinant unchanged, while the middle one multiplies by $\epsilon(\pi)$.

COROLLARY 11.8

If A has two equal rows or two equal columns, then $\det A = 0$.

Proof

The transposition interchanging these rows (or columns) has no effect, but on the other hand it multiplies the determinant by -1.

THEOREM 11.9

$$\det AB = \det A \det B.$$

Proof

Let $A = \{a_{ij}\}$, $B = \{b_{ij}\}$, and $AB = \{c_{ij}\}$, so

$$c_{ij} = \sum_k a_{ik} b_{kj}, \tag{3}$$

and

$$\det AB = \sum_\mu \epsilon(\mu) c_{1\mu(1)} \cdots c_{n\mu(n)}. \tag{4}$$

Now we shall substitute (3) into (4), but for each i we shall have to use a different index of summation. We get

$$\det AB = \sum_\mu \sum_{k_1} \cdots \sum_{k_n} \epsilon(\mu) a_{1k_1} b_{k_1\mu(1)} \cdots a_{nk_n} b_{k_n\mu(n)}.$$

Fix $k_1, \ldots, k_n$ and consider the sum with respect to μ. The sum

$$d_{k_1,\ldots,k_n} = \sum_\mu \epsilon(\mu) b_{k_1\mu(1)} \cdots b_{k_n\mu(n)}$$

is just the determinant of the matrix in which the first row is the k_1st row of B, and in general the jth row is the k_jth row of B. If $k_1, \ldots, k_n$ are not all distinct, then this is 0 by Corollary 11.8. If $k_1, \ldots, k_n$ are all distinct, then the function π defined by $\pi(j) = k_j$ is a permutation, and by Corollary 11.7 we have

$$d_{k_1,\ldots,k_n} = \epsilon(\pi) \det B.$$

Thus,

$$\det AB = \det B \sum_{\pi} \epsilon(\pi)a_{1\pi(1)} \cdots a_{n\pi(n)} = \det B \det A.$$

THEOREM 11.10

If A has $\lambda_1, \ldots, \lambda_n$ along the diagonal and 0's everywhere else, then

$$\det A = \lambda_1 \cdots \lambda_n.$$

Proof

This is clear from the definition. If μ is not the identity permutation in formula (2), then the corresponding term in the sum is 0, for it contains an a_{ij} off the diagonal.

Theorems 11.9 and 11.10 make it possible to define the determinant of a linear transformation.

DEFINITION 11.11

If T is a linear transformation from $\mathbf{R}^n$ to $\mathbf{R}^n$, then $\det T$ is the determinant of the matrix of T relative to any basis.

For the definition to make sense it must be true that the determinant of one matrix of T is the same as the determinant of any other matrix of T. If A and B are matrices of T relative to different bases, then, according to Exercise 10 of Section 5, we have $B = CAC^{-1}$, where C is some invertible matrix. First we notice that if I is the identity matrix, then by Theorem 11.10, $\det I = 1$. Next we notice that by Theorem 11.9 we have $\det C \det C^{-1} = 1$, because $CC^{-1} = I$. Finally, again by Theorem 11.9, we have $\det B = \det C \det A \det C^{-1} = \det A$.

THEOREM 11.12

The function $d(T) = \det T$ is characterized by the following properties:
(a) *$d(T)$ is continuous from $\mathcal{L}_{nn}$ to $\mathbf{R}^1$.*
(b) *$d(T^*) = d(T)$.*
(c) *$d(ST) = d(S)d(T)$.*
(d) *If H is self-adjoint with eigenvalues $\lambda_1, \ldots, \lambda_n$, then $d(H) = \lambda_1 \cdots \lambda_n$.*

Proof

To see (a), identify $\mathcal{L}_{nn}$ with $\mathbf{R}^{n^2}$ by identifying each linear transformation with its matrix relative to the standard orthonormal basis of $\mathbf{R}^n$. The absolute value on $\mathcal{L}_{nn}$ is equivalent to the absolute value on $\mathbf{R}^{n^2}$, for any two absolute values on $\mathbf{R}^{n^2}$ are equivalent. It is plain that the determinant is continuous on $\mathbf{R}^{n^2}$, for it is a polynomial in the n^2 coordinates. Parts (b) and (c) come from Theorems 11.6 and 11.9. Part (d) comes from Theorem 11.10. (Use the orthonormal basis of eigenvectors.)

Now we have to show that if d is any function with the properties listed, then $d(T)$ must be the determinant of T. In the course of the proof we shall find a number of additional basic properties of the determinant, which we shall list as theorems as we go along. Notice first that again we have

$$d(I) = 1 \qquad \text{and} \qquad d(J) = -1,$$

for I has n eigenvalues all equal to 1, and J has $n - 1$ eigenvalues equal to 1 and one equal to -1. In both cases the standard basis of $\mathbf{R}^n$ is an orthonormal basis of eigenvectors.

THEOREM 11.13

T is invertible if and only if $d(T) \neq 0$.

Proof

If T is invertible, then $I = TT^{-1}$, so $1 = d(T)d(T^{-1})$, which certainly means that $d(T) \neq 0$. If T is not invertible, then it has a nontrivial null space N. If P is the projection on $N^{\perp}$ (see Exercise 12 of Section 9), then $T = TP$. Indeed, if $x \in N$, then Tx and Px are both 0, while if $x \in N^{\perp}$, then $Px = x$. Thus, $d(T) = d(T)d(P)$. But $d(P) = 0$, for P is self-adjoint and 0 is an eigenvalue (with any nonzero vector in N as a corresponding eigenvector).

THEOREM 11.14

The space $\mathscr{G}$ of invertible transformations from $\mathbf{R}^n$ to $\mathbf{R}^n$ has exactly two connected components.

Proof

According to Theorem 10.6, $\mathscr{G}$ has at most two components and $\mathscr{G}$ cannot be connected, for the continuous function $d(T)$ takes the value 1 at I and the value -1 at J, but does not take the value 0 anywhere on $\mathscr{G}$.

THEOREM 11.15

The space $\mathbf{O}_n$ of orthogonal transformations has exactly two connected components. If U is orthogonal, then $d(U) = 1$ if U is in the component of I, and $d(U) = -1$ if U is in the component of J.

Proof

If U is orthogonal, then $I = U^*U$; so

$$1 = d(U^*)d(U) = d(U)^2,$$

which shows that $d(U) = \pm 1$. A continuous function that takes only the values ± 1 must be constant on any connected set. Hence, $d(U) = 1$ on the component of I and $d(U) = -1$ on the component of J. Theorem 10.5 shows that there are no other components.

These theorems characterize d as the determinant. If T is not invertible, then $d(T) = 0$. If T is invertible, then $T = UH$, where U is orthogonal and H is positive definite; then $d(T) = d(U)d(H)$. The value of $d(U)$ is determined by Theorem 11.15 and the value of $d(H)$ is determined by property (d) of Theorem 11.12. There are various ways to characterize determinants. This one happens to be natural if one has in mind the applications to volumes and areas, although in that case it is really the absolute value of the determinant that is relevant.

THEOREM 11.16 *The function $J(T) = |\det T|$ is characterized by the following properties:*
(a) $J(T^*) = J(T)$.
(b) $J(ST) = J(S)J(T)$.
(c) *If $Te_i = \lambda_i e_i$, where $e_1, \ldots, e_n$ is the standard orthonormal basis of $\mathbf{R}^n$ and $\lambda_i \geq 0$, then $J(T) = \lambda_1 \cdots \lambda_n$.*

Exercise 5 Prove the theorem. [You will have a little work to do because property (c) in this theorem appears to be quite a bit weaker than the corresponding property (d) of Theorem 11.12. The idea is to deal first with orthogonal transformations and then to get from the weaker property to the stronger by an orthogonal transformation. It is too bad to have two J's around, but in fact you do not need the reflection at all. The present J is called the *Jacobian* and J is the traditional letter for it.]

Exercise 6 If $Te_i = \lambda_i e_i$, where $e_1, \ldots, e_n$ is any basis of $\mathbf{R}^n$ (not necessarily orthonormal), then $\det T = \lambda_1 \cdots \lambda_n$.

Exercise 7 Let $T:\mathbf{R}^n \to \mathbf{R}^n$ and set $p(\lambda) = \det(T - \lambda I)$. Show that p is a polynomial of degree n whose real zeros are the eigenvalues of T.

From the fact that T is invertible if and only if $\det T \neq 0$, one might hope that there is a formula for T^{-1} in terms of the determinant. In fact, there is a very nice formula which we can discover by calculating the determinant in a particular way. If A is a matrix, then according to the definition we have

$$\det A = \Sigma \, \epsilon(\pi) a_{1\pi(1)} \cdots a_{n\pi(n)}.$$

Fix an index j and sum first over the permutations π with $\pi(n) = j$, and then sum over j to get

$$\det A = \sum_j a_{nj} \sum_{\pi(n)=j} \epsilon(\pi) a_{1\pi(1)} \cdots a_{n-1\pi(n-1)}.$$

If π_j is the transposition that interchanges n and j and if $\mu = \pi_j\pi$, then $\mu(n) = n$, so μ is a permutation of $\{1, \ldots, n-1\}$. Since $\pi = \pi_j\mu$, we have

$$\det A = -\sum_j a_{nj} \sum_\mu \epsilon(\mu) a_{1\pi_j\mu(1)} \cdots a_{n-1\pi_j\mu(n-1)}.$$

Let B_j be the $(n-1)$ by $(n-1)$ matrix defined by

$$b_{km} = a_{k\pi_j(m)} \qquad \text{for } 1 \leq k, m \leq n-1.$$

Then

$$\det A = -\sum_j a_{nj} \det B_j.$$

At this point we have almost proved the following lemma.

LEMMA
11.17

If A_{nj} is the matrix obtained by crossing out the nth row and jth column of A, then

$$\det A = \sum_j (-1)^{n+j} a_{nj} \det A_{nj}.$$

To finish the proof simply note that

$$\det B_j = (-1)^{n-j-1} \det A_{nj},$$

which follows from the fact that we get A_{nj} from B_j by moving the jth column of B_j successively past the $n - j - 1$ columns that follow it.

It is customary to write δ_{ik} for the number in the ith row and kth column of the identity matrix. Thus, δ_{ik} is 1 if $i = k$ and is 0 if $i \neq k$.

THEOREM
11.18

If A_{ij} is the matrix obtained by crossing out the ith row and jth column of A, then

$$\sum_j (-1)^{k+j} a_{ij} \det A_{kj} = \delta_{ik} \det A. \qquad (5)$$

Proof

First suppose that $k = i$. Then formula (5) follows from Lemma 11.17 if we move the ith row of A successively past the $n - i$ rows that follow it. Now suppose that $k \neq i$. If we replace the kth row of A by the ith row of A, we get a new matrix B with determinant 0 (because two rows are the same). Now note that $A_{kj} = B_{ij}$ and use formula (5) to calculate the determinant of B.

Theorem 11.18 gives the following explicit formula for the inverse matrix.

THEOREM
11.19

If $\det A \neq 0$, then the element in the jth row and kth column of A^{-1} is $(-1)^{k+j} \det A_{kj}/\det A$.

Exercise 8   Use adjoints to show that

$$\sum_i (-1)^{k+i} a_{ij} \det A_{ik} = \delta_{jk} \det A. \qquad (6)$$

Exercise 9
Cramer's Rule

Let $T : R^n \to R^n$ be a linear transformation with matrix A. Show that if $Tx = y$, then

$$x_k \det A = \det A_k,$$

where A_k is the matrix obtained from A by replacing the kth column by the vector y.

[*Hint:* Multiply both sides of (6) by x_j and sum on j.]

10 { Linear Approximation

1 DIRECTIONAL DERIVATIVES AND PARTIAL DERIVATIVES

The derivative of a function from $\mathbf{R}^m$ to $\mathbf{R}^n$ cannot be defined as it was for functions from $\mathbf{R}^1$ to $\mathbf{R}^n$. The difference quotient $f(x) - f(a)/x - a$ is meaningless because $x - a$ is a vector. Nevertheless, the derivative in any direction can be defined.

DEFINITION 1.1

The function $f\colon \mathbf{R}^m \to \mathbf{R}^n$ is differentiable at the point a in the direction θ if the function $\varphi(t) = f(a + t\theta)$ is differentiable at $t = 0$. If this is the case, then $\varphi'(0)$ is called the directional derivative of f at the point a in the direction θ and is written $D_\theta f(a)$.

Equivalently,

$$D_\theta f(a) = \lim_{\substack{t \to 0 \\ t \neq 0}} \frac{f(a + t\theta) - f(a)}{t}. \tag{1}$$

It should be clear that a is a point in $\mathbf{R}^m$ and θ is a direction in $\mathbf{R}^m$. It is not necessary, of course, that f be defined on all of $\mathbf{R}^m$. All points near a are enough. Quite often we shall write $f\colon \mathbf{R}^m \to \mathbf{R}^n$ when f is only defined on a suitable subset of $\mathbf{R}^m$, and in most cases we shall leave it to the reader to make the correction.

The directions along the coordinate axes are particularly important.

DEFINITION 1.2

Let $e_1, \ldots, e_m$ be the natural basis of $\mathbf{R}^m$, and let $f\colon \mathbf{R}^m \to \mathbf{R}^n$. The directional derivative in the direction e_j is called the partial derivative of f with respect to x_j and is written $D_j f(a)$, or $\partial f(a)/\partial x_j$, or $f_{x_j}(a)$.

It will be shown in the next section that under suitable hypotheses on the function f the derivative in any direction θ is expressed in terms of the partial derivatives by the formula

$$D_\theta f(a) = \sum_{j=1}^{m} \theta_j D_j f(a). \tag{2}$$

Thus, the partial derivatives determine the derivatives in all directions. Furthermore, the partial derivatives can be calculated as easily as ordinary derivatives.

Exercise 1

$$D_1 f(a) = \lim_{\substack{x_1 \to a_1 \\ x_1 \neq a_1}} \frac{f(x_1, a_2, \ldots, a_n) - f(a_1, a_2, \ldots, a_n)}{x_1 - a_1}.$$

What the exercise means is this. To get the partial derivative with respect to x_1, fix all the other variables and differentiate the resulting function of x_1.

Example   Calculate the partial derivatives of

$$f(x, y) = e^{xy} \sin x$$

To get the partial derivative with respect to x, we consider y constant:

$$\frac{\partial f}{\partial x} = e^{xy} \cos x + y e^{xy} \sin x.$$

To get the derivative with respect to y, we consider x constant:

$$\frac{\partial f}{\partial y} = x e^{xy} \sin x.$$

To some extent the directional derivatives or partial derivatives serve the purpose that the ordinary derivative served for functions of one variable. For example:

THEOREM 1.3   *Let $f : \mathbf{R}^m \to \mathbf{R}^1$ have a maximum or minimum at the point a. If f is differentiable at a in some direction θ, then $D_\theta f(a) = 0$.*

Proof   It is evident that if f has a maximum or minimum at a, then the function φ in Definition 1.1 has a maximum or minimum at 0, and the old theorem applies to φ.

The theorem is most useful if all the partial derivatives exist at a. It gives the m equations

$$\frac{\partial f}{\partial x_1} = 0, \ldots, \frac{\partial f}{\partial x_m} = 0 \tag{3}$$

for the m unknowns $x_1, \ldots, x_m$. In general, of course, these equations are not linear and are hard to solve. Moreover, the equations (3) only give the possibilities. It always has to be proved that a solution of (3) really is a maximum or minimum.

Exercise 2 Find the maxima and minima of the function

$$f(x, y) = 2x^2 - 2xy + y^2 - 2x + 2y + 1.$$

(It may be helpful to look back at Example 2, Section 3 of Chapter 3.)

Exercise 3 Use the definition of the directional derivative and Theorem 2.2 of Chapter 8 to prove the following theorem.

THEOREM 1.4 *The function $f : \mathbf{R}^m \to \mathbf{R}^n$ is differentiable at the point a in the direction θ if and only if each coordinate function is differentiable; if this is the case, then*

$$D_\theta f(a) = \big(D_\theta f_1(a), \ldots, D_\theta f_n(a)\big).$$

2 THE DIFFERENTIAL

While the partial derivatives are convenient in calculations, they are not at all convenient in theoretical questions. For one thing they are too unwieldy, and for another their existence implies almost nothing.

Exercise 1 Give an example where the partial derivatives exist, but certain directional derivatives do not.

Exercise 2 Give an example where all directional derivatives exist at a point, but the function is not continuous there.

What is needed in order to prove theorems is the possibility of approximating the given function by a linear transformation in the following sense.

DEFINITION 2.1 *The function $f : \mathbf{R}^m \to \mathbf{R}^n$ is differentiable at the point a if there is a linear transformation $T : \mathbf{R}^m \to \mathbf{R}^n$ such that*

$$\lim_{\substack{h \to 0 \\ h \neq 0}} \frac{f(a + h) - f(a) - Th}{|h|} = 0. \tag{1}$$

The linear transformation T is called the differential or derivative of f at a, and is written $df(a)$, or $f'(a)$.

Remark The term differential is the traditional one, and the traditional symbol is $df(a)$; but the term derivative and the symbol $f'(a)$ are becoming more common now. Note that differentiability requires that the function be defined on some ball with center a, but not, of course, on all of $\mathbf{R}^m$.

The differential and the directional derivatives are related by the following formula—which, incidentally, shows that the differential is uniquely determined by formula (1).

THEOREM 2.2 *If f is differentiable at a, then f is differentiable in every direction, and*

$$D_\theta f(a) = df(a)\theta. \tag{2}$$

Proof Let θ be given and take $h = t\theta$, $t > 0$, in formula (1) to get

$$\lim_{\substack{t \to 0 \\ t > 0}} \frac{f(a + t\theta) - f(a)}{t} - T\theta = 0. \tag{3}$$

Exercise 3 Show that the same formula holds for $t < 0$.

Formula (3) shows that f is differentiable in the direction θ and that $D_\theta f(a) = T\theta$.

Recall that the matrix of a linear transformation T relative to bases e_1, . . . , e_m and f_1, . . . , f_n has the coordinates of Te_j relative to f_1, . . . , f_n down the jth column. When f_1, . . . , f_n is the natural basis of $\mathbf{R}^n$, a vector *is* its n-tuple of coordinates. Moreover, in the present case, where T is the differential, Theorem 2.2 shows that $Te_j = D_{e_j}f(a)$, and the latter is just the partial derivative. Therefore, we have a theorem about the matrix of the differential.

THEOREM 2.3 *If f is differentiable at a, then the matrix of $df(a)$ is*

$$\frac{\partial f}{\partial x} = \begin{pmatrix} \dfrac{\partial f_1}{\partial x_1} & \dfrac{\partial f_1}{\partial x_2} & \cdots & \dfrac{\partial f_1}{\partial x_m} \\ \cdot & \cdot & & \cdot \\ \cdot & \cdot & & \cdot \\ \cdot & \cdot & & \cdot \\ \dfrac{\partial f_n}{\partial x_1} & \dfrac{\partial f_n}{\partial x_2} & \cdots & \dfrac{\partial f_n}{\partial x_m} \end{pmatrix}, \tag{4}$$

where the partial derivatives are calculated at a. Consequently,

$$(df(a)h)_i = \sum_{j=1}^{m} \frac{\partial f_i}{\partial x_j} h_j. \qquad (5)$$

The matrix $\partial f/\partial x$ is called the Jacobi matrix of f.

One way to remember the formula for the matrix is that to get the jth column you differentiate with respect to x_j.

Consider formula (5). If f is a real-valued function, that is, $f: \mathbf{R}^m \to \mathbf{R}^1$, then the right-hand side is an inner product—the inner product of h with the vector

$$\nabla f = \left(\frac{\partial f}{\partial x_1}, \cdots, \frac{\partial f}{\partial x_n} \right) \qquad (6)$$

(calculated, of course, at a).

THEOREM 2.4

If f is differentiable at a, then

$$df(a)h = \langle \nabla f(a), h \rangle, \qquad (7)$$

and, in particular,

$$D_\theta f(a) = \langle \nabla f(a), \theta \rangle. \qquad (8)$$

The vector ∇f defined by (6) is called the *gradient* of f. One convenient way to remember formula (5) is to use (6) as a formal definition even in the general case where f goes from $\mathbf{R}^m$ to $\mathbf{R}^n$. In this case ∇f is not really a vector—each component $\partial f/\partial x_j$ is a vector. However, the formal expression for $\langle \nabla f, h \rangle$, which is

$$\langle \nabla f(a), h \rangle = \sum_{j=1}^{m} \frac{\partial f}{\partial x_j} h_j, \qquad (9)$$

makes perfectly good sense, because h_j is a number. With these definitions, formula (5) is identical with formula (7).

Another way to remember the formulas (and probably the best one for calculations) is like this. Write $y = f(x)$ and then dy instead of df. Then formula (5) becomes

$$dy_i = \sum_{j=1}^{m} \frac{\partial y_i}{\partial x_j} dx_j. \qquad (10)$$

Notice that ∂x_j and dx_j "cancel" in each term. Two things are involved in proving formula (10). In writing dx_j we are, of course, thinking of x_j as a function; that is, x_j is the jth coordinate of x. Thus, x_j is a function from $\mathbf{R}^m$ to $\mathbf{R}^1$, so dx_j is a linear function from $\mathbf{R}^m$ to $\mathbf{R}^1$. It is clear that

$$dx_j(a)h = h_j. \tag{11}$$

Therefore, the right-hand side of (10) is a linear function from $\mathbf{R}^m$ to $\mathbf{R}^1$, and its value at h is the right-hand side of (5). On the other hand, it is shown in the next section [formula (1)] that $(dyh)_i = dy_ih$. Hence, (10) is the same formula as (5).

Example 1   Let $f(r, \theta) = (r \cos \theta, r \sin \theta)$; that is,

$$x = r \cos \theta,$$
$$y = r \sin \theta.$$

Then

$$dx = \frac{\partial x}{\partial r} dr + \frac{\partial x}{\partial \theta} d\theta = \cos \theta \, dr - r \sin \theta \, d\theta,$$

$$dy = \frac{\partial y}{\partial r} dr + \frac{\partial y}{\partial \theta} d\theta = \sin \theta \, dr + r \cos \theta \, d\theta,$$

and the Jacobi matrix is

$$\frac{\partial f}{\partial (r, \theta)} = \begin{pmatrix} \cos \theta & -r \sin \theta \\ \sin \theta & r \cos \theta \end{pmatrix}.$$

Exercise 4   Calculate the Jacobi matrix of $f(x, y) = (x^2 + y^2, xy)$.

Exercise 5   Calculate the Jacobi matrix of

$$x = \cos u, \qquad y = \sin u, \qquad z = v,$$

that is, of $f(u, v) = (\cos u, \sin u, v)$.

3   EXISTENCE OF THE DIFFERENTIAL

It is the existence of the differential that is essential in proving theorems, but it is the partial derivatives that one can get his hands on. What is needed is a way to tell that a function is differentiable by looking at the partial derivatives.

THEOREM 3.1   *If the partial derivatives of f exist on a ball with center a and are continuous at a, then f is differentiable at a.*

The proof will make use of the mean-value theorem—which is false for functions with values in $\mathbf{R}^n$. Therefore, the first step is to reduce to the case $n = 1$.

THEOREM 3.2

The function $f: \mathbf{R}^m \to \mathbf{R}^n$ is differentiable at a if and only if each coordinate function is.

Proof

Assume that f is differentiable at a, with differential T. We shall show that if f_j and T_j are the jth coordinate functions of f and T, then

$$df_j(a) = T_j. \tag{1}$$

In other words, the differential of the jth coordinate function is the jth coordinate function of the differential. In fact, this is obvious simply because the absolute value of any coordinate of a vector is less than or equal to the absolute value of the vector itself. That is,

$$\left| \frac{f_j(a + h) - f_j(a) - T_j h}{|h|} \right| \leq \left| \frac{f(a + h) - f(a) - Th}{|h|} \right|,$$

and by hypothesis the right side goes to 0 as h goes to 0.

Exercise 1 It was tacitly assumed that T_j is linear. Prove it.

Assume that each f_j is differentiable, let $T_j = df_j(a)$, and let T be the linear transformation with coordinate functions T_j. [That is, $Tx = (T_1 x, \ldots, T_n x)$; prove that T is linear.] If $\epsilon > 0$ is given, we can find $\delta > 0$ such that if $|h| < \delta$, then

$$\left| \frac{f_j(a + h) - f_j(a) - T_j h}{|h|} \right| < \epsilon \qquad \text{for each } j.$$

If each coordinate of a vector is smaller than ϵ, then the absolute value of the vector itself is smaller than $\sqrt{n}\,\epsilon$. Therefore, if $|h| < \delta$, then

$$\left| \frac{f(a + h) - f(a) - Th}{|h|} \right| < \sqrt{n}\,\epsilon.$$

Proof of Theorem 3.1

From Theorem 1.4 it follows that the partial derivatives of each coordinate function exist on a ball with center a and are continuous at a; and according to Theorem 3.2, we can deal with the coordinate functions separately. Therefore, we can suppose from now on that f is a function from $\mathbf{R}^m$ to $\mathbf{R}^1$.

In order to use the partial derivatives, we want to express the difference $f(x) - f(a)$ as a sum of terms in each of which only one coordinate varies.

If we put

$$a^j = (a_1, \ldots, a_j, x_{j+1}, \ldots, x_m),$$

then $a^0 = x$ and $a^m = a$, and we have

$$f(x) - f(a) = \sum_{j=1}^{m} f(a^{j-1}) - f(a^j) \tag{2}$$

because of the cancellation between each term and the next. Geometrically, in two dimensions we go from a to x along the path shown in Figure 1.

We want to apply the mean-value theorem to the function

$$g_j(t) = f(a_1, \ldots, a_{j-1}, t, x_{j+1}, \ldots, x_m)$$

in order to evaluate $f(a^{j-1}) - f(a^j) = g_j(x_j) - g_j(a_j)$. If we can do so, we shall obtain

$$f(a^{j-1}) - f(a^j) = g_j'(\xi_j)(x_j - a_j) = D_j f(\xi^j)(x_j - a_j), \tag{3}$$

where ξ_j is some point between a_j and x_j, and

$$\xi^j = (a_1, \ldots, a_{j-1}, \xi_j, x_{j+1}, \ldots, x_m).$$

This use of the mean-value theorem is justified if g_j is differentiable on the open interval from a_j to x_j and continuous on the corresponding closed interval. Now, if x is a point of a ball $B(a, r)$, then each a^j is also a point of the ball; for it is clear that $|a^j - a| \leq |x - a|$. Moreover, a ball has the property that if it contains two points, then it contains the entire line segment joining them. Therefore, if $B(a, r)$ is a ball on which the partial derivatives of f exist, then each of the line segments $[a^{j-1}, a^j]$ belongs to $B(a, r)$; so the partial derivatives of f exist on each such line

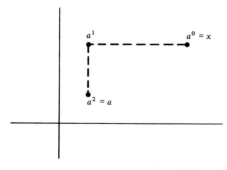

Figure 1

segment—and the existence of $D_j f$ is (by definition) what is needed for the differentiability of g_j.

Formulas (2) and (3) with $x = a + h$ give

$$f(a + h) - f(a) = \sum_{j=1}^{m} D_j f(\xi^j) h_j \qquad \text{if } |h| < r. \tag{4}$$

Let

$$Th = \sum_{j=1}^{m} D_j f(a) h_j \tag{5}$$

and let $\epsilon > 0$ be given. Use the continuity of the partial derivatives at the point a to find $\delta < r$ such that if $|y - a| < \delta$, then

$$|D_j f(y) - D_j f(a)| < \epsilon \qquad \text{for each } j.$$

If $|h| < \delta$, this can be applied to $y = \xi^j$ to give

$$|D_j f(\xi^j) - D_j f(a)| < \epsilon \qquad \text{for each } j.$$

Subtracting formulas (4) and (5) and making use of this, we find

$$|f(a + h) - f(a) - Th| \leq \sqrt{n}\, \epsilon |h|,$$

which implies that f is differentiable with differential T.

Remark   Differentiability does not imply continuity of the partial derivatives. It is somewhere between continuity of the partial derivatives and simple existence.

Exercise 2   Write out the proof of the theorem for $m = 2$.

Exercise 3   A function $f: \mathbf{R}^1 \to \mathbf{R}^n$ is differentiable in the present sense if and only if it is differentiable in the sense of Chapter 8. If this is the case, then

$$df(a)h = hf'(a).$$

4   COMPOSITE FUNCTIONS

The formula for the differential of a composite function is absolutely fundamental.

THEOREM 4.1   *Let $\mathbf{R}^l \xrightarrow{f} \mathbf{R}^m \xrightarrow{g} \mathbf{R}^n$. If f is differentiable at a, and g is differentiable at $b = f(a)$, then $g \circ f$ is differentiable at a, and*

$$d(g \circ f)(a) = dg(b) \circ df(a). \tag{1}$$

In other words, the differential of the composite is the composite of the differentials.

In proving the theorem it is handy to write the condition of differentiability in a slightly different, but obviously equivalent, form: f is differentiable at a with differential S if and only if

$$f(a + h) = f(a) + Sh + |h|\epsilon(h), \qquad \text{where } \epsilon(h) \to 0 \text{ as } h \to 0. \qquad (2)$$

Proof of the Theorem

If S is the differential of f at a, and T is the differential of g at b, then we have formula (2); similarly,

$$g(b + k) = g(b) + Tk + |k|\eta(k), \qquad \text{where } \eta(k) \to 0 \text{ as } k \to 0. \quad (3)$$

Setting $k = f(a + h) - f(a)$, we get

$$\begin{aligned}
g \circ f(a + h) - g \circ f(a) = g(b + k) - g(b) &= Tk + |k|\eta(k) \\
&= T(Sh + |h|\epsilon(h)) + |k|\eta(k) \\
&= T \circ Sh + |h|T\epsilon(h) + |k|\eta(k). \qquad (4)
\end{aligned}$$

In accordance with the interpretation of differentiability by formula (2), what has to be proved is that

$$T\epsilon(h) \to 0 \text{ as } h \to 0, \qquad \text{and} \qquad \frac{|k|\eta(k)}{|h|} \to 0 \text{ as } h \to 0. \qquad (5)$$

The first part is clear, for

$$|T\epsilon(h)| \leq \|T\| \, |\epsilon(h)| \qquad \text{and} \qquad \epsilon(h) \to 0.$$

As for the second, we have $k = Sh + |h|\epsilon(h)$, so that

$$|k| \leq |h|(\|S\| + |\epsilon(h)|),$$

which shows first that $k \to 0$ as $h \to 0$, and then that

$$\frac{|k|\eta(k)}{|h|} \leq (\|S\| + |\epsilon(h)|)|\eta(k)| \to 0.$$

Since the matrix of the differential is made up of the partial derivatives of the coordinate functions, Theorem 4.1 includes the formula for the partial derivatives of a composite function. Using Theorem 8.1 of Chapter 9 on the matrix of a product of linear transformations, we get

THEOREM 4.2

Let $\mathbf{R}^l \xrightarrow{f} \mathbf{R}^m \xrightarrow{g} \mathbf{R}^n$. If f is differentiable at a, and g is differentiable at $b = f(a)$, then

$$D_j(g \circ f)_i = \sum_{k=1}^{m} (D_k g_i)(D_j f_k), \qquad (6)$$

where the partial derivatives of f and $g \circ f$ are calculated at a, and those of g are calculated at b.

The theorem is obviously impossible to remember, but in the right notation it becomes easy. With $y = f(x)$ and $z = g(y)$ it becomes

$$\frac{\partial z_i}{\partial x_j} = \sum_{k=1}^{m} \frac{\partial z_i}{\partial y_k} \frac{\partial y_k}{\partial x_j}. \tag{7}$$

The notation is arranged so that if we think of the things that look like fractions as fractions, then the tops and bottoms cancel out nicely.

Exercise 1   Prove the following theorem:

THEOREM 4.3   *Let* $\mathbf{R}^1 \xrightarrow{\varphi} \mathbf{R}^m \xrightarrow{f} \mathbf{R}^n$. *If* φ *is differentiable at* a, *and* f *is differentiable at* $b = \varphi(a)$, *then* $(f \circ \varphi)' = \langle \nabla f, \varphi' \rangle$.

Example   One way to define an "$(n - 1)$-dimensional surface" in $\mathbf{R}^n$ is by an equation $f(x) = 0$, where $f : \mathbf{R}^n \to \mathbf{R}^1$. For instance, $f(x, y) = 0$ defines a curve in the plane, $f(x, y, z) = 0$ a two-dimensional surface in $\mathbf{R}^3$, and so on. There are sticky problems in this, but let us play with the idea a little.

To say that a path $\varphi : \mathbf{R}^1 \to \mathbf{R}^n$ lies on the surface means that $f(\varphi(t)) = 0$ for each t; that is, that $f \circ \varphi = 0$. If this is so, then Theorem 4.3 says that $\varphi'(a)$ is perpendicular to $\nabla f(b)$. But $\varphi'(a)$ is the tangent vector to the path at b.

THEOREM 4.4   *Let* $f : \mathbf{R}^n \to \mathbf{R}^1$ *be differentiable at the point* b *on the surface* $f(x) = 0$.
The tangent vector to any path on the surface at the point b *is perpendicular to* $\nabla f(b)$.

This makes it very tempting to define the tangent subspace to the surface to be the orthogonal complement of $\nabla f(b)$—which, as we know, is an $(n - 1)$-dimensional subspace of $\mathbf{R}^n$ as long as $\nabla f(b) \neq 0$. Only one thing is lacking, and that is to know that every vector in $\nabla f(b)^\perp$ is tangent to some path on the surface at b. Then the tangent subspace would have a pretty description as the set of all tangent vectors to paths on the surface at b. This is absolutely all right (as long as $\nabla f(b) \neq 0$), but it will require the implicit-function theorem of Section 8 to prove it.

Exercise 2   Prove the result just discussed in the case that $f(x) = x_n - g(x_1, \ldots, x_{n-1})$. [This would mean in the plane, for instance, that the curve is given by an equation $y = g(x)$, instead of $f(x, y) = 0$; and in the three-space that the surface is given by an equation $z = g(x, y)$, instead of $f(x, y, z) = 0$.] Obviously, this is already an important and interesting case.

5   THE MEAN-VALUE THEOREM

Although the mean-value theorem is false for functions with values in $\mathbf{R}^n$, there is nothing wrong with it for functions from $\mathbf{R}^m$ to $\mathbf{R}^1$.

**THEOREM
5.1**

Let $f: \mathbf{R}^m \to \mathbf{R}^1$ be continuous on the closed segment from a to x and differentiable on the open segment.   Then there is a point ξ between a and x such that

$$f(x) - f(a) = df(\xi)(x - a) = \langle \nabla f(\xi), x - a \rangle. \tag{1}$$

Proof

Set $h = x - a$ and $g(t) = f(a + th)$.   Then $g = f \circ \varphi$, where

$$\varphi(t) = a + th; \qquad \text{hence } \varphi'(t) = h.$$

Assuming there is no trouble with differentiability, we can apply the ordinary mean-value theorem to g to get

$$f(x) - f(a) = g(1) - g(0) = g'(\tau) = \langle \nabla f(a + \tau h), \varphi'(\tau) \rangle$$
$$= \langle \nabla f(\xi), h \rangle,$$

and $\xi = a + \tau h$ is between a and $x = a + h$ because τ is between 0 and 1.

Exercise 1   Establish the differentiability needed in the above calculation and write out the proof in full.

Exercise 2   If the partial derivatives of $f: \mathbf{R}^m \to \mathbf{R}^n$ exist and are identically 0 on a connected open set G, then f is constant on G.

**DEFINITION
5.2**

A function $f: \mathbf{R}^m \to \mathbf{R}^n$ is continuously differentiable, or C^1, at a point a if it is differentiable at every point near a, and the partial derivatives are continuous at a.   It is C^1 on an open set if it is C^1 at each point of the open set.

Exercise 3   Let $f: \mathbf{R}^m \to \mathbf{R}^n$ be differentiable on a set D, and think of df as a function from D into the space $\mathfrak{L}_{mn}$ of linear transformations.   Show that f is C^1 at a if and only if D includes a ball with center a and df is continuous at a.

Remark   Usage varies.   In some books "C^1 at a" means "C^1 on a neighborhood of a."

**THEOREM
5.3**

Let $f: \mathbf{R}^m \to \mathbf{R}^n$ be C^1 at a.   For every $\epsilon > 0$ there is a $\delta > 0$ such that if $|x - a| < \delta$ and $|y - a| < \delta$, then

$$|f(x) - f(y) - df(a)(x - y)| \leq \epsilon |x - y|. \tag{2}$$

Proof It is enough to prove the theorem for each coordinate function, so we can suppose that $f : \mathbf{R}^m \to \mathbf{R}^1$. Let $\epsilon > 0$ be given. Since the partial derivatives are continuous at a, we can find $\delta > 0$ such that

$$|\nabla f(z) - \nabla f(a)| < \epsilon \qquad \text{if } |z - a| < \delta.$$

The mean-value theorem gives $f(x) - f(y) = \langle \nabla f(\xi), x - y \rangle$ with ξ between x and y; therefore, $|\xi - a| < \delta$. Hence

$$|f(x) - f(y) - df(a)(x - y)| = |\langle \nabla f(\xi) - \nabla f(a), x - y \rangle| \le \epsilon |x - y|,$$

by virtue of Cauchy–Schwarz.

Exercise 4 Show that it is really enough to prove the theorem for each coordinate function.

The following theorem is a good example of how simple properties of the differential reflect deep properties of the function itself. The first half can be proved easily now. The other will have to wait until Section 7.

THEOREM 5.4 *Let $f : \mathbf{R}^m \to \mathbf{R}^n$ be C^1 at a. If $df(a)$ is one to one, then f itself is one to one on some ball with center a. If $df(a)$ is onto, then f maps each neighborhood of a onto a neighborhood of $b = f(a)$; that is, for every $r > 0$ there is an $r' > 0$ such that*

$$f(B(a; r)) \supset B(b; r').$$

Proof of the First Half According to Theorem 8.5 of Chapter 9, there is a number $m > 0$ such that

$$|df(a)h| \ge m|h|.$$

Combined with formula (2), this gives

$$|f(x) - f(y)| \ge (m - \epsilon)|x - y| \qquad \text{if } |x - a| < \delta \text{ and } |y - a| < \delta, \quad (3)$$

and we have only to take $\epsilon < m$.

Exercise 5 In the first half of the theorem it must be true that $m \le n$, and in the second half that $m \ge n$.

Exercise 6 What does a function like the one in Figure 2 show about the necessity of the hypothesis that f is C^1 at a in Theorem 5.4?

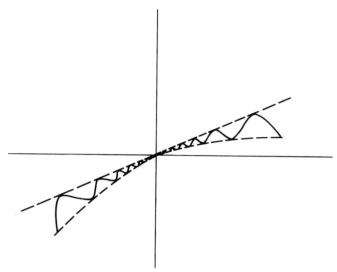

Figure 2

6   A FIXED-POINT THEOREM

A fixed point of a function $F: X \to X$ is a point $x \in X$ such that $F(x) = x$. There are two or three fundamental theorems that assert that fixed points exist under suitable conditions. One of them (the easy one) is proved here. It is not immediately apparent why such theorems are fundamental, but we shall give an application of obvious interest in Section 7.

THEOREM 6.1

Let X be a complete metric space, and let $F: X \to X$ have the property that

$$d\big(F(x), F(y)\big) \leq Md(x, y) \qquad \text{with } M < 1. \tag{1}$$

Then there is one and only one point $x \in X$ such that $F(x) = x$.

Proof

It is apparent that there cannot be more than one fixed point, for if x and y are both fixed points, then formula (1) gives $d(x, y) \leq Md(x, y)$ with $M < 1$. The problem is to show that there is at least one.

Let x_0 be any point of X. Set $x_1 = F(x_0)$, $x_2 = F(x_1)$, and in general $x_n = F(x_{n-1})$. If the sequence $\{x_n\}$ converges, say $x_n \to x$, then it follows that

$$F(x_n) \to F(x) \qquad \text{and} \qquad F(x_n) = x_{n+1} \to x. \tag{2}$$

Since a sequence can converge to at most one point, it must be true that $F(x) = x$, and the theorem is proved.

Exercise 1   The first part of formula (2) depends on the fact that F is continuous.   Indeed, F is uniformly continuous; given $\epsilon > 0$, one can take $\delta = \epsilon$.

The problem that remains is to show that the sequence $\{x_n\}$ does converge. Notice that from property (1) it follows that

$$d(x_1, x_2) = d\big(F(x_0), F(x_1)\big) \leq Md(x_0, x_1),$$
$$d(x_2, x_3) = d\big(F(x_1), F(x_2)\big) \leq Md(x_1, x_2) \leq M^2d(x_0, x_1),$$
$$d(x_3, x_4) = d\big(F(x_2), F(x_3)\big) \leq Md(x_2, x_3) \leq M^3d(x_0, x_1),$$

and in general that

$$d(x_k, x_{k+1}) \leq M^k d(x_0, x_1).$$

If $n > m$, then by the triangle inequality we have

$$d(x_m, x_n) \leq \sum_{k=m}^{n-1} d(x_k, x_{k+1}) \leq d(x_0, x_1) \sum_{k=m}^{n-1} M^k. \qquad (3)$$

If m and n are large enough, the term on the right is as small as we please, for the series

$$\sum_{k=0}^{\infty} M^k$$

converges, when $M < 1$.   This shows that the sequence $\{x_n\}$ is Cauchy, so it must converge, as X is complete.

Exercise 2   Show that the sum in (3) is equal to

$$\frac{M^m - M^n}{1 - M}$$

and hence that

$$d(x_m, x) \leq \frac{M^m}{1 - M} d(x_0, x_1). \qquad (4)$$

Formula (4) is interesting in numerical work.   If the point x_m, which is obtained by explicit calculation, is considered as an approximation to the fixed point x, then formula (4) gives the error.

7   THE INVERSE-FUNCTION THEOREM

The inverse-function theorem is one of the prettiest and most important theorems of calculus.   It says roughly that to decide whether a function $f : \mathbf{R}^n \to \mathbf{R}^n$ has an inverse, you just look at the differential.   If the linear function $df(a)$ has an inverse, then f itself has an inverse at least on a neighborhood of $b = f(a)$. Equivalently, the equations $f(x) = y$ have a unique solution for each y near b if the linear equations $df(a)x = y$ do.

THEOREM
7.1

(Inverse-Function Theorem) *Let $f : \mathbf{R}^n \to \mathbf{R}^n$ be C^1 at a. If $df(a)$ is invertible, then f itself is locally invertible in the sense that there is a function φ which is defined on a neighborhood of $b = f(a)$, is differentiable at b, and satisfies*

$$f \circ \varphi = I \quad and \quad \varphi \circ f = I.$$

If f is C^1 on a neighborhood of a, then φ is C^1 on a neighborhood of b.

Before turning to the proof, let us discuss the theorem. The hypothesis that $df(a)$ is invertible is plainly necessary, for if $f \circ \varphi = I$ and $\varphi \circ f = I$, then the chain rule for composite functions shows that

$$df(a) \circ d\varphi(b) = I \quad and \quad d\varphi(b) \circ df(a) = I.$$

Incidentally, this shows why we consider functions from $\mathbf{R}^n$ to $\mathbf{R}^n$ rather than functions from $\mathbf{R}^m$ to $\mathbf{R}^n$ with $m \neq n$. A linear transformation from $\mathbf{R}^m$ to $\mathbf{R}^n$ can never be both one to one and onto if m and n are different. (There is still something that can be said, though, as we shall see later.)

It is natural to ask whether there is a reason for singling out a particular point a and claiming only that f is locally invertible on a neighborhood of $b = f(a)$. Suppose that f is C^1 on an open set G and that $df(a)$ is invertible for every $a \in G$. Is it then true that f has an inverse that is defined on the whole set $f(G)$? Equivalently, is it true that f is one to one on the whole set G? In dimension 1 we know that the answer is "yes" if G is an interval (why?); but in higher dimensions the answer is "no." Consider the function $f : R^2 \to R^2$ defined by

$$f(z) = z^2 \quad complex\ multiplication.$$

It is easily checked that

$$df(a)h = 2ah,$$

so $df(a)$ is invertible for each $a \neq 0$, that is, for each a in the open set $G = R^2 - \{0\}$. On the other hand, f is not one to one on G, for every complex number $w \neq 0$ has two square roots—if one of them is z, then the other is $-z$.

To look at things a little more closely, let H_+ and H_- be the open half-planes bounded by a line through 0. It is plain that if z lies in H_+, then $-z$ lies in H_-, and therefore that f is one to one on both H_+ and H_- and that

$$f(H_+) = f(H_-).$$

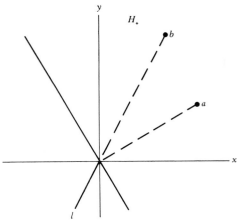

Figure 3

Calling this set K, we have two local inverses:

$$\varphi_+:K \to H_+ \quad \text{and} \quad \varphi_-:K \to H_- \quad (\varphi_- = -\varphi_+).$$

To be a little more explicit, let a be a given point $\neq 0$, and let $b = f(a)$. Let H_+ be the half-space that contains a and is bounded by the line through 0 that is perpendicular to a (Figure 3). Then $f(H_+) = \mathbf{R}^2 - l$, where l is the half-line determined by $-b$, so φ_+ is defined on $\mathbf{R}^2 - l$. φ_+ is the continuous square-root function that is defined on this set and takes the value a at the point b. Note that if w is close to $-b$ and on one side of l, then $\varphi_+(w)$ is close to ia; while if w is close to $-b$ and on the other side of l, then $\varphi_+(w)$ is close to $-ia$. This suggests (correctly) that there is no continuous square-root function on a ring around the origin (Figure 4). Nevertheless, there is a continuous

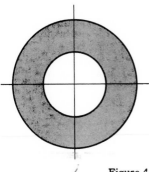

Figure 4

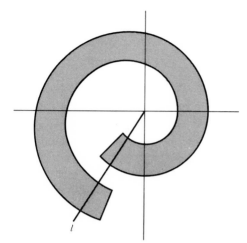

Figure 5

square-root function on the set shown in Figure 5. Try to give a formula in terms of φ_+ and φ_-.

The question of what are the "biggest" sets on which the local inverses are defined is a complicated one. The inverse-function theorem does not attempt to answer it, but merely maintains that the restriction of f to some neighborhood of a has an inverse that is defined on some neighborhood of $b = f(a)$ and has suitable differentiability properties. We shall see that even this weaker kind of result is very powerful.

Remark   A few years ago some questions of this kind came up in my own research, and I thought it might be helpful to see what Jacobi himself had to say on the subject of Jacobians and the inverse-function theorem. I dug back into the dusty old pages, and found to my considerable disgust that the paper was written in Latin. However, I had fancied myself pretty good in high school Latin, and also I thought that the formulas should provide plenty of clues. Whether I was good in high school Latin, I am no longer at all sure. As for the formulas, neatly displayed in the center of each page was

$$y = f(x)$$

and no other. So I have no idea what Jacobi himself had to say on the subject of Jacobians and the inverse-function theorem (and I did not succeed in solving the problem).

Karl Jacobi

Now let us turn to the proof.   It is convenient to begin with the special case in which $df(a) = I$.   The calculations are a little easier, and they give more explicit information that is useful for other purposes (e.g., changes of variable in multiple integrals).   This more explicit information is contained in the following theorem.

THEOREM 7.2   *Let $g : \mathbf{R}^n \to \mathbf{R}^n$ be C^1 at a, and let $g(a) = b$ and $dg(a) = I$.   For every ϵ, $0 < \epsilon < 1$, there is a $\delta > 0$ such that if $r < \delta$, then*

$$B(b; (1 + \epsilon)r) \supset g(B(a; r)) \supset B(b; (1 - \epsilon)r). \tag{1}$$

Proof   If ϵ is given, recall that $dg(a) = I$, and choose $\delta > 0$ so that (Theorem 5.3)

$$|g(x) - g(y) - (x - y)| \leq \epsilon |x - y| \qquad \text{if } x, y \in B(a; \delta). \tag{2}$$

Then, in particular,

$$(1 - \epsilon)|x - y| \leq |g(x) - g(y)| \leq (1 + \epsilon)|x - y| \text{ if } x, y \in B(a; \delta). \tag{3}$$

The left side of this inequality shows that g is one to one on $B(a; \delta)$. And the right side, with $y = a$, shows that

$$g(B(a; r)) \subset B(b; (1 + \epsilon)r),$$

which is the first half of formula (1).

To prove the second half of formula (1), we shall use the fixed-point theorem with

$$X = \bar{B}(a; r) \qquad \text{and} \qquad F(x) = -g(x) + x + z,$$

where z is any point of $B(b; (1 - \epsilon)r)$. It is plain that

$$F(x) = x \qquad \text{if and only if } g(x) = z,$$

so all we have to prove is that the fixed-point theorem is applicable—that is, that X is complete (and we do know that a closed ball is complete), and that

$$F : \bar{B}(a; r) \to \bar{B}(a; r) \tag{4}$$

and

$$|F(x) - F(y)| \le M|x - y| \qquad \text{with } M < 1. \tag{5}$$

Formula (5) is the same as (2), with $M = \epsilon$, for z cancels in the difference. And formula (4) follows from (2), with $y = a$, for

$$|F(x) - a| = |-g(x) + b + x - a + z - b|$$
$$\le \epsilon|x - a| + |z - b| < \epsilon r + (1 - \epsilon)r = r.$$

Note that the strict inequality in the middle shows that $F : \bar{B}(a; r) \to B(a; r)$, so the fixed point x lies in the open ball, as required.

Now let us do the differentiability of the inverse of g. The function $h = g^{-1}$ is defined on the ball $B(b; (1 - \epsilon)\delta)$, and if z and w are any two points of this ball, then formulas (2) and (3) [with $x = h(z)$ and $y = h(w)$] give

$$\left| z - w - \bigl(h(z) - h(w) \bigr) \right| \le \epsilon|x - y| \le \frac{\epsilon}{1 - \epsilon}|z - w|$$
$$\text{if } z, w \in B(b; (1 - \epsilon)\delta). \tag{6}$$

This obviously implies that h is differentiable at b.

Exercise 1   Prove the last statement—and be careful that your proof does not show that h is differentiable at an arbitrary $w \in B(b; (1 - \epsilon)\delta)$.

Proof of Theorem 7.1   Let $T = df(a)$ and $g = T^{-1} \circ f$. The composite-function theorem shows that $dg(a) = I$, so everything proved above is applicable. The function $\varphi = h \circ T^{-1}$ is the one sought in the theorem. Indeed,

$$f \circ \varphi = T \circ g \circ h \circ T^{-1} = T \circ I \circ T^{-1} = I,$$
$$\varphi \circ f = h \circ T^{-1} \circ T \circ g = h \circ I \circ g = I,$$

and φ is differentiable because both h and T^{-1} are.

It remains to prove the last part of the theorem to the effect that if f is C^1 not just at a, but on a neighborhood of a, then φ is C^1 on a neighborhood of b. First of all, it is clear that φ is differentiable on a neighborhood of b, because what has been proved can be applied at each point of a neighborhood of a. Now, look at df and $d\varphi$ as functions from $\mathbf{R}^n$ to the space $\mathcal{L}_{nn}$ of linear transformations. From this point of view the composite function theorem says

$$d\varphi = \mathcal{R} \circ df \circ \varphi, \tag{7}$$

where $\mathcal{R}$ is the inverse function; that is, $\mathcal{R}(T) = T^{-1}$ for any invertible $T \in \mathcal{L}_{nn}$. Now, φ is continuous because it is differentiable, df is continuous by hypothesis, and $\mathcal{R}$ is continuous by Theorem 8.8 of Chapter 9. Consequently, $d\varphi$ is continuous.

Exercise 2 Check formula (7). In chasing around composite functions it is convenient to make a diagram of the following sort:

$$\begin{array}{ccc} \mathcal{L}_{nn} & \xrightarrow{\mathcal{R}} & \mathcal{L}_{nn} \\ df \uparrow & \varphi & \uparrow d\varphi \\ \mathbf{R}^n & \leftrightarrows & \mathbf{R}^n \\ & f & \end{array}$$

The assertion of formula (7) is that the arrows can be followed around in either direction with the same result.

Example Consider the function $f : \mathbf{R}^2 \to \mathbf{R}^2$ given by

$$x = r \cos \theta, \qquad y = r \sin \theta. \tag{8}$$

Ordinarily, these are interpreted geometrically as the equations between rectangular and polar coordinates in the same geometric plane, but they can also be interpreted as defining a function from $\mathbf{R}^2$ to $\mathbf{R}^2$ which fits into the present scheme. The Jacobi matrix is

$$\begin{pmatrix} \dfrac{\partial x}{\partial r} & \dfrac{\partial x}{\partial \theta} \\ \dfrac{\partial y}{\partial r} & \dfrac{\partial y}{\partial \theta} \end{pmatrix} = \begin{pmatrix} \cos \theta & -r \sin \theta \\ \sin \theta & r \cos \theta \end{pmatrix}.$$

The determinant of this matrix is r, so the differential is invertible if and only if $r \neq 0$. The inverse-function theorem gives the following: Take any (r_0, θ_0) with $r_0 \neq 0$ and the corresponding (x_0, y_0). If (x, y) is close enough to (x_0, y_0), then the equations (8) have a unique solution $r = \varphi(x, y)$, $\theta = \psi(x, y)$ which is close to (r_0, θ_0), and the functions φ and ψ are C^1 on a neighborhood of (x_0, y_0).

Exercise 3 In this polar coordinate example, analyze the "global" existence of the local inverse as was done in the example $f(z) = z^2$.

Exercise 4 Discuss the equations $x = u^2 - v^2$, $y = uv$, and also the equations $x = u^2 + v^2$, $y = uv$.

Exercise 5 It is often convenient to study a surface

$$x_n = \varphi(x_1, \ldots, x_{n-1})$$

by flattening it out with the transformation

$$y_i = x_i \quad \text{for } i < n \qquad \text{and} \qquad y_n = x_n - \varphi(x_1, \ldots, x_{n-1}).$$

Show that the differential is invertible and calculate the Jacobi matrix. Discuss the transformation. What happens to the surface?

One part of the inverse function theorem is all right for functions from $\mathbf{R}^m$ to $\mathbf{R}^n$ with $m > n$. The result was stated initially as Theorem 5.4.

THEOREM 7.3 *Let $f: \mathbf{R}^m \to \mathbf{R}^n$ be C^1 at a. If $df(a)$ maps $\mathbf{R}^m$ onto $\mathbf{R}^n$, then f itself maps every neighborhood of a onto a neighborhood of $b = f(a)$. In fact, there is a function $\varphi: \mathbf{R}^n \to \mathbf{R}^m$ which is defined on a neighborhood of b, is differentiable at b, and satisfies $f \circ \varphi = I$. If f is C^1 on a neighborhood of a, then φ is C^1 on a neighborhood of b.*

Proof Since the dimension of the range of a linear transformation is the number of linearly independent columns in its matrix, it follows that $\partial f/\partial x$ has n linearly independent columns. Suppose for simplicity of notation that they are the first n, and define $g: \mathbf{R}^n \to \mathbf{R}^n$ by

$$g(x) = f(x, a_{n+1}, \ldots, a_m).$$

Now apply the inverse-function theorem to g.

Exercise 6 The function φ of the theorem is not precisely the local inverse of g. What is it? Is there just one function φ that satisfies the conditions of the theorem? Draw a picture.

Exercise 7 It is tempting to conjecture a corresponding theorem on the existence of a "left inverse"—a function φ such that $\varphi \circ f = I$. State the conjecture, and disprove it by using a figure six.

Exercise 8 Let $f: \mathbf{R}^m \to \mathbf{R}^n$ be C^1 on an open set G. If at each point of G, df maps $\mathbf{R}^m$ onto $\mathbf{R}^n$, then $f(G)$ is open in $\mathbf{R}^n$.

8   THE IMPLICIT-FUNCTION THEOREM

The inverse-function theorem involves the solution of equations $y = f(x)$, or equivalently $y - f(x) = 0$. The implicit-function theorem involves the solution of equations that appear more general at first, that is, $F(x, y) = 0$, but reduce quite easily to the former. Here $x \in \mathbf{R}^m$ and $y \in \mathbf{R}^n$, and the point is to solve for y when x is given. In order to have a decent expectation of a unique solution, there should be as many equations as unknowns, so F should be a function from $\mathbf{R}^{m+n}$ to $\mathbf{R}^n$. If the equation $F(x, y) = 0$ is written out in full in terms of all the coordinates, it looks like this:

$$
\begin{aligned}
F_1(x_1, \ldots, x_m, y_1, \ldots, y_n) &= 0 \\
F_2(x_1, \ldots, x_m, y_1, \ldots, y_n) &= 0 \\
\cdot \\
\cdot \\
\cdot \\
F_n(x_1, \ldots, x_m, y_1, \ldots, y_n) &= 0
\end{aligned}
\tag{1}
$$

which shows mainly that there is good reason for writing $F(x, y) = 0$ instead, and for introducing some suitable additional notation.

DEFINITION 8.1

If $F:\mathbf{R}^{m+n} \to \mathbf{R}^n$, then $d_y F$ is the differential of the function from $\mathbf{R}^n$ to $\mathbf{R}^n$ obtained by fixing x, and $\partial F/\partial y$ is the corresponding Jacobi matrix. Thus,

$$
\frac{\partial F}{\partial y} = \begin{pmatrix} \dfrac{\partial F_1}{\partial y_1} & \cdots & \dfrac{\partial F_1}{\partial y_n} \\ \dfrac{\partial F_n}{\partial y_1} & \cdots & \dfrac{\partial F_n}{\partial y_n} \end{pmatrix}.
$$

THEOREM 8.2

*(**Implicit-Function Theorem**)   Let $F:\mathbf{R}^{m+n} \to \mathbf{R}^n$ be C^1 at a point (a, b) with $F(a, b) = 0$. If $d_y F(a, b)$ is invertible, there are positive numbers ϵ and δ such that*

(a) If $|x - a| < \delta$, then there is one and only one point $y = \varphi(x)$ satisfying

$$
|y - b| < \epsilon \qquad and \qquad F(x, y) = 0.
$$

(b) The function φ is differentiable at a and satisfies

$$
\frac{\partial F}{\partial x} + \frac{\partial F}{\partial y}\frac{\partial \varphi}{\partial x} = 0 \qquad or \qquad \frac{\partial F_i}{\partial x_j} + \sum_{k=1}^{n} \frac{\partial F_i}{\partial y_k}\frac{\partial \varphi_k}{\partial x_j} = 0.
\tag{2}
$$

(c) If F is C^1 on a neighborhood of (a, b), then φ is C^1 on a neighborhood of a.

The partial derivatives of φ are evaluated at a, those of F at (a, b). Formula (2) results from differentiating the equation $F(x, \varphi(x)) = 0$. Note that this formula gives the means to calculate the partial derivatives of φ, since the matrix $\partial F/\partial y$ is invertible.

Exercise 1 Verify formula (2).

Proof   The theorem is proved by using the inverse-function theorem on the function $f:\mathbf{R}^{m+n} \to \mathbf{R}^{m+n}$ defined by

$$f(x, y) = (x, F(x, y)). \tag{3}$$

The equation $F(x, y) = 0$ is equivalent to the equation $f(x, y) = (x, 0)$. Assume for the moment that $df(a, b)$ is invertible, so f has a local inverse ψ defined on some neighborhood of $(a, 0)$. If x is near a, then there is one and only one solution of the equation $f(x, y) = (x, 0)$ which is near (a, b), namely $\psi(x, 0)$. In other words, if J and P are the functions

$$J:\mathbf{R}^m \to \mathbf{R}^{m+n}, \qquad Jx = (x, 0),$$
$$P:\mathbf{R}^{m+n} \to \mathbf{R}^n, \qquad P(x, y) = y,$$

then

$$\varphi = P \circ \psi \circ J \tag{4}$$

is the function sought in the theorem. Since J and P are undoubtedly differentiable—and by Theorem 7.1, ψ is, too—it follows that φ is differentiable. So all depends upon showing that $df(a, b)$ is invertible.

The matrix of $df(a, b)$ is

$$\begin{pmatrix} I & 0 \\ \dfrac{\partial F}{\partial x} & \dfrac{\partial F}{\partial y} \end{pmatrix},$$

where I stands for the m by m identity matrix, and 0 stands for the m by n 0 matrix. Therefore,

$$df(a, b)(h, k) = (h, d_xFh + d_yFk).$$

If follows that if $df(a, b)(h, k) = 0$, then first $h = 0$, and then $d_yFk = 0$. But then also $k = 0$, since d_yF is one to one. Thus, $df(a, b)$ is invertible.

Exercise 2 There are ϵ's and δ's in the statement of Theorem 8.2, but not in the proof. Write out the proof in enough detail to produce them.

The following version of the implicit-function theorem can be regarded as an improvement of Theorem 7.3.

**THEOREM
8.3**

Let $F:\mathbf{R}^{m+n} \to \mathbf{R}^n$ be C^1 at (a, b), and let $F(a, b) = c$. If $d_y F(a, b)$ is invertible, then there are positive numbers ϵ and δ such that

(a) If $|x - a| < \delta$ and $|z - c| < \delta$, then there is one and only one point $y = \varphi(x, z)$ satisfying

$$|y - b| < \epsilon \qquad and \qquad F(x, y) = z.$$

(b) The function φ is differentiable at (a, b) and satisfies

$$\frac{\partial F}{\partial x} + \frac{\partial F}{\partial y}\frac{\partial \varphi}{\partial x} = 0 \qquad and \qquad \frac{\partial F}{\partial y}\frac{\partial \varphi}{\partial z} = I.$$

(c) If F is C^1 on a neighborhood of (a, b), then φ is C^1 on a neighborhood of (a, c).

The theorem is proved by applying the implicit-function theorem to the function $G:\mathbf{R}^{m+n+n} \to \mathbf{R}^n$ defined by

$$G(x, y, z) = F(x, y) - z.$$

The point (x, z) plays the role that x played before and (a, c) the role that a played.

Exercise 3 Instead of deducing Theorem 8.3 from the implicit-function theorem, prove it directly by going back over the proof of the implicit-function theorem.

How is Theorem 8.3 an improvement on Theorem 7.3? It shows not only that F maps every neighborhood of (a, b) onto a neighborhood of c, but also that the points that map onto a given point z form an "m-dimensional surface" given by an equation $y = \varphi(x, z)$ and that the surface varies differentiably as z varies. [This refers, however, not to all the points that map onto z, but to those in a neighborhood of (a, b).]

What about the hypotheses? In Theorem 7.3 the hypothesis is that dF maps $\mathbf{R}^{m+n}$ onto $\mathbf{R}^n$, while in Theorem 8.3 it is that $d_y F$ is invertible. The two look different, but the difference is more one of appearance than of substance. If dF maps $\mathbf{R}^{m+n}$ onto $\mathbf{R}^n$, then its matrix must have n linearly independent columns. If the coordinates are relabeled so that these columns correspond to the y's and the others to the x's, then the two hypotheses are exactly the same.

Example Let $F(x, y) = x^2 + y^2 - 1$, so that $F(x, y) = 0$ is the equation of the unit circle in the plane. We have

$$\frac{\partial F}{\partial x} = 2x, \qquad \frac{\partial F}{\partial y} = 2y,$$

and, consequently,

$$d_x F(a, b)h = 2ah, \qquad d_y F(a, b)k = 2bk, \qquad dF(a, b)(h, k) = 2ah + 2bk.$$

The implicit-function theorem says that the equation $x^2 + y^2 = 1$ can be solved for y as a function of x in a neighborhood of any point (a, b) with $a^2 + b^2 = 1$ and $b \neq 0$. A glance at the graphs shows that it cannot be solved for y in any neighborhood of either of the points $(\pm 1, 0)$. The equation can be solved for x in a neighborhood of any (a, b) with $a^2 + b^2 = 1$ and $a \neq 0$.

From the point of view of Theorem 8.3, we see that the points which satisfy $F(x, y) = z$ form a nice curve—the circle with center 0 and radius $1 + z$—as long as $z > -1$. For $z < -1$ there are no such points, and for $z = -1$ there is just the one point $(0, 0)$, which is the one point at which dF does not map $\mathbf{R}^2$ onto $\mathbf{R}^1$.

Exercise 4   Discuss the solution of the equation $xy^2 - 2y + 1 = 0$. Draw a picture of the set of points in the plane that satisfy the equation.

Exercise 5   Discuss the solution of the equation $y^2 - x^3 = 0$ and draw the picture. Notice that near the point $(0, 0)$ the equation can be solved for x but not for y. The solution for x is not differentiable. At this point both $d_x F$ and $d_y F$ are 0.

11 { Surfaces

1 ALGEBRAIC CURVES

A (real) algebraic curve is the set of points in the plane satisfying an equation $F(x, y) = 0$, where F is a polynomial:

$$F(x, y) = \sum_{j,k=0}^{n} a_{jk} x^j y^k. \tag{1}$$

It is a good exercise to try to get some information about such curves by using the implicit-function theorem and some of the other basic results of calculus.

There are two kinds of points $x = a$ that cause mischief. The first is obvious from the implicit-function theorem. If the equations

$$F(a, y) = 0 \qquad \text{and} \qquad \frac{\partial F(a, y)}{\partial y} = 0 \tag{2}$$

have a common solution $y = b$, then on the one hand the point (a, b) is on the curve, but on the other we cannot expect to be able to solve for y as a function of x.

The second kind is less obvious. In formula (1) collect together all the terms with a given power of y and write

$$F(x, y) = \sum_{k=0}^{n} A_k(x) y^k. \tag{3}$$

Each A_k is then a polynomial in x. The second points that cause mischief are the points a such that

$$A_n(a) = 0, \tag{4}$$

that is, the points where the coefficient of the highest power of y vanishes. The reason these must be avoided will be apparent in the proofs.

In order to start out, we shall need two theorems from algebra to show that these troublemakers are only finite in number.

THEOREM 1.1   *The polynomial*

$$p(x) = \sum_{j=0}^{n} a_j x^j, \qquad a_n \neq 0, \tag{5}$$

does not vanish at more than n points.

Proof   For any α we can write

$$p(x) = \sum_{j=0}^{n} \alpha_j (x - \alpha)^j \tag{6}$$

simply by putting $x = (x - \alpha) + \alpha$ in the original formula and multiplying out. (Taylor's formula for polynomials!) It is clear that $\alpha_0 = p(\alpha)$; so if $p(\alpha) = 0$, then each term in the sum in (6) contains $x - \alpha$, and

$$p(x) = (x - \alpha)q(x) = (x - \alpha) \sum_{j=1}^{n} \alpha_j (x - \alpha)^{j-1}.$$

We can use induction on the number n, which is called the degree of p. Since q has degree $n - 1$, it cannot vanish at more than $n - 1$ points, and p cannot vanish except at these and at α.

This takes care of the points that satsify equation (4). The other theorem from algebra is not so easy, and we shall simply assume it. The polynomial $F(x, y)$ is said to have a square factor if there exist polynomials G and H such that $F = G^2H$. If this is the case, then $F_1 = GH$ vanishes at exactly the same set of points, so in studying algebraic curves we can assume that F has no square factor. The second theorem from algebra is the following:

THEOREM 1.2   *If the polynomial $F(x, y)$ has no square factor, then there exist polynomials $A(x, y)$, $B(x, y)$, and $R(x)$ such that*

$$R = AF + B \frac{\partial F}{\partial y}, \qquad R \neq 0. \tag{7}$$

THEOREM 1.3   *If the polynomial $F(x, y)$ has no square factor, then there are only a finite number of points α such that the equations*

$$F(\alpha, y) = 0 \qquad and \qquad \frac{\partial F}{\partial y}(\alpha, y) = 0 \tag{8}$$

have a common solution.

Proof If the equations (8) do have a common solution, then by (7) we have $R(\alpha) = 0$, and by Theorem 1.1 this can happen for only a finite number of α.

Exercise 1 Show that both theorems are false if F has a square factor.

Exercise 2 Calculate the A, B, and R for the polynomials $x^2 + y^2 - 1$, $xy^2 - 2y + 1$, and $y^2 - x^3$, which appear in the exercises at the end of the last section.

Henceforth, we shall assume that F contains no square factor and will let N be the finite set of points a such that either $A_n(a) = 0$ or the equations $F(a, y) = 0$ and $\partial F(a, y)/\partial y = 0$ have a common solution. The main theorem that can be proved is as follows.

THEOREM 1.4 *Let $I = (\alpha, \beta)$ be any open interval that contains no point of the finite set N. If for some $a \in I$ the equation $F(a, y) = 0$ has exactly k distinct solutions, then for every $x \in I$ the equation $F(x, y) = 0$ has exactly k distinct solutions, and there are functions $\varphi_1, \ldots, \varphi_k$ on I such that*
(a) $F(x, y) = 0$ if and only if $y = \varphi_j(x)$ for some j.
(b) Each φ_j is C^1 on I.
(c) φ_j has a limit at α and at β, provided $\pm \infty$ are allowed.

In two respects the theorem provides more precise information than the implicit-function theorem. It shows that the local solutions given by the latter actually exist on the whole interval I and that the limits exist at the end points. A typical picture of an algebraic curve would be as in Figure 1.

Proof If b is one of the solutions of $F(a, y) = 0$, then since $a \in I$, it follows that $\partial F(a, b)/\partial y \neq 0$; the implicit-function theorem provides a function φ that is C^1 on an interval $(a - \delta, a + \delta)$ and satisfies $\varphi(a) = b$ and $F(x, \varphi(x)) = 0$.

Let d be the upper bound of the numbers d' such that there is a function ψ that is C^1 on $(a - \delta, d')$ and satisfies $\psi(a) = b$ and $F(x, \psi(x)) = 0$.

LEMMA 1.5 *Any two continuous functions ψ_1 and ψ_2 on an interval $[a, d')$ which satisfy the conditions*
$$\psi(a) = b \quad and \quad F(x, \psi(x)) = 0$$
must be identical.

Proof Since ψ_1 and ψ_2 are continuous, the subset of $[a, d')$ on which $\psi_1 = \psi_2$ is a closed subset. By the uniqueness in the implicit-function theorem, it is also an open subset—hence, the whole interval, as an interval, is connected.

By Lemma 1.5 there is a unique function, which we shall call φ again, which is continuous on $(a - \delta, d)$ and satisfies $\varphi(a) = b$ and $F(x, \varphi(x)) = 0$,

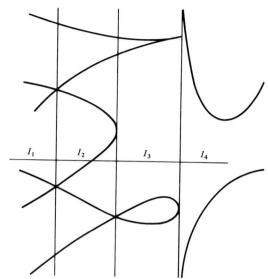

I_1 I_2 I_3 I_4

Figure 1

and this φ is automatically C^1.   It is all right to call the new function φ, because Lemma 1.5 shows that it coincides with the original φ on $(a - \delta, a + \delta)$.   We have to show, of course, that $d = \beta$, but some other steps come first.

LEMMA 1.6

φ has a limit at d.

Proof

Let l_1 be the limit inferior of $\varphi(x)$ as x approaches d from the left, and let l_2 be the limit superior.   If the two are different, let l be any number between them.   There must be points x' as close to d as we please with $\varphi(x') < l$, and points x'' as close to d as we please with $\varphi(x'') > l$, and therefore points x as close to d as we please with $\varphi(x) = l$.   But this implies that the equation $F(x, l) = 0$ has infinitely many solutions, which is impossible by Theorem 1.1.

LEMMA 1.7

If $d < \beta$, then the limit of φ at d must be finite.

Proof

Here is where the fact that the leading coefficient $A_n(d)$ does not vanish comes in.   Suppose that $p(y) = 0$, where

$$p(y) = \sum_{k=0}^{n} A_k y^k. \tag{9}$$

Then

$$A_n y^n = - \sum_{k=0}^{n-1} A_k y^k,$$

so if $|y| \geq 1$, then

$$|y| \leq \frac{1}{|A_n|} \sum_{k=0}^{n-1} |A_k|. \tag{10}$$

Now, $y = \varphi(x)$ does satisfy the equation $p(y) = 0$, and each A_k is bounded on some interval with center d, while A_n is bounded away from 0. Hence (10) gives a bound for $\varphi(x)$.

Now we can show that $d = \beta$. If not, simply let l be the limit of $\varphi(x)$ as x approaches d and apply the implicit-function theorem at the point (d, l) to extend φ beyond d.

At this point the theorem is effectively proved. The same argument to the left of a instead of to the right pushes the interval on which φ is defined out to the whole interval I. Using each of the solutions $b_1, \ldots, b_k$ of $F(a, y) = 0$, we get solutions $\varphi_1, \ldots, \varphi_k$, which are C^1 on I, and, according to Lemma 1.6, have limits (possibly $\pm\infty$) at α and β. If at some other point c there were additional solutions to $F(c, y) = 0$, then we could start at c and perform the same construction to obtain an additional solution at a. (Note Lemma 1.5, which says that no two solutions can be equal at a single point unless they are identical!)

This proof is an example of a fundamental process called *analytic continuation.*

Exercise 3   Go through the exercises at the end of the last section again in the light of the general theorem. Find the points in N, examine their significance, and so on.

Remark   The results of this section, which depend partly on the unproved algebraic Theorem 1.2, are included for their intrinsic interest. They are not results that will be needed in the sequel.

2   MANIFOLDS

In the next few sections we shall study various features of smooth surfaces in $\mathbf{R}^n$. There are several ways to define such surfaces. A curve in the plane (one-dimensional surface in $\mathbf{R}^2$) has appeared in the following guises:

(a) $S = \{(x, y): y = f(x)\}$.
(b) $S = \{(x, y): F(x, y) = 0\}$.
(c) $S = \varphi(I)$, where $\varphi: I \to \mathbf{R}^2$.

In (a), S is the graph of the function f.   In (b), S is perhaps an algebraic curve of the kind discussed in Section 1.   In (c), S is a parametric curve or a path.

Consider the analogs in $\mathbf{R}^3$.

(a) $S = \{(x, y, z): z = f(x, y)\}$.
(b) $S = \{(x, y, z): F(x, y, z) = 0\}$.
(c) $S = \varphi(I)$, where $\varphi: I \to \mathbf{R}^3$.

In (a), S is the graph of the function f.   It is not a curve, but a two-dimensional surface.   For instance, the equations

$$z = 2x - 3y + 1 \qquad \text{and} \qquad z = \sqrt{1 - x^2 - y^2}$$

define respectively a plane and the top half of the unit sphere.   In (b), S is again a two-dimensional surface.   For instance, the equation $x^2 + y^2 + z^2 - 1 = 0$ defines the unit sphere.   In (c), however, if I is an interval, then S is a parametric curve or path.   The way to get a two-dimensional surface out of (c) is to take I to be a square (or disk or some other two-dimensional figure).   The way to get a curve out of (b) is to intersect a pair of surfaces, that is, to take a pair of equations $F(x, y, z) = 0$ and $G(x, y, z) = 0$—or, equivalently, to take a single equation $F(x) = 0$, where F is a function from $\mathbf{R}^3$ to $\mathbf{R}^2$.

Exercise 1   What is the curve in $\mathbf{R}^3$ described by the pair of equations $2x - 3y - z + 1 = 0$ and $x^2 + y^2 + z^2 - 1 = 0$?

The way to get a curve out of (a) is also to intersect a pair of surfaces, but in this case the surfaces are usually taken to be special ones:

$$S = \{(x, y, z): y = f(x) \text{ and } z = g(x)\}. \qquad (\text{a}')$$

Exercise 2   What is special about these surfaces?

We are going to use the n-dimensional analog of (b) for the basic definition of a surface in $\mathbf{R}^n$, and then, of course, there will be a basic problem of showing how to pass back and forth between this and the analogs of (a) and (c).   The analog of (b) is to define an m-dimensional surface in $\mathbf{R}^n$ to be a set of points satisfying $n - m$ equations $F_j(x) = 0$, or, equivalently, a single equation $F(x) = 0$, where F is a function from $\mathbf{R}^n$ to $\mathbf{R}^{n-m}$.   In order to obtain a set with some resemblance to an intuitive m-dimensional surface, we shall have to put some restrictions on F, for we have the following result.

Exercise 3   If S is any closed set whatever in $\mathbf{R}^n$, let $F(x) = d(x, S)$ be the distance from x to S. Show that $F:\mathbf{R}^n \to \mathbf{R}^1$ is continuous and that $S = \{x:F(x) = 0\}$.

It is even possible to modify F so that it is of class C^1 on $\mathbf{R}^n$ (in fact of class C^∞, although we have not defined this yet). (See Exercise 18.) According to the tentative definition above, this would mean that every closed set in $\mathbf{R}^n$ is a surface of dimension $n - 1$, which certainly is not desirable. What we shall do is impose a condition on F that makes the surface smooth and of the right dimension.

DEFINITION 2.1   *A point $a \in \mathbf{R}^p$ is a regular point of a function $f:\mathbf{R}^p \to \mathbf{R}^q$ iff f is of class C^1 on a neighborhood of a and $df(a)$ has the maximum possible rank. If $p \leq q$, this means that the rank is p, and hence that $df(a)$ is one to one. If $p \geq q$, this means that the rank is q, and hence that $df(a)$ maps $\mathbf{R}^p$ onto $\mathbf{R}^q$.*

DEFINITION 2.2   *A smooth surface or smooth manifold of dimension m in $\mathbf{R}^n$ is a set $\mathbf{M}$ with the following property: For each point $a \in \mathbf{M}$ there is a function $F:\mathbf{R}^n \to \mathbf{R}^{n-m}$ which is regular on an open set G containing a and is such that*

$$\mathbf{M} \cap G = \{x:F(x) = 0\} \cap G.$$

Within the open set G, $\mathbf{M}$ is exactly the set of points satisfying the equation $F(x) = 0$. Ordinarily we shall abbreviate this by saying simply that $\mathbf{M} = \{x:F(x) = 0\}$ *on a neighborhood of a.*

Example   If $f:\mathbf{R}^m \to \mathbf{R}^{n-m}$ is of class C^1 on an open set $G \subset \mathbf{R}^m$, then the graph of f is a smooth m-dimensional manifold in $\mathbf{R}^n$.

In discussing the example we shall introduce some notation that will be standard through the next few sections. If x is a point of $\mathbf{R}^n$, then $\bar{x}$ will denote the first m coordinates of x. If y is a point of $\mathbf{R}^{n-m}$, the coordinates of y will be written $(y_{m+1}, \ldots, y_n)$, rather than $(y_1, \ldots, y_{n-m})$. In this notation the graph of f is the set

$$\mathbf{M} = \{x \in \mathbf{R}^n : x_i = f_i(\bar{x}), i > m, \bar{x} \in G\}.$$

The notation allows the use of matching indices on both sides of the equation $x_i = f_i(\bar{x})$ and makes it easier to keep track of the indices.

To show that the graph of f is a smooth m-dimensional manifold, let $F_i(x) = x_i - f_i(\bar{x})$, and let $F:\mathbf{R}^n \to \mathbf{R}^{n-m}$ be the function with coordinate functions F_i, $i > m$.   It is plain that

$$\mathbf{M} = \{x : F(x) = 0, \bar{x} \in G\},$$

.so the problem is to show that F is regular at each point.   The Jacobi matrix of F looks like this:

$$\left(-\frac{\partial f}{\partial \bar{x}} \quad I\right),$$

where I is the $(n - m)$ by $(n - m)$ identity matrix.   The last $n - m$ columns are plainly independent, so dF has rank $n - m$ and F is regular.

THEOREM 2.3   *Every smooth m-dimensional manifold* $\mathbf{M}$ *in* $\mathbf{R}^n$ *is locally the graph of a* C^1 *function.   More precisely, if a is a point of* $\mathbf{M}$, *then with a suitable relabeling of the coordinates there is a* C^1 *function* $f : \mathbf{R}^m \to \mathbf{R}^{n-m}$ *on a neighborhood of* $\bar{a}$ *such that*

$$\mathbf{M} = \{x : x_i = f_i(\bar{x}), i > m\} \qquad \textit{on a neighborhood of a.}$$

Exercise 4   Prove the theorem by using the implicit-function theorem.

Theorem 2.3 provides a local parametric representation of a smooth manifold.

THEOREM 2.4   *Let a be a point of a smooth m-dimensional manifold* $\mathbf{M}$ *in* $\mathbf{R}^n$. *There is a function* $\varphi : \mathbf{R}^m \to \mathbf{R}^n$ *which is regular on an open set* G *in* $\mathbf{R}^m$ *and such that* $\varphi(G)$ *is a neighborhood of a in* $\mathbf{M}$. *There is an inverse function* $P : \mathbf{R}^n \to \mathbf{R}^m$ *which is* C^1 *on a neighborhood of a in* $\mathbf{R}^n$ *and satisfies* $P \circ \varphi = I$ *on a neighborhood of Pa in* $\mathbf{R}^m$, *and* $\varphi \circ P = I$ *on a neighborhood of a in* $\mathbf{M}$.

Such a function φ is called a *local parametric representation* of $\mathbf{M}$ at a.   Note that $\varphi(G)$ is not a neighborhood of a in $\mathbf{R}^n$, but rather there is such a neighborhood G_n such that $\varphi(G) = \mathbf{M} \cap G_n$.   The inverse function P is defined on the full neighborhood G_n and is C^1 there, but it is the restriction of P to $\mathbf{M}$ that is the inverse of φ.

Proof   Use Theorem 2.3 to write $\mathbf{M}$ as the graph of a function f in a neighborhood of a, and set $\varphi(t) = \big(t, f(t)\big)$.   It is immediately checked that this does the job and that the inverse function is just the projection $Px = \bar{x}$.

COROLLARY 2.5   *The dimension of a smooth manifold is uniquely determined—a given set* $\mathbf{M}$ *cannot be a smooth manifold of two different dimensions.*

Proof   Suppose that $\mathbf{M} \subset \mathbf{R}^n$ is a smooth manifold of dimension m and also of dimension k.   Use the theorem to find local parametric representations

$\varphi : \mathbf{R}^m \to \mathbf{R}^n$ and $\psi : \mathbf{R}^k \to \mathbf{R}^n$.   Let P and Q be the inverse functions, and set $f = Q \circ \varphi$ and $g = P \circ \psi$.   Then $f \circ g = I$ on a neighborhood of Qa and $g \circ f = I$ on a neighborhood of Pa.   The chain rule gives the same equations for the differentials and shows that $df(Pa) : \mathbf{R}^m \to \mathbf{R}^k$ is both one to one and onto, which implies that $m = k$.

Exercise 5   Draw a picture to illustrate the proof.

Exercise 6   Let $\mathbf{M} \subset \mathbf{R}^n \subset \mathbf{R}^l$.   Show that if $\mathbf{M}$ is a smooth m-dimensional manifold in $\mathbf{R}^n$, then it is also a smooth m-dimensional manifold in $\mathbf{R}^l$.   (This was tacitly used in the proof of Corollary 2.5.)

THEOREM 2.6   *Let $\mathbf{M}$ be a smooth m-dimensional manifold in $\mathbf{R}^n$.   Let $\psi : \mathbf{R}^m \to \mathbf{M}$ be regular at the point t_0.   Then ψ is a local parametric representation of $\mathbf{M}$ at the point $a = \psi(t_0)$.*

The theorem says that if G is any open set containing t_0, then $\psi(G)$ is a neighborhood of $a = \psi(t_0)$ in $\mathbf{M}$, and there is a C^1 inverse function Q on some neighborhood of a in $\mathbf{R}^n$.   Before turning to the proof, consider the example of a figure six in the plane.   It is obvious that there is a one-to-one regular function $\psi : (0, 1) \to \mathbf{R}^2$ that traces out the figure six.   It is equally obvious that the inverse function is not even continuous at the point where the six comes together.   Consequently, the theorem shows that the figure six is not a smooth one-dimensional manifold, or even contained in one.

Proof   Use Theorem 2.3 to write $\mathbf{M}$ as the graph of a function f in a neighborhood of a, and let P be the projection on the first m coordinates.   The first step is to show that the function $g = P \circ \psi$ is regular at t_0.   Since P is linear, we have

$$dg(t_0) = P\, d\psi(t_0) = d\psi(t_0).$$

Since $\psi(t)$ lies on the graph of f, we have $\psi_i(t) = f_i(\bar{\psi}(t))$ for $i > m$; hence $d\psi_i(t_0) = df(\bar{a})d\bar{\psi}(t_0)$.   From this it follows that if $d\psi(t_0)h = 0$, then $d\psi(t_0)h = 0$, which is possible only if $h = 0$, because ψ is regular.   Thus, g is regular.

According to the inverse-function theorem, g has a local inverse h that is defined on a neighborhood of Pa.   The function we are looking for to invert ψ is just $Q = h \circ P$:

Q is plainly of class C^1 on a neighborhood of a and satisfies

$$Q \circ \psi = h \circ P \circ \psi = h \circ g = I$$

on a neighborhood of t_0.   If φ is the local parametric representation of **M** coming from the graph [that is, $\varphi(t) = (t, f(t))$ as in the proof of Theorem 2.4], then P is the inverse function to φ; so we have $\varphi \circ g = \varphi \circ P \circ \psi = \psi$, hence

$$\psi \circ Q = \varphi \circ g \circ Q = \varphi \circ g \circ h \circ P = I$$

on a neighborhood of a in **M**.

Exercise 7   Does the last equation imply that if G is open, then $\psi(G)$ is a neighborhood of a in **M**?

THEOREM 2.7   *Let $\varphi: \mathbf{R}^m \to \mathbf{R}^n$, $m \leq n$, be regular at the point t_0.   If r is small enough, then $\mathbf{M} = \varphi(B(t_0, r))$ is a smooth m-dimensional manifold in $\mathbf{R}^n$, and φ is a local parametric representation.*

Proof   Let $a = \varphi(t_0)$.   The idea is to relabel the x coordinates so as to be able to solve the equations

$$0 = G_i(x, t) = x_i - \varphi_i(t), \qquad i = 1, \ldots, n$$

for $x_{m+1}, \ldots, x_n, t_1, \ldots, t_m$ on a neighborhood of (a, t_0).   If we can find solutions

$$x_i = f_i(\bar{x}) \qquad \text{for } i > m, \, t = g(\bar{x}),$$

then we shall have expressed **M** as the graph of f in a neighborhood of a, and so will know that it is a smooth m-dimensional manifold.

The Jacobi matrix of the function $G: \mathbf{R}^{n+m} \to \mathbf{R}^n$ is the matrix

$$\left(I \left(-\frac{\partial \varphi}{\partial t} \right) \right),$$

where I is the n by n identity matrix.   This matrix has rank n, and the regularity of φ means that the last m columns are linearly independent. Therefore, by rearranging the first n columns, we can ensure that the last n columns are linearly independent.   Now use the implicit-function theorem to solve.   There are positive numbers ϵ and δ such that if $|\bar{x} - \bar{a}| < \delta$, then there is one and only one point (x, t), $x_i = f_i(\bar{x})$ for $i > m$ and $t = g(\bar{x})$, satisfying $G(x, t) = 0$, $|x - a| < \epsilon$, and $|t - t_0| < \epsilon$.

Exercise 8   Check that if r is sufficiently small, then $\varphi(B(t_0, r))$ is the graph of f in a neighborhood of a.   (Note that the example of the figure six shows that the restriction to small r is absolutely necessary.)

Knowing that **M** is a smooth m-dimensional manifold, we can conclude from Theorem 2.6 that φ is a local parametric representation.

Remark   In the theorems of this section we have used the term smooth manifold in preference to the term smooth surface, and we shall continue to do so.   The term smooth manifold is more in current favor, and also we shall want to use the term surface (though never smooth surface) in a slightly vague and nontechnical way.   We shall speak of the surface $F = 0$ or the parametric surface φ in cases where we do not assume a priori that F or φ is regular.   In such cases we shall have to realize that the surface may not be a surface at all from the intuitive point of view—but still the language is convenient.   Moreover, it is usually true that F or φ is regular at "most" points, so if we remove a "small" set from the surface, then what is left is actually a smooth manifold.   In the case of an algebraic curve, for example, the results of the last section show that if we remove a finite number of points, then what is left is a smooth one-dimensional manifold.

Exercise 9   Draw a picture of the surface in $\mathbf{R}^3$ defined by $z^3 = x^2 + y^2$.   Show that if the origin is removed, then what is left is a smooth two-dimensional manifold.

Exercise 10   The surface in $\mathbf{R}^3$ defined by $x^2 - y^2 = 0$ consists of two planes whose intersection is the z axis.   If the z axis is removed, then what is left is a smooth two-dimensional manifold.

Exercise 11   Discuss the curve in $\mathbf{R}^2$ that is defined in polar coordinates by the equation $r = \sin 2\theta$.   Draw a picture.   Show that the same curve is defined by the equation $0 = F(x, y) = (x^2 + y^2)^3 - 4x^2y^2$.   Show that F is regular at every point except 0, and hence that if 0 is removed, then what is left is a smooth one-dimensional manifold.   Show that the same curve is defined parametrically by the equations

$$x = 2 \sin \theta \cos^2 \theta, \qquad y = 2 \sin^2 \theta \cos \theta.$$

Which points are regular for the corresponding function φ?   Discuss this in the light of Theorem 2.7.

Exercise 12   Show that the torus (= hollow doughnut) obtained by revolving the circle $x^2 + (y - 2)^2 = 1$ around the x axis has the equation

$$0 = F(x, y, z) = (x^2 + y^2 + z^2 + 3)^2 - 16(y^2 + z^2).$$

Show that F is regular at each point of the torus, and hence that the torus is a two-dimensional manifold.

Exercise 13 If **M** is a smooth manifold, then each point $a \in \mathbf{M}$ has a neighborhood in **M** that is connected.   [*Hint:* Theorems 2.4 and 2.7 show that each point has a neighborhood of the form $\varphi\big(B(t_0, r)\big)$, where φ is a local parametric representation.]

Exercise 14 A compact smooth manifold has only a finite number of connected components. (*Hint:* If there is an infinite number of components, pick a sequence $\{x_k\}$ in distinct components and apply Exercise 13 at a point a which is a limit of some subsequence.)

The two extreme cases—zero-dimensional manifolds in $\mathbf{R}^n$ and n-dimensional manifolds in $\mathbf{R}^n$—are easy to describe.

THEOREM 2.8   *A set $\mathbf{M} \subset \mathbf{R}^n$ is a smooth zero-dimensional manifold if and only if $\mathbf{M}$ is isolated—that is, for each point $a \in \mathbf{M}$ there is an $r > 0$ such that $\mathbf{M} \cap B(a; r) = \{a\}$.*

Exercise 15 Prove the theorem by using the inverse-function theorem.

Exercise 16 A compact smooth zero-dimensional manifold consists of a finite number of points.   (*Hint:* Use Exercise 14 and Theorem 2.8.)

THEOREM 2.9   *A set $\mathbf{M} \subset \mathbf{R}^n$ is a smooth n-dimensional manifold if and only if $\mathbf{M}$ is open in $\mathbf{R}^n$.*

Exercise 17 Prove the theorem.   (You should interpret $\mathbf{R}^0$ as a zero-dimensional space; that is, $\mathbf{R}^0 = \{0\}$.   In this case, the function F in Definition 2.2 is the function identically 0.)

In Section 3 we shall also describe the connected one-dimensional manifolds.

Exercise 18 Every closed set in $\mathbf{R}^n$ is the set of zeros of a C^1 function.   (*Hint:* (a) If B is an open ball, then there is a nonnegative C^1 function f such that $B = \{x : f(x) > 0\}$. (b) Every open set in $\mathbf{R}^n$ is the union of a sequence of open balls.   (c) Now let F be the given closed set, write the complement as the union of a sequence of open balls B_k, and choose f_k as in (a).   If M_k is the maximum of f_k and its first derivatives, and $\alpha_k = 1/2^k M_k$, then

$$f = \Sigma \alpha_k f_k$$

has the desired property.   [You can also use cubes instead of balls, in which case it is enough to do (a) in dimension 1 and then take a product of the resulting functions.].)

3 TANGENT SPACES

**DEFINITION
3.1**

Let **M** *be a smooth m-dimensional manifold in* $\mathbf{R}^n$ *given by an equation* $F = 0$ *in a neighborhood of a point a.   If a is a regular point of F, then the tangent space to* **M** *at the point a, written* $T_a(\mathbf{M})$, *is the null space of* $dF(a)$.   *The tangent plane is the parallel plane through a.*

According to Corollary 2.5, F must be a function from $\mathbf{R}^n$ to $\mathbf{R}^{n-m}$.   Since F is regular, the dimension of the range of $dF(a)$ must be $n - m$; therefore, the dimension of the null space must be m.   Thus, the tangent space is a subspace of the same dimension m as the manifold.   It is not immediately apparent that the definition makes sense, however, for a given manifold can always be described by many different equations.

Exercise 1   If **M** is described by $F = 0$ in a neighborhood of a, and $G = \Sigma F_i^2$, then **M** is also described by $G = 0$.   In this case, however, G is not regular at any point of **M**, so it is of no use in Definition 3.1.

Exercise 2   If **M** is described by $F = 0$ in a neighborhood of a, and $G = pF$, where p is a positive real-valued C^1 function on a neighborhood of a, then **M** is also described by $G = 0$.   If F is regular at a, then so is G.

**THEOREM
3.2**

If φ *is a local parametric representation of the smooth manifold* **M**, *then the tangent space to* **M** *at the point* $a = \varphi(t_0)$ *is the range of* $d\varphi(t_0)$.

Proof

Let **M** be m-dimensional and let it be given by the equation $F = 0$ in a neighborhood of a, where F is regular at a.   For every t close to t_0, $\varphi(t)$ lies on **M**, so $F(\varphi(t)) = 0$; that is, $F \circ \varphi = 0$.   From the chain rule it follows that $dF(a)\, d\varphi(t_0) = 0$, which shows that the range of $d\varphi(t_0)$ is contained in the tangent space.   Since φ is regular, the range of $d\varphi(t_0)$ has dimension m, the same dimension as the tangent space.   Therefore, the range of $d\varphi(t_0)$ is equal to the tangent space.

This theorem shows that Definition 3.1 does make sense.   The tangent space is determined by the manifold **M**, not by the equation $F = 0$ that is used to define it (as long as F is regular).

Exercise 3   Prove this last statement.

**THEOREM
3.3**

If **M** *and* **N** *are smooth manifolds with* $\mathbf{M} \subset \mathbf{N}$ *in a neighborhood of* $a \in \mathbf{M}$, *then the tangent space to* **M** *at a is contained in the tangent space to* **N** *at a.*

Proof

To say that $\mathbf{M} \subset \mathbf{N}$ in a neighborhood of a means, of course, that there is a ball $B(a; r)$ such that $\mathbf{M} \cap B(a; r) \subset \mathbf{N}$. Let φ be a local parametric representation of $\mathbf{M}$ at a with $\varphi(t_0) = a$, and let $\mathbf{N}$ be defined by $G = 0$ in a neighborhood of a with G regular. Then $G \circ \varphi = 0$, so that $dG(a) \, d\varphi(t_0) = 0$. Theorem 3.2 finishes the job.

Now we shall turn to more geometrical characterizations of the tangent space. In Chapter 8 we defined the tangent vector to a path φ in $\mathbf{R}^n$ at a regular point t_0 to be the vector $\varphi'(t_0)$. This is related to the differential $d\varphi(t_0)$ by the formula

$$d\varphi(t_0)h = h\varphi'(t_0),$$

so the range of the differential is simply the line determined by the tangent vector. If B is a small-enough ball with center t_0, then we know (Theorem 2.7) that $\varphi(B)$ is a smooth one-dimensional manifold, and Theorem 3.2 shows that the tangent line to $\varphi(B)$ at $\varphi(t_0)$ is simply the line determined by the tangent vector $\varphi'(t_0)$. If it happens that the path φ lies on a smooth manifold $\mathbf{M}$, then Theorem 3.3 shows that the tangent vector $\varphi'(t_0)$ lies in the tangent space to the manifold. This proves half of the following nice characterization of the tangent space.

THEOREM 3.4

Let a be a point of a smooth manifold $\mathbf{M}$. The tangent space to $\mathbf{M}$ at a consists of all tangent vectors at a to paths on $\mathbf{M}$ that pass through a.

Proof

Let $\mathbf{M}$ be m-dimensional in $\mathbf{R}^n$ and choose the coordinates so that in a neighborhood of a it is the graph of a function f, that is, is given by equations

$$0 = F_i(x) = x_i - f_i(\bar{x}), \qquad i > m. \tag{1}$$

To prove the remaining half of the theorem we have to start with an arbitrary vector h in the tangent space $T_a(\mathbf{M})$ and produce a path φ on $\mathbf{M}$ that is defined (for example) on a neighborhood of $0 \in \mathbf{R}^1$ and satisfies $\varphi(0) = a$ and $\varphi'(0) = h$. The idea is to project h down on the space of the first m coordinates, to take the line in that space with direction $\bar{h}$, and then to pull the line back up to the surface. This is accomplished by defining $\varphi(t) = \left(\bar{a} + t\bar{h}, f(\bar{a} + t\bar{h})\right)$; that is,

$$\bar{\varphi}(t) = \bar{a} + t\bar{h} \quad \text{and} \quad \varphi_i(t) = f_i(\bar{a} + t\bar{h}) \qquad \text{for } i > m. \tag{2}$$

Straight from the definition it follows that $\varphi(t)$ does lie on $\mathbf{M}$ if t is small and that $\varphi(0) = a$. Differentiation of (2) gives

$$\bar{\varphi}'(0) = \bar{h} \quad \text{and} \quad \varphi_i'(0) = \sum_{j=1}^{m} \frac{\partial f_i(\bar{a})}{\partial x_j} h_j \qquad \text{for } i > m. \tag{3}$$

On the other hand, the fact that h is in the tangent space gives

$$0 = dF_i(a)h = h_i - \sum_{j=1}^{m} \frac{\partial f_i(\bar{a})}{\partial x_j} h_j \qquad \text{for } i > m. \tag{4}$$

Comparison of (3) and (4) shows that $\varphi'(0) = h$.

The tangent plane can also be described as the set of limits of chords.

THEOREM 3.5
Let a be a point of the smooth manifold $\mathbf{M}$. A unit vector θ lies in the tangent space to $\mathbf{M}$ at a if and only if there is a sequence $\{x_k\}$ in $\mathbf{M}$ such that

$$x_k \to a \qquad and \qquad \frac{x_k - a}{|x_k - a|} \to \theta. \tag{5}$$

Proof
If θ is in the tangent space, choose a path φ on $\mathbf{M}$ with $\varphi(0) = a$ and $\varphi'(0) = \theta$, take any sequence $t_k \to 0$, $t_k > 0$, and set $x_k = \varphi(t_k)$.

Exercise 4
In Chapter 8 we already checked that such a sequence has the property (5), but check it again.

On the other hand, if $\{x_k\}$ is a sequence in $\mathbf{M}$ with the property (5), and $\mathbf{M}$ is defined in a neighborhood of a by $F = 0$, then we have

$$0 = F(x_k) - F(a) = dF(a)(x_k - a) + \epsilon(x_k - a)|x_k - a|,$$

so

$$dF(a) \frac{x_k - a}{|x_k - a|} = -\epsilon(x_k - a) \to 0.$$

From this it follows that $dF(a)\theta = 0$, so θ is in the tangent space.

Exercise 5
What are the equations for the tangent space to a manifold in graph form, $x_i = f_i(\bar{x})$, $i > m$?

DEFINITION 3.6
The normal to a smooth manifold $\mathbf{M} \subset \mathbf{R}^n$ at a point $a \in \mathbf{M}$ is the orthogonal complement of the tangent space.

THEOREM 3.7
If $\mathbf{M}$ is given by $F = 0$ in a neighborhood of a, $F:\mathbf{R}^n \to \mathbf{R}^{n-m}$ regular at a, then the normal to $\mathbf{M}$ at a is the space spanned by $\nabla F_{m+1}(a), \ldots, \nabla F_n(a)$.

Proof
For any $h \in \mathbf{R}^n$, we have $dF_i(a)h = \langle \nabla F_i(a), h \rangle$, so that $dF(a)h = 0$ if and only if h is orthogonal to each $\nabla F_i(a)$.

THEOREM 3.8

If $\mathbf{M}$ is a smooth manifold with local parametric representation $\varphi : \mathbf{R}^m \to \mathbf{R}^n$, then the normal to $\mathbf{M}$ at $\varphi(t_0)$ is the set of vectors $h \in \mathbf{R}^n$ satisfying

$$\sum_{i=1}^{n} \frac{\partial \varphi_i(t_0)}{\partial t_j} h_i = 0 \qquad for \; j = 1, \ldots, m.$$

Exercise 6   Prove the theorem.   (*Hint:* The orthogonal complement of the range is the null space of the adjoint.)

Exercise 7   What is the normal to the surface $x_n = f(x_1, \ldots, x_{n-1})$?

Remark   If $\mathbf{M} \subset \mathbf{R}^n \subset \mathbf{R}^l$ is a smooth manifold, then the tangent space to $\mathbf{M}$ at a point is the same whether we look at $\mathbf{M}$ as a subset of $\mathbf{R}^n$ or of $\mathbf{R}^l$.   The normal, on the contrary, is quite different.   On the one hand, we take the orthogonal complement relative to $\mathbf{R}^n$ and on the other relative to $\mathbf{R}^l$.   The normal relative to $\mathbf{R}^n$ is the intersection of $\mathbf{R}^n$ with the normal relative to $\mathbf{R}^l$.

Exercise 8   Prove the assertions in the remark.

It is interesting (and the results will be useful later) to analyze the one-dimensional manifolds, particularly the connected ones.   In this case the arc length provides local parametric representations that can be pieced together to give a parametric representation of the whole manifold.

DEFINITION 3.9

Let $\mathbf{M}$ be a smooth one-dimensional manifold.   A local parametric representation by arc length on an open interval I is a C^1 function $\varphi : I \to \mathbf{M}$ such that $|\varphi'(t)| = 1$ for each $t \in I$.

Since the arc length of the path φ on an interval $[\sigma, \tau]$ is just

$$\int_{\sigma}^{\tau} |\varphi'(t)| \, dt,$$

it follows that the condition $|\varphi'(t)| = 1$ means simply that the arc length on any interval $[\sigma, \tau]$ is $\tau - \sigma$.   In particular, we have

$$|\varphi(t) - \varphi(s)| \le |t - s| \qquad \text{for any } s, t \in I \tag{6}$$

(which can be proved in many other ways too).   From the results of Section 6 of Chapter 8 it follows that if a is a given point of the smooth one-dimensional manifold $\mathbf{M}$ and t_0 is a given point of $\mathbf{R}^1$, then $\mathbf{M}$ has a local parametric representation by arc length that is defined on a neighborhood of t_0 and satisfies $\varphi(t_0) = a$.   Our problem is to piece together these local parametric representations by arc length.

LEMMA 3.10

Let φ, $\psi:I \to \mathbf{M}$ be local parametric representations by arc length. If $\varphi(t_0) = \psi(t_0)$ and $\varphi'(t_0) = \psi'(t_0)$ for some point t_0 in I, then $\varphi(t) = \psi(t)$ for all $t \in I$.

Proof

It is enough to show that $\varphi(t) = \psi(t)$ on a neighborhood of t_0, for this will show that the set of points t with $\varphi(t) = \psi(t)$ and $\varphi'(t) = \psi'(t)$ is both open and closed in I, and hence equal to all of I. Let P be the usual local inverse of φ and set $f = P \circ \psi$, so $\psi = \varphi \circ f$, and hence

$$\psi'(t) = \varphi'(f(t))f'(t).$$

Since both φ' and ψ' have absolute value 1, it follows that f' has absolute value 1, and therefore that f' is either identically $+1$ or identically -1. Now, $f(t_0) = t_0$ and $\varphi'(t_0) = \psi'(t_0)$, so it must be that f' is identically $+1$, and therefore that $f(t) = t$ for all t near t_0.

LEMMA 3.11

Let $\varphi_1:I_1 \to \mathbf{M}$ and $\varphi_2:I_2 \to \mathbf{M}$ be local parametric representations by arc length. If $\varphi_1(I_1) \cap \varphi_2(I_2) \neq \varnothing$, then φ_1 has an extension $\varphi:I \to \mathbf{M}$, which is a local parametric representation by arc length satisfying $\varphi(I) = \varphi_1(I_1) \cup \varphi_2(I_2)$.

Proof

Let $a = \varphi_1(t_1) = \varphi_2(t_2)$, and set

$$\psi_2(t) = \varphi_2(t - t_1 + t_2) \qquad \text{on } I_2 + t_1 - t_2.$$

It is plain that ψ_2 is again a local parametric representation by arc length and that $\psi_2(t_1) = \varphi_2(t_2) = \varphi_1(t_1)$. Since the tangent space to $\mathbf{M}$ at a is one-dimensional, there are just two possibilities for $\psi_2'(t_1)$: It is either $\varphi_1'(t_1)$ or $-\varphi_1'(t_1)$. In the first case put $\psi = \psi_2$, and in the second case put $\psi(t) = \psi_2(2t_1 - t)$, which just reverses the direction. By Lemma 3.10, $\psi = \varphi_1$ on the interval where both are defined; so we can put $\varphi = \varphi_1$ on I_1 and $\varphi = \psi$ on the interval where ψ is defined (which is $t_1 - t_2 + I_2$ in the first case and $t_1 + t_2 - I_2$ in the second). This function φ does the job.

THEOREM 3.12

Let $\mathbf{M}$ be a connected smooth one-dimensional manifold. There is a local parametric representation by arc length $\varphi:I \to \mathbf{M}$ with $\varphi(I) = \mathbf{M}$.

Proof

Choose a point $a \in \mathbf{M}$ and a unit tangent vector θ at a. Let I be the union of all intervals J containing 0 such that there is a local parametric representation by arc length $\varphi_J:J \to \mathbf{M}$ with $\varphi_J(0) = a$ and $\varphi_J'(0) = \theta$. If J is small enough, then φ_J certainly exists (why?); and if J and K are any two such intervals, then by Lemma 3.10, $\varphi_J = \varphi_K$ on $J \cap K$. Consequently, the φ_J determine a φ that is defined on the whole interval I. By the construction there is no extension of φ to a local parametric representation by arc length on a larger interval.

From the definition of a local parametric representation it follows that $\varphi(I)$ is a neighborhood in $\mathbf{M}$ of each point $\varphi(t_0)$. In other words, $\varphi(I)$ is open in $\mathbf{M}$. If we can show that it is also closed in $\mathbf{M}$, then by connectedness we shall have $\varphi(I) = \mathbf{M}$. If $\varphi(I)$ is not closed in $\mathbf{M}$, let $b \in \mathbf{M}$ be a point in the closure, but not in $\varphi(I)$ itself. Let $\psi : J \to \mathbf{M}$ be a local parametric representation by arc length in a neighborhood of b. Since $\psi(J)$ is open in $\mathbf{M}$, we must have $\varphi(I) \cap \psi(J) \neq \varnothing$, so we can use Lemma 3.11 to extend φ. Since this is impossible, our assumption that $\varphi(I)$ is not closed in $\mathbf{M}$ does not hold up.

Now let us check what happens in this construction when φ is not one to one. If $\varphi(t_0) = \varphi(t_1)$, then (one-dimensional tangent space, as usual) either $\varphi'(t_0) = \varphi'(t_1)$ or $\varphi'(t_0) = -\varphi'(t_1)$. First let us eliminate the second possibility. In this case Lemma 3.10 shows that we must have $\varphi(t_0 + t_1 - t) = \varphi(t)$ on $I \cap t_0 + t_1 - I$, for the two functions and their derivatives are the same at t_0. Differentiation gives $-\varphi'(t_0 + t_1 - t) = \varphi'(t)$, which is obviously impossible at the point $t = t_0 + t_1/2$.

Consequently, we must have $\varphi'(t_0) = \varphi'(t_1)$, and in this case Lemma 3.10 gives

$$\varphi(t) = \begin{cases} \varphi(t - t_0 + t_1) & \text{on } I \cap t_0 - t_1 + I, \\ \varphi(t - t_1 + t_0) & \text{on } I \cap t_1 - t_0 + I. \end{cases}$$

If I is not the whole line, then these two formulas permit the extension of φ either to the right or the left. For instance, if $t_0 < t_1$, then the first formula extends φ to the left to the interval $t_0 - t_1 + I$, and the second one extends φ to the right. Since φ cannot be extended, it must be true that $I = \mathbf{R}^1$ and that $\varphi(t) = \varphi(t - t_0 + t_1)$ for all t. In this case, $\mathbf{M}$ must of course be compact, for the range of φ is the same as the range of its restriction to the interval $[0, t_1 - t_0]$. With this additional information we can refine Theorem 3.12 as follows.

THEOREM 3.13

Let $\mathbf{M}$ be a connected noncompact smooth one-dimensional manifold. There is a local parametric representation by arc length $\varphi : I \to \mathbf{M}$ with $\varphi(I) = \mathbf{M}$ and φ one to one.

THEOREM 3.14

Let $\mathbf{M}$ be a connected compact smooth one-dimensional manifold. There is a local parametric representation by arc length $\varphi : \mathbf{R}^1 \to \mathbf{M}$ with $\varphi(\mathbf{R}^1) = \mathbf{M}$. Moreover, there is a positive number α such that φ is one to one on $(0, \alpha)$ and $\varphi(t + \alpha) = \varphi(t)$ for each t.

Proof

Theorem 3.13 is already proved, and so is the first part of Theorem 3.14. The number α is defined to be the lower bound of the numbers $t > 0$ with $\varphi(t) = \varphi(0)$. Note that when $\mathbf{M}$ is compact, φ cannot be one to one, for then the inverse would have to be continuous, while the range of the inverse is $\mathbf{R}^1$, which is not compact. If $\varphi(t_0) = \varphi(t_1)$, then the discussion above shows that $\varphi(t - t_0 + t_1) = \varphi(t)$, so in particular $\varphi(t_1 - t_0) = \varphi(0)$. This shows that the set of numbers $t > 0$ with $\varphi(t) = \varphi(0)$ is not empty, so the definition of α makes sense. It also shows that if $t_1 > t_0$, then $t_1 - t_0 \geq \alpha$, so φ is one to one on the interval $(0, \alpha)$. Finally, by continuity $\varphi(\alpha) = \varphi(0)$, so we can take $t_0 = 0$ and $t_1 = \alpha$ to obtain $\varphi(t + \alpha) = \varphi(t)$ for each t.

Exercise 9   Show that $\alpha \neq 0$.   What is the geometric meaning of α?

Exercise 10   Invent converses of Theorems 3.13 and 3.14; that is, if $\varphi : I \to \mathbf{R}^n$ is regular at each point and is one to one and something else, then $\varphi(I)$ is a connected noncompact smooth one-dimensional manifold, and so on.

4   FUNCTIONS ON MANIFOLDS

Consider the problem of finding the maximum of the function $f(x, y, z) = 3x - 2y + z$ on the ball $x^2 + y^2 + z^2 \leq 1$. We know that if a function $f : \mathbf{R}^n \to \mathbf{R}^1$ has a local maximum or minimum at a point a and if f is differentiable at a, then $df(a) = 0$, or, equivalently, $\nabla f(a) = 0$. In this case $\nabla f(a) = (3, -2, 1) \neq 0$. The conclusion is not that there is no maximum, for the closed ball is compact, and there certainly is one. The proper conclusion is that f is not differentiable at the point where the maximum occurs. The function $f(x, y, z) = 3x - 2y + z$ certainly looks differentiable, but the point is that we are looking at the restriction of this function to the closed ball, and the restriction is not differentiable at any boundary point, that is, at any point of the sphere $S^2 = \{(x, y, z) : x^2 + y^2 + z^2 = 1\}$. The maximum does lie on the sphere, and at the moment we are helpless to locate it; so what we are going to do is discuss the idea of the differential of a function $f : \mathbf{M} \to \mathbf{R}^q$ which is defined on a smooth m-dimensional manifold $\mathbf{M} \subset \mathbf{R}^n$. (In this case $\mathbf{M} = S^2$.)

Exercise 1   By accident, we are not really helpless in this particular problem.   Use Cauchy–Schwarz to show that the maximum is at $(3, -2, 1)/\sqrt{14}$.

Throughout the section it is assumed that $\mathbf{M}$ is a smooth m-dimensional manifold in $\mathbf{R}^n$ and that $f : \mathbf{M} \to \mathbf{R}^q$ is defined on a neighborhood in $\mathbf{M}$ of some point $a \in \mathbf{M}$ (not on a neighborhood of a in $\mathbf{R}^n$, but on a neighborhood of a in $\mathbf{M}$, which is the intersection of $\mathbf{M}$ with a neighborhood of a in $\mathbf{R}^n$). If φ is a local parametric representation of $\mathbf{M}$ in a neighborhood of a with $\varphi(t_0) = a$, we shall call $f_\varphi = f \circ \varphi$ a local representation of f at (t_0, a). Note that f_φ is defined on a neighborhood of t_0.

LEMMA 4.1

Let f_φ and f_ψ be local representations of f at (t_0, a) and at (s_0, a). If f_φ is differentiable at t_0, is of class C^1 at t_0, or is of class C^1 in a neighborhood of t_0, then the same is true of f_ψ at s_0.

Proof

Let P be a local inverse of φ, and set $g = P \circ \psi$, so that $\psi = g \circ \varphi$; hence

$$f_\psi = f_\varphi \circ g. \tag{1}$$

Since g is of class C^1 on a neighborhood of t_0, the theorem follows.

The lemma makes it reasonable to make the following definition.

DEFINITION 4.2

The function $f : \mathbf{M} \to \mathbf{R}^q$ is differentiable at a, or of class C^1 at a, or of class C^1 on a neighborhood of a, if some (and hence any) local representation f_φ at (t_0, a) has the same property at t_0.

There is another equally reasonable but quite different way to make the definition, which the following theorem shows is exactly the same.

THEOREM 4.3

The function $f : \mathbf{M} \to \mathbf{R}^q$ is differentiable at a, or of class C^1 at a, or of class C^1 on a neighborhood of a, if and only if there is an extension $\tilde{f}$ of f to a full neighborhood of a in $\mathbf{R}^n$ with the same property.

Proof

If there is an extension $\tilde{f}$ with one of the properties listed, and φ is a local parametric representation, then $f_\varphi = f \circ \varphi = \tilde{f} \circ \varphi$; so it is apparent that f_φ has the same property. On the other hand, if f_φ has one of the properties, and P is the local inverse of φ, then $\tilde{f} = f_\varphi \circ P = f \circ \varphi \circ P$ is the required extension.

DEFINITION 4.4

If $f : \mathbf{M} \to \mathbf{R}^q$ is differentiable at a, then $df(a)$ is the restriction of $d\tilde{f}(a)$ to the tangent space $T_a(\mathbf{M})$, where $\tilde{f}$ is any differentiable extension of f.

A given function f has various extensions $\tilde{f}$, each with its own differential. It must be shown that all these have the same restriction to the tangent space $T_a(\mathbf{M})$. If φ is any local parametric representation, then $f_\varphi = \tilde{f} \circ \varphi$; therefore,

$$df_\varphi(t_0) = d\tilde{f}(a) \, d\varphi(t_0). \tag{2}$$

Consequently,

$$df(a) = d\tilde{f}(a) = df_\varphi(t_0) \, d\varphi(t_0)^{-1} \qquad on \ T_a(\mathbf{M}). \tag{3}$$

This makes sense because $d\varphi(t_0)$ is one to one and maps $\mathbf{R}^m$ onto the tangent space $T_a(\mathbf{M})$. It shows that any two extensions do have differentials with the same restriction to $T_a(\mathbf{M})$ because the same φ can be used with all extensions.

Exercise 2 Start with formula (3) [without the $d\tilde{f}(a)$] as the definition of $df(a)$, and show that any two local parametric representations give the same result without using the extension $\tilde{f}$.

THEOREM 4.5 *If $f:\mathbf{M} \to \mathbf{R}^1$ is differentiable at a, then there is a unique vector $\nabla f(a)$ in $T_a(\mathbf{M})$ such that*

$$df(a)h = \langle \nabla f(a), h \rangle \qquad \text{for all } h \in T_a(\mathbf{M}).$$

$\nabla f(a)$, which is called the gradient of f at a, is the projection of $\nabla\tilde{f}(a)$ on $T_a(\mathbf{M})$, where $\tilde{f}$ is any differentiable extension of f.

Proof If P is the projection on $T_a(\mathbf{M})$ and $h \in T_a(\mathbf{M})$, then

$$df(a)h = d\tilde{f}(a)h = \langle \nabla\tilde{f}(a), h \rangle = \langle \nabla\tilde{f}(a), Ph \rangle = \langle P\nabla\tilde{f}(a), h \rangle.$$

The uniqueness comes as usual from the fact that two distinct vectors in $T_a(\mathbf{M})$ cannot have the same inner product with every vector in $T_a(\mathbf{M})$.

THEOREM 4.6 *If $f:\mathbf{M} \to \mathbf{R}^1$ has a local maximum or minimum at a, and f is differentiable at a, then $df(a) = 0$.*

Proof If f has a local maximum or minimum at a, then f_φ has the same at t_0, so by the original theorem of this kind, $df_\varphi(t_0) = 0$. Formula (3) shows that $df(a) = 0$.

In order to use the theorem for actual calculations, we shall have to know the equations of $\mathbf{M}$, and we shall want to deal with some specific extension of f. So as not to multiply the notations, we shall call the extension f. Then we shall be interested in the local maxima and minima of the restriction of f to $\mathbf{M}$, which is written $f|_\mathbf{M}$. If $f|_\mathbf{M}$ has a local maximum or minimum at a, and f is differentiable at a, then, according to Theorems 4.5 and 4.6, $\nabla f(a)$ is normal to $\mathbf{M}$ at a (for its projection on the tangent space is 0). Now if $\mathbf{M}$ is given by the equation $F = 0$ in a neighborhood of a, where $F:\mathbf{R}^n \to \mathbf{R}^{n-m}$ is regular at a, then the normal is the space spanned by $\nabla F_{m+1}(a), \ldots, \nabla F_n(a)$; so there exist numbers $\lambda_{m+1}, \ldots, \lambda_n$ such that

$$\nabla f(a) = \sum_{j=m+1}^{n} \lambda_j \nabla F_j(a). \tag{4}$$

The numbers $\lambda_{m+1}, \ldots, \lambda_n$ are called *Lagrange multipliers*.

THEOREM
4.7

(*Lagrange Multipliers*)   *Let* $\mathbf{M}$ *be a smooth manifold given by the equation* $F = 0$ *in a neighborhood of the point* a, F *regular at* a.   *Let* $f : \mathbf{R}^n \to \mathbf{R}^1$ *be differentiable at* a, *and set*

$$g(x, \lambda) = f(x) - \sum_{j=m+1}^{n} \lambda_j F_j.$$

If $f|_\mathbf{M}$ *has a local maximum or minimum at* a, *then* $\nabla g(a, \lambda) = 0$ *for some* $\lambda \in \mathbf{R}^{n-m}$.

Proof

We are considering g as a function of all the $2n - m$ variables x and λ, and ∇g refers to the gradient in all these variables.   Since $\partial g / \partial \lambda_j = F_j$, the equations $\partial g / \partial \lambda_j = 0$ say simply that a lies on M.   The equations $\partial g / \partial x_j = 0$ are just another way of writing (4).

Example

Find the maximum and minimum of the function $f(x, y, z) = 3x - 2y + z$ on the ball $x^2 + y^2 + z^2 \le 1$.

At the beginning of the section we saw that both the maximum and minimum exist and that both must occur on the boundary $\mathbf{M} = S^2 = \{(x, y, z) : x^2 + y^2 + z^2 - 1 = 0\}$, so we consider Lagrange multipliers and the function

$$g(x, y, z, \lambda) = 3x - 2y + z - \lambda(x^2 + y^2 + z^2 - 1).$$

We have

$$\frac{\partial g}{\partial x} = 3 - 2\lambda x, \quad \frac{\partial g}{\partial y} = -2 - 2\lambda y, \quad \frac{\partial g}{\partial z} = 1 - 2\lambda z, \quad \frac{\partial g}{\partial \lambda} = -(x^2 + y^2 + z^2 - 1).$$

When we put $\nabla g = 0$, the first three equations give

$$x = \frac{3}{2\lambda}, \qquad y = \frac{-1}{\lambda}, \qquad z = \frac{1}{2\lambda}, \tag{5}$$

and then substitution in the last equation gives

$$\lambda = \frac{\pm\sqrt{14}}{2}.$$

Since the closed ball is compact, there must be both a maximum and a minimum.   One must be given by (5) with $\lambda = \sqrt{14}/2$ and the other by (5) with $\lambda = -\sqrt{14}/2$.   Calculation of the value of f at the two points shows that the former is the maximum and the latter the minimum.

Exercise 3   Find the distance from the point $(-1, 1)$ to the curve $xy = 1$, $x > 0$.

Exercise 4   Let $b \in \mathbf{R}^n$ and let a be a point of a smooth manifold $\mathbf{M}$ that is closest to b. Show that $b - a$ is normal to $\mathbf{M}$ at a.   Does such a closest point always exist?

Exercise 5   Show that a self-adjoint linear $H: \mathbf{R}^n \to \mathbf{R}^n$ must have an eigenvalue (Theorem 9.3 of Chapter 9) by using Lagrange multipliers.

Exercise 6   Let $\mathbf{M}$ be the torus obtained by revolving the circle $x^2 + (y - 2)^2 = 1$ around the x axis.   Use Lagrange multipliers to find the maximum and minimum of the function $f(x, y, z) = z$ on $\mathbf{M}$.

Let us consider this last exercise from the geometric point of view.   According to the discussion preceding Theorem 4.7, we should look for the points where the gradient of f is normal to $\mathbf{M}$.   The gradient of f is

$$\nabla f = (0, 0, 1)$$

so we should look for points where the tangent plane to $\mathbf{M}$ is horizontal.   The torus $\mathbf{M}$ looks as shown in Figure 2.   The tangent plane is horizontal at the four points $(0, 0, \pm 3)$ and $(0, 0, \pm 1)$, and these are the four points the method of Lagrange multipliers will produce for you when you do the exercise.   It is plain that $(0, 0, 3)$ is the maximum point and $(0, 0, -3)$ is the minimum point, and that the other two are neither maximum nor minimum points.   In other

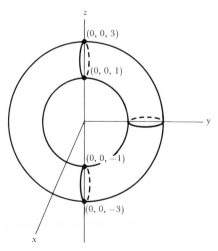

Figure 2

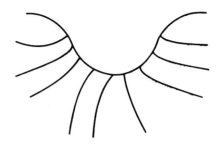

Figure 3

words, the method of Lagrange multipliers turns up the possibilities for the maximum and minimum points, but these possibilities have to be checked. The points $(0, 0, 1)$ and $(0, 0, -1)$ are called *saddle points* because the surface looks something like a saddle (or upside-down saddle) in a neighborhood of these points (Figure 3).

Exercise 7   Let $f:\mathbf{M} \to \mathbf{R}^q$ be differentiable at the point $a \in \mathbf{M}$. Let $\mathbf{N}$ be a smooth manifold in $\mathbf{R}^q$. If $f(\mathbf{M}) \subset \mathbf{N}$, then $df(a): T_a(\mathbf{M}) \to T_b(\mathbf{N})$, $b = f(a)$.

5   QUADRATIC FORMS AND QUADRIC SURFACES

The quadratic forms are the simplest functions after the linear ones, and the quadric surfaces are the simplest surfaces after the planes. They can be analyzed in detail with the aid of the results of Section 9 of Chapter 9.

DEFINITION 5.1   *A quadratic form on $\mathbf{R}^n$ is a function $Q:\mathbf{R}^n \to \mathbf{R}^1$ of the form*

$$Q(x) = \langle Tx, x \rangle, \tag{1}$$

where T is a linear transformation from $\mathbf{R}^n$ to $\mathbf{R}^n$.

If $\{a_{ij}\}$ is the matrix of T relative to any orthonormal basis of $\mathbf{R}^n$, then (1) becomes

$$Q(x) = \sum_{i,j=1}^{n} a_{ij}x_i x_j. \tag{2}$$

Equation (1) does not determine the linear transformation T uniquely. Indeed,

$$\langle Tx, x \rangle = \langle x, T^*x \rangle = \langle T^*x, x \rangle,$$

so T and T^* determine the same quadratic form Q, and so does the self-adjoint transformation

$$H = \frac{T + T^*}{2}.$$

THEOREM 5.2

The equation

$$Q(x) = \langle Hx, x \rangle \tag{3}$$

determines a one-to-one correspondence between the quadratic forms and the self-adjoint linear transformations.

Proof   What has to be shown is that equation (3) determines H uniquely. If $\langle Hx, y \rangle$ can be expressed in terms of Q, then this will do the trick, for two distinct vectors cannot have the same inner product with every vector y. $\langle Hx, y \rangle$ is expressed in terms of Q by the identity

$$4\langle Hx, y \rangle = Q(x + y) - Q(x - y). \tag{4}$$

Exercise 1   Prove this identity when H is self-adjoint and satisfies (3) by simply writing out the right-hand side.

THEOREM 5.3

Let Q be a quadratic form with corresponding self-adjoint transformation H. Let $\lambda_1, \ldots, \lambda_n$ be the eigenvalues of H and let $e_1, \ldots, e_n$ be an ortho-normal basis of eigenvectors. In coordinates relative to $e_1, \ldots, e_n$ we have

$$Q(x) = \sum_{i=1}^{n} \lambda_i x_i^2.$$

Exercise 2   Prove the theorem.

DEFINITION 5.4

A quadric surface in $\mathbf{R}^n$ is the set of points that satisfy a quadratic equation

$$\langle Hx, x \rangle + \langle x, b \rangle + c = 0, \tag{5}$$

where $H:\mathbf{R}^n \to \mathbf{R}^n$ is self-adjoint, $b \in \mathbf{R}^n$, and $c \in \mathbf{R}^1$.

If $\{a_{ij}\}$ is the matrix of H relative to any orthonormal basis, and $b = (b_1, \ldots, b_n)$ relative to this basis, then (5) becomes

$$\sum_{i,j=1}^{n} a_{ij}x_i x_j + \sum_{i=1}^{n} b_i x_i + c = 0. \tag{6}$$

In terms of coordinates relative to an orthonormal basis of eigenvectors of H, this equation simplifies to

$$\sum_{i=1}^{n} \lambda_i y_i^2 + \sum_{i=1}^{n} \mu_i y_i + c = 0. \tag{7}$$

Exercise 3   What are the λ_i and μ_i in terms of the original data in (5)?

Exercise 4   If Q is a quadratic form with corresponding self-adjoint H, then $\nabla Q(a) = 2Ha$.

Exercise 5   Show that the set of singular points (= nonregular points) of the quadric surface defined by (5) is either empty or is a plane contained in the plane parallel to the null space of H.

Formula (7) can be simplified still further by combining the quadratic and linear terms by completing the square.   If $\lambda_i \neq 0$, then

$$\lambda_i y_i^2 + \mu_i y_i = \lambda_i \left(y_i + \frac{\mu_i}{2\lambda_i} \right)^2 - \frac{\mu_i^2}{4\lambda_i}.$$

Thus, if $\lambda_1, \ldots, \lambda_k$ are $\neq 0$ and the rest are 0, then equation (7) becomes

$$\sum_{i=1}^{k} \lambda_i (y_i + \alpha_i)^2 + \sum_{i=k+1}^{n} \mu_i y_i + d = 0. \tag{8}$$

A final coordinate change,

$$z_i = y_i + \alpha_i \quad \text{for } i \le k, \qquad z_i = y_i \quad \text{for } i > k, \tag{9}$$

gives

$$\sum_{i=1}^{k} \lambda_i z_i^2 + \sum_{i=k+1}^{n} \mu_i z_i + d = 0. \tag{10}$$

Note that the coordinate change (9) is not of the kind that we have been considering.   The new axes are parallel to the old, but the origin is at the point $0 = z_i = y_i + \alpha_i$; that is, $y_i = -\alpha_i$, for $i \le k$.   Such a coordinate change is called a *translation*.   Note that it is not linear.

THEOREM 5.5   *By means of a suitable choice of an orthonormal basis in $\mathbf{R}^n$, and a translation, the equation (5) can be put in the form (10), where the λ_i are the nonzero eigenvalues of H.*

Study of formula (10) shows that the nature of the quadric surface depends on the number of positive λ_i, the number of negative λ_i, the number of nonzero

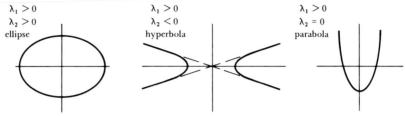

Figure 4

μ_i, and the sign of the number d.   For example, if there are n positive λ_i, then there can be no μ_i; and the surface is empty if $d > 0$, consists of a single point if $d = 0$, and is an ellipsoid if $d < 0$.

In two dimensions the complete analysis is immediate (Figure 4).   When $d = 0$ the ellipse becomes a single point, and the hyperbola becomes a pair of lines.   When $\mu_2 = 0$, the parabola becomes either empty, or a single line, or a pair of parallel lines, according as $d > 0$, $d = 0$, or $d < 0$.

Exercise 6   Discuss the various quadric surfaces in $\mathbf{R}^3$.

The actual reduction of a given equation (5) to the form (10) is easy enough in two dimensions, but beyond that it is pretty cumbersome.

Example   Discuss $x^2 - 2xy - y^2 + x - 6 = 0$.

The matrices of H and of $H - \lambda I$ are

$$\begin{pmatrix} 1 & -1 \\ -1 & -1 \end{pmatrix} \quad \text{and} \quad \begin{pmatrix} 1 - \lambda & -1 \\ -1 & -1 - \lambda \end{pmatrix}.$$

The eigenvalues are the numbers λ for which the latter is not one to one, that is, for which the determinant

$$D(\lambda) = (1 - \lambda)(-1 - \lambda) - 1 = \lambda^2 - 2$$

is zero.   These are $\lambda_1 = \sqrt{2}$ and $\lambda_2 = -\sqrt{2}$.   From this we can tell already that the curve is a hyperbola because one of the eigenvalues is positive and the other negative; and we can tell that the quadratic part of the new equation will be $\sqrt{2}\, z^2 - \sqrt{2}\, w^2$.   But to get the rest of the equation and to see where the new axes go, we have to push on with the calculation.

To get the eigenvector e_1 corresponding to $\lambda_1 = \sqrt{2}$ we have to solve the equations

$$h - k = \sqrt{2}\, h \quad \text{and} \quad -h - k = \sqrt{2}\, k,$$

but it is sufficient to solve just one of them, and then the other will be automatically satisfied.

Exercise 7   Why?

Taking $h = 1$, we get $k = 1 - \sqrt{2}$, but to get e_1 we should normalize this by dividing by its length.   Thus,

$$e_1 = \frac{1}{\sqrt{4 - 2\sqrt{2}}}\,(1, 1 - \sqrt{2}), \qquad e_2 = \frac{1}{\sqrt{4 - 2\sqrt{2}}}\,(-1 + \sqrt{2}, 1).$$

Note that e_2 comes free because it is perpendicular to e_1 and has length one.

If a given vector v has coordinates (x, y) relative to the initial basis f_1, f_2 of $\mathbf{R}^2$, and coordinates (z, w) relative to the new basis, then

$$x = \langle v, f_1 \rangle = z\langle e_1, f_1 \rangle + w\langle e_2, f_1 \rangle,$$
$$y = \langle v, f_2 \rangle = z\langle e_1, f_2 \rangle + w\langle e_2, f_2 \rangle.$$

Hence,

$$
\begin{aligned}
x &= \frac{1}{\sqrt{4 - 2\sqrt{2}}}\,(z + (-1 + \sqrt{2})w), \\
y &= \frac{1}{\sqrt{4 - 2\sqrt{2}}}\,((1 - \sqrt{2})z + w).
\end{aligned}
\tag{11}
$$

Now we go back to the original equation and substitute these values of x and y to get

$$\sqrt{2}\,z^2 - \sqrt{2}\,w^2 + \frac{1}{\sqrt{4 - 2\sqrt{2}}}\,z + \frac{-1 + \sqrt{2}}{\sqrt{4 - 2\sqrt{2}}}\,w - 6 = 0. \tag{12}$$

(This part of the calculation looks fearsome, but remember that we already know the quadratic part and can ignore it completely.)   At this point the equation is in the form (7), and all that remains is to complete the square to arrive at (8) and then to make the substitution (9).

Exercise 8   Carry out the rest of the calculation.

Consider the general equation

$$ax^2 + bxy + cy^2 + dx + ey + f = 0 \tag{13}$$

in dimension 2.   The matrices of H and $H - \lambda I$ are

$$\begin{pmatrix} a & \dfrac{b}{2} \\[2mm] \dfrac{b}{2} & c \end{pmatrix} \qquad \text{and} \qquad \begin{pmatrix} a - \lambda & \dfrac{b}{2} \\[2mm] \dfrac{b}{2} & c - \lambda \end{pmatrix}.$$

The eigenvalues are the numbers λ_1 and λ_2 for which

$$D(\lambda) = (a - \lambda)(c - \lambda) - \frac{b^2}{4} = \lambda^2 - (a + c)\lambda + ac - \frac{b^2}{4}$$

is 0.   Hence, also $D(\lambda) = (\lambda - \lambda_1)(\lambda - \lambda_2)$; therefore,

$$\lambda_1 + \lambda_2 = a + c \qquad \text{and} \qquad 4\lambda_1\lambda_2 = 4ac - b^2. \tag{14}$$

The curve is an ellipse exactly when both eigenvalues have the same sign, that is, when $4ac - b^2 > 0$.   It is a hyperbola when they have different signs, that is, when $4ac - b^2 < 0$.   And it is a parabola when one is 0, that is, when $4ac - b^2 = 0$.

THEOREM
5.6

If a, b, and c are not all three 0 in equation (13), then the equation represents an ellipse if $b^2 - 4ac < 0$, a hyperbola if $b^2 - 4ac > 0$, and a parabola if $b^2 - 4ac = 0$.   The number $b^2 - 4ac$ is called the discriminant of the equation.

(Of course, allowance must be made for the degenerate cases when the ellipse collapses to a point, and so on.)

12 { Higher Derivatives

1 SECOND DERIVATIVES

The partial derivatives of a function $f:\mathbf{R}^m \to \mathbf{R}^n$ are again functions from $\mathbf{R}^m$ to $\mathbf{R}^n$, which may have partial derivatives of their own. If so, the latter are called the *second partial derivatives:*

$$\frac{\partial^2 f(a)}{\partial x_i\, \partial x_j} = D_{ij}f(a) = D_i(D_jf)(a). \tag{1}$$

The formula says that to get $D_{ij}f(a)$, you take first D_jf, which must exist on a neighborhood of a, and differentiate it with respect to x_i. It appears that this would be quite different from $D_{ji}f(a)$, which is formed by differentiating first with respect to x_i and then with respect to x_j. In fact, the two are usually the same.

THEOREM 1.1 *If D_if and D_jf are both differentiable at a, then $D_{ij}f(a) = D_{ji}f(a)$.*

Exercise 1 Show that it is enough to treat the case where $f:\mathbf{R}^2 \to \mathbf{R}^1$.

Proof Let $f:\mathbf{R}^2 \to \mathbf{R}^1$, and write (a, b) instead of (a_1, a_2), and so on. Consider the quantity

$$F(h) = f(a + h, b + h) - f(a + h, b) - f(a, b + h) + f(a, b).$$

If $\varphi(x) = f(x, b + h) - f(x, b)$, then $F(h) = \varphi(a + h) - \varphi(a)$, and application of the mean-value theorem to φ gives

$$F(h) = \varphi'(\xi)h = [D_1f(\xi, b + h) - D_1f(\xi, b)]h,$$

where ξ is between a and $a + h$. The fact that $D_1 f$ is differentiable at (a, b) gives

$$D_1 f(\xi, b + h) = D_1 f(a, b) + D_{11} f(a, b)(\xi - a) + D_{21} f(a, b)h + \epsilon_1(h)|h|,$$
$$D_1 f(\xi, b) = D_1 f(a, b) + D_{11} f(a, b)(\xi - a) + \epsilon_2(h)|h|.$$

Therefore,

$$F(h) = D_{21} f(a, b)h^2 + \epsilon(h)|h|^2. \qquad (2)$$

On the other hand, if $\psi(y) = f(a + h, y) - f(a, y)$, then $F(h) = \psi(b + h) - \psi(b)$, and a repetition of the above argument gives

$$F(h) = D_{12} f(a, b)h^2 + \epsilon(h)|h|^2. \qquad (3)$$

From (2) and (3) it follows that

$$D_{21} f(a, b) = \lim_{h \to 0} \frac{F(h)}{h^2} = D_{12} f(a, b).$$

Exercise 2 Give an example of an $f : \mathbf{R}^2 \to \mathbf{R}^1$ for which $D_{12} f(a)$ and $D_{21} f(a)$ both exist but are different.

Exercise 3 If $D_2 f$ and $D_{21} f$ exist on a neighborhood of a, and $D_{21} f$ is continuous at a, then $D_{12} f$ exists at a and is equal to $D_{21} f$.

2 HIGHER DERIVATIVES

The third derivatives are the derivatives of the second, the fourth are the derivatives of the third, and so on. The general definition is best put inductively. The notion to be defined is the derivative $D_i f$, where i is now not a single integer between 1 and n but a finite sequence $i = (i_1, \ldots, i_r)$. The number r is called the *order of the derivative* and is written $r = |i|$.

DEFINITION 2.1   *If $f : \mathbf{R}^m \to \mathbf{R}^n$, and $i = (i_1, \ldots, i_r)$, then*

$$D_i f = D_{i'}(D_{i_r}) f$$

where $i' = (i_1, \ldots, i_{r-1})$.

For technical reasons it is convenient to write 0 for the empty sequence i and to set $D_0 f = f$. This avoids exceptional cases in many formulas.

A function is of class C^r at a point if the partial derivatives of orders $\leq r$ all exist on a neighborhood of the point and are continuous at the point itself. Again the definition is best put inductively.

**DEFINITION
2.2**

A function $f: \mathbf{R}^m \to \mathbf{R}^n$ is of class C^r at a point a if f and df are of class C^{r-1} at a. It is of class C^r on an open set if it is of class C^r at each point of the open set.

This determines C^r for every $r \geq 1$, because C^1 is already known. Again, however, it is convenient to invent something for $r = 0$. The useful notion is that f is of class C^0 at a if it is defined on a neighborhood of a and is continuous at a.

Exercise 1   Show that Definition 2.2 gives back the original notion of C^1, if C^0 is defined as above.

From the inductive nature of the definitions it is clear that most proofs should rest on induction too.

**THEOREM
2.3**

If $f, g: \mathbf{R}^m \to \mathbf{R}^n$ are of class C^r at a, and α is a real number, then $f + g$ and αf are of class C^r at a, and

$$D_i(f + g) = D_i f + D_i g \quad and \quad D_i(\alpha f) = \alpha D_i f \quad if \ |i| \leq r.$$

Exercise 2   Prove the theorem.

**THEOREM
2.4**

If $f: \mathbf{R}^m \to \mathbf{R}^1$ and $g: \mathbf{R}^m \to \mathbf{R}^n$ are of class C^r at a, then the product $fg: \mathbf{R}^m \to \mathbf{R}^n$ is of class C^r at a.

Exercise 3   Prove the theorem.

**THEOREM
2.5**

Let $\mathbf{R}^l \xrightarrow{f} \mathbf{R}^m \xrightarrow{g} \mathbf{R}^n$. If f is of class C^r at a and g is of class C^r at $b = f(a)$, then $g \circ f$ is of class C^r at a.

Exercise 4   Prove the theorem.

One consequence of Theorem 2.5 is that the notion of class C^r does not depend on the coordinate system in $\mathbf{R}^n$. Let F be a given function on $\mathbf{R}^n$, and suppose that a given vector v has coordinates x with respect to one basis and coordinates $y = \varphi(x)$ with respect to another. In terms of the first coordinates, the function F determines a function f by $f(x) = F(v)$, and in terms of the second it determines a function g by $g(y) = F(v)$. Thus,

$$g(y) = f(x) \quad if \quad y = \varphi(x)$$

or, in other words, $f = g \circ \varphi$. In this case φ is linear, so it is clearly C^r for every r, and the same is true of its inverse. Therefore, the theorem shows that f is C^r if and only if g is.

Exercise 5   Let $f: \mathbf{R}^m \to \mathbf{R}^n$. In the discussion above we made a change of coordinates in $\mathbf{R}^m$. Now make a change of coordinates in $\mathbf{R}^n$ and show that the notion of class C^r remains the same.

Consider the space $\mathcal{L}_{mn}$ of linear transformations from $\mathbf{R}^m$ to $\mathbf{R}^n$. As soon as bases are fixed in $\mathbf{R}^m$ and $\mathbf{R}^n$, $\mathcal{L}_{mn}$ can be identified with the space of matrices, and this in turn with $\mathbf{R}^{mn}$. This amounts to choosing a basis in $\mathcal{L}_{mn}$ in the following way. If $e_1, \ldots, e_m$ is the basis in $\mathbf{R}^m$, and $f_1, \ldots, f_n$ is the basis in $\mathbf{R}^n$, let T_{ij} be the linear transformation such that

$$T_{ij}e_j = f_i, \qquad T_{ij}e_k = 0 \qquad \text{if } k \neq j.$$

This is the linear transformation whose matrix has 1 in the ith row, jth column, and 0 everywhere else. It is plain that

$$T = \sum_{i,j} \alpha_{ij} T_{ij}$$

if and only if T has matrix $\{\alpha_{ij}\}$. In other words, the T_{ij} form a basis of $\mathcal{L}_{mn}$, and the matrix of a linear transformation is nothing but its set of coordinates relative to the basis.

Once $\mathcal{L}_{mn}$ is identified with $\mathbf{R}^{mn}$ it makes perfectly good sense to speak of functions of class C^r from $\mathbf{R}^k$ to $\mathcal{L}_{mn}$, from $\mathcal{L}_{mn}$ to $\mathbf{R}^k$, from $\mathcal{L}_{mn}$ to $\mathcal{L}_{pq}$, and so on. The above discussion shows that how the bases are chosen in the various spaces is immaterial. To be sure the ideas are fixed clearly, let us state the definition explicitly in one case.

DEFINITION 2.6   *Let bases be fixed in $\mathbf{R}^m$ and $\mathbf{R}^n$. A function $f: \mathbf{R}^k \to \mathcal{L}_{mn}$ is of class C^r at a point a if each element of its matrix is of class C^r at a.*

Of course, the elements of the matrix are functions $f_{ij}: \mathbf{R}^k \to \mathbf{R}^1$. If the bases are changed, then these functions are changed; but the discussion above shows that if they are of class C^r for one pair of bases, then they are of class C^r for any other.

THEOREM 2.7   *Let $f: \mathbf{R}^m \to \mathbf{R}^n$ be differentiable at each point of an open set G. Then f is of class C^r on G if and only if $df: \mathbf{R}^m \to \mathcal{L}_{mn}$ is of class C^{r-1} on G.*

Proof   The elements of the matrix are the first derivatives $D_j f_i$, so the result follows straight from Definitions 2.6 and 2.2. The theorem looks like nothing more than a ponderous tautology, which it is, but watch what happens in Section 3.

Exercise 6   Let $f, g : \mathbf{R}^k \to \mathfrak{L}_{nn}$. If f and g are of class C^r at a point a, then $h(x) = f(x)g(x)$ is also of class C^r at a.

Remark   Other notations for the derivative $D_i f$ are

$$\frac{\partial^r f}{\partial x_{i_1}\, \partial x_{i_2}\, \cdots\, \partial x_{i_r}} \quad \text{and} \quad f_{x_{i_1} x_{i_2} \cdots x_{i_r}}.$$

If f is of class C^r, then all the differentiations with respect to x_1 can be lumped together, all those with respect to x_2 can be lumped together, and so on. In this case the notation is often

$$\frac{\partial^r f}{\partial x_1^{k_1}\, \partial x_2^{k_2}\, \cdots\, \partial x_n^{k_n}} \quad \text{where } r = \sum_{j=1}^{n} k_j.$$

Exercise 7   Every closed set in $\mathbf{R}^n$ is the set of zeros of a C^∞ function. (*Hint:* This is the extension of Exercise 18, Section 2, Chapter 11, to the C^∞ case, and the proof is almost the same. All that is necessary is to take each f_k of class C^∞ and the α_k so small that the series can be differentiated any number of times without losing the uniform convergence. This can be accomplished by redefining M_k to be the maximum of f_k and all its derivatives through order k.)

3   THE INVERSE- AND IMPLICIT-FUNCTION THEOREMS

THEOREM 3.1   *Suppose in the inverse-function theorem that the function f is of class C^r on a neighborhood of a, with $r \geq 1$. Then the inverse φ is of class C^r on a neighborhood of $b = f(a)$.*

THEOREM 3.2   *Suppose in the implicit-function theorem that the function F is of class C^r on a neighborhood of (a, b), with $r \geq 1$. Then the solution φ is of class C^r on a neighborhood of a.*

It is enough to treat the inverse-function theorem because the other was derived from it. If $\mathfrak{R} : \mathfrak{L}_{nn} \to \mathfrak{L}_{nn}$ is the inverse map, that is, $\mathfrak{R}(T) = T^{-1}$, then the composite-function formula gives

$$d\varphi = \mathfrak{R} \circ df \circ \varphi. \tag{1}$$

From the inverse-function theorem itself we know that φ is of class C^0, so we are in a position to use induction. (Actually we know that φ is of class C^1, since we already carried out the first step of the induction in Section 7 of Chapter 10.)

Assume for the moment that $\mathcal{R}$ is of class C^{r-1}. If we take as an inductive hypothesis that φ is of class C^{r-1}, then formula (1) and Theorem 2.5 show that $d\varphi$ is of class C^{r-1}, and then Theorem 2.7 shows that φ is of class C^r, and we are done.

LEMMA 3.3

$\mathcal{R}$ *is of class* C^r *for every* r.

Proof

We shall calculate $d\mathcal{R}$! If A and B are invertible linear transformations, then

$$A^{-1} - B^{-1} = A^{-1}(B - A)B^{-1}.$$

Therefore, $A^{-1} = B^{-1} + A^{-1}(B - A)B^{-1}$, and if this is put back into the first formula, we get

$$A^{-1} - B^{-1} = B^{-1}(B - A)B^{-1} + A^{-1}(B - A)B^{-1}(B - A)B^{-1}.$$

Taking $A = X + H$ and $B = X$, we find

$$(X + H)^{-1} - X^{-1} = -X^{-1}HX^{-1} + (X + H)^{-1}HX^{-1}HX^{-1}. \quad (2)$$

From this we shall conclude that

$$d\mathcal{R}(X)H = -X^{-1}HX^{-1}. \quad (3)$$

First of all, the function $F(H) = -X^{-1}HX^{-1}$ is certainly linear in H. (X is fixed here.) Therefore, what has to be shown is that

$$\frac{(X + H)^{-1}HX^{-1}HX^{-1}}{\|H\|} \to 0 \qquad \text{as } H \to 0. \quad (4)$$

Exercise 1 Show that if $\|H\| \leq \frac{1}{2}\|X^{-1}\|$, then

$$\|(X + H)^{-1}HX^{-1}HX^{-1}\| \leq 2\|X^{-1}\|^3\|H\|^2. \quad (5)$$

(*Hint:* If necessary, look back at Theorem 8.8 of Chapter 9.)

From formula (3) we can read off the lemma by induction. If we take as an inductive hypothesis that $\mathcal{R}$ is of class C^{r-1}, then formula (3) and Exercise 6 of Section 2 show that $d\mathcal{R}$ is of class C^{r-1}, and therefore that $\mathcal{R}$ is of class C^r.

Exercise 2 Does this induction begin all right?

Exercise 3 $d\mathcal{R}$ is a function from where to where?

Exercise 4 Use formula (6), Section 11 of Chapter 9, to prove Theorem 3.1 directly without using $\mathcal{R}$ at all.

The basic tool in the study of manifolds in the last chapter was simply the implicit-function theorem. Now that we have this theorem for class C^r, we can review the theory of the last chapter from the point of view of manifolds of class C^r.

DEFINITION 3.4

A set $\mathbf{M} \subset \mathbf{R}^n$ *is a smooth m-dimensional manifold of class* C^r *if for each point* $a \in \mathbf{M}$ *there is a function* $F:\mathbf{R}^n \to \mathbf{R}^{n-m}$ *that is regular and of class* C^r *on a neighborhood of a and such that* $\mathbf{M} = \{x : F(x) = 0\}$ *in a neighborhood of a.*

In terms of this definition, the original smooth manifolds are simply the smooth manifolds of class C^1.

All the original theorems remain valid in this setting—a smooth manifold of class C^r is locally the graph of a C^r function, it has local parametric representations that are of class C^r and have local inverses of class C^r, and so on. The original proofs remain valid, too. The only additional fact that is needed is the fact that the implicit-function theorem works for C^r.

Exercise 5 Go through the definitions and theorems of Chapter 11 and restate them for smooth manifolds of class C^r.

Since the theory of smooth manifolds of class C^r appears to involve nothing new, the question arises as to why they should be considered at all. There are many interesting problems that do depend on higher derivatives. We simply have not encountered them.

Exercise 6 Let $f:\mathbf{R}^1 \to \mathbf{R}^1$ be a function that is of class C^1 but not C^2; for example, $f(x) = x^2$ for $x \geq 0$ and $f(x) = -x^2$ for $x < 0$. Show that the graph of f is a manifold of class C^1 but not of class C^2.

4 TAYLOR'S FORMULA

Once again Taylor's formula shows how to approximate general functions by polynomials. The idea is to apply the ordinary Taylor's formula to the function

$$g(t) = f(a + th),$$

where f is a given function from $\mathbf{R}^n$ to $\mathbf{R}^1$, and a and h are given points in $\mathbf{R}^n$. Let us do it and see what happens.

$$f(a + h) = g(1) = \sum_{k=0}^{r} \frac{1}{k!} g^k(0)(1 - 0)^k + \frac{g^{r+1}(\tau)(1 - 0)^{r+1}}{(r + 1)!}, \qquad (1)$$

where τ is between 0 and 1. Therefore, we have to calculate the derivatives of g. Note that $g = f \circ \varphi$, with $\varphi(t) = a + th$, so

$$g'(t) = \sum_{j=1}^{n} D_j f(a + th) \varphi_j'(t) = \sum_{j=1}^{n} D_j f(a + th) h_j.$$

By the same argument

$$g''(t) = \sum_{i,j=1}^{n} D_{ij} f(a + th) h_i h_j,$$

and, in general,

$$g^k(t) = \sum D_{i_1 i_2 \ldots i_k} f(a + th) h_{i_1} h_{i_2} \cdots h_{i_k},$$

where $i_1, \ldots, i_k$ all vary from 1 to n. If we write

$$h^i = h_{i_1} h_{i_2} \cdots h_{i_k}, \tag{2}$$

the formula becomes simply

$$g^k(t) = \sum_{|i|=k} D_i f(a + th) h^i; \tag{3}$$

and formula (1) becomes

$$f(a + h) = \sum_{|i| \leq r} \frac{1}{|i|!} D_i f(a) h^i + \frac{1}{(r+1)!} \sum_{|i|=r+1} D_i f(a + \tau h) h^i,$$

$$\text{where } 0 < \tau < 1. \tag{4}$$

THEOREM 4.1   **(*Taylor's Formula*)** *If $f : \mathbf{R}^n \to \mathbf{R}^1$ is of class C^{r+1} at each point of the line segment from a to x, then there is a point ξ on this segment such that*

$$f(x) = \sum_{|i| \leq r} \frac{1}{|i|!} D_i f(a)(x - a)^i + \frac{1}{(r+1)!} \sum_{|i|=r+1} D_i f(\xi)(x - a)^i. \tag{5}$$

The first sum is the Taylor polynomial $T_a^r f$ of f of order r at the point a. The second sum is the remainder.

Exercise 1   Show that the hypothesis that f is of class C^{r+1} justifies the calculations that went into the proof of the theorem. (Actually, a little less will do, for example, class C^r plus differentiability of the derivatives of order r.)

5   LOCAL MAXIMA AND MINIMA

In the case of functions of one variable we were able to use Taylor's formula to obtain a rather complete test for local maxima and minima. The only situations left open were those in which the derivatives fail to exist, or else all vanish. The same argument gives interesting results for functions of several variables; but they are less complete, and the calculations are likely to be fearsome.

Let $f : \mathbf{R}^n \to \mathbf{R}^1$ be of class C^r on a neighborhood of a. Suppose that all derivatives of order > 0 and $< r$ vanish at a, but at least one of order r does not vanish, and define

$$Q_r(h) = \frac{1}{r!} \sum_{|i|=r} D_i f(a) h^i. \tag{1}$$

From Taylor's formula in the form (5) of Section 4, and the fact that the derivatives of order r are continuous at a, it results that for every positive ϵ there is a positive δ such that

$$|f(a+h) - f(a) - Q_r(h)| \le \epsilon |h|^r \qquad \text{if } |h| < \delta;$$

hence

$$Q_r(h) - \epsilon |h|^r \le f(a+h) - f(a) \le Q_r(h) + \epsilon |h|^r \qquad \text{if } |h| < \delta. \tag{2}$$

This formula suggests that the sign of $f(a+h) - f(a)$ (i.e., whether f has a maximum or minimum at a) depends on the sign of Q_r.

THEOREM 5.1   *f has a local minimum at a if $Q_r(h) > 0$ for every $h \ne 0$, a local maximum at a if $Q_r(h) < 0$ for every $h \ne 0$, and neither one if Q_r is positive at some h and negative at others.*

Remark   What the theorem does is to reduce the study of f to the study of the much simpler function Q_r. But it leaves open the case where Q_r is nonnegative for all h, but not actually positive, and also the case where Q_r is nonpositive for all h, but not actually negative. For example, both functions $f(x, y) = x^2 + y^4$ and $g(x, y) = x^2 + y^3$ vanish together with the first derivatives at the origin, and they have the same Q_2, that is, $Q_2(h, k) = 2h^2$, which is ≥ 0 but not > 0 for all (h, k). Obviously, f has a minimum at the origin, and g does not.

Proof   The important fact to notice is that Q_r is homogeneous of degree r, that is,

$$Q_r(th) = t^r Q_r(h), \tag{3}$$

which is perfectly obvious from the definition in formula (1).

First suppose that $Q_r(h) > 0$ for all $h \ne 0$. Since the unit sphere $S(0; 1)$ is compact, Q_r has a minimum m on this set, and m must be positive, because it is a minimum and not just a lower bound. For every $h \ne 0$,

$h/|h|$ is on the unit sphere, so we have $Q_r(h/|h|) \geq m$. Then formula (3) gives

$$Q_r(h) \geq m|h|^r, \tag{4}$$

and this holds for all h, since both sides are 0 for $h = 0$. Taking $\epsilon < m$ and using formula (2), we get

$$f(a + h) - f(a) \geq (m - \epsilon)|h|^r \qquad \text{if } |h| < \delta, \tag{5}$$

which shows that f has a strict local minimum at a.

Exercise 1 Write out the proof of the fact that f has a strict local maximum at a if $Q_r(h) < 0$ for all $h \neq 0$.

Suppose that $Q_r(h_1) > 0$ and $Q_r(h_2) < 0$. According to formulas (2) and (3), we have

$$f(a + th_1) - f(a) \geq t^r(Q_r(h_1) - \epsilon|h_1|^r) \qquad \text{if } |th_1| < \delta.$$

First choose ϵ so that $Q_r(h_1) - \epsilon|h_1|^r > 0$ and find the corresponding δ. If $t > 0$ and small enough so that $|th_1| < \delta$, then $f(a + th_1) - f(a) > 0$; this means that f cannot have a local maximum at a.

Exercise 2 Use h_2 in the same way to show that f cannot have a local minimum at a.

Exercise 3 If r is odd, then there are always points where Q_r is positive and others where it is negative, so there is never a local maximum or minimum.

Although Q_r is generally simpler than f, it is still a difficult job to decide whether such a function is always positive. The case $r = 2$ is interesting because Q_2 is a quadratic form. A quadratic form Q is positive for all $h \neq 0$ if and only if the corresponding self-adjoint linear transformation H is strictly positive definite, and this is true if and only if all the eigenvalues are positive. Calculation of the eigenvalues is still hard, though, so it is worthwhile to have a theorem that is easier to manage, at least in low dimensions.

THEOREM 5.2 *Let $H: \mathbf{R}^n \to \mathbf{R}^n$ be self-adjoint with matrix A relative to an orthonormal basis. Let A_k be the k by k matrix in the upper left-hand corner of A. Then H is strictly positive definite if and only if $\det A_k > 0$ for $k = 1, \ldots, n$.*

Proof Let V_k be the space spanned by the first k basis vectors, let P_k be the projection on V_k, and let K_k be the restriction of $P_k H P_k$ to V_k. Then A_k is nothing but the matrix of K_k, and the determinant of A_k is the product of the eigenvalues of K_k.

Suppose first that H is strictly positive definite. For $x \in V_k$, $x \neq 0$, we have

$$\langle K_k x, x \rangle = \langle P_k H P_k x, x \rangle = \langle H P_k x, P_k x \rangle = \langle H x, x \rangle > 0.$$

This shows that K_k is strictly positive definite, so all its eigenvalues are positive; therefore, their product is positive.

To go the other way we shall use induction and Exercise 13, Section 9 of Chapter 9. Let $\lambda_1 \geq \cdots \geq \lambda_n$ be the eigenvalues of H, and let $\mu_1 \geq \cdots \geq \mu_n$ be the eigenvalues of $P_{n-1} H P_{n-1}$. By the exercise cited we have

$$\lambda_k \geq \mu_{k+1} \qquad \text{for } k = 1, \ldots, n - 1.$$

Now $\mu_1, \ldots, \mu_{n-1}$ are just the eigenvalues of K_{n-1}, and by the induction hypothesis these are all positive, while clearly $\mu_n = 0$. Thus, we have $\lambda_k \geq 0$ for $k = 1, \ldots, n - 1$. But we also have

$$0 < \det A_n = \det A = \lambda_1 \cdots \lambda_n,$$

and the two together give $\lambda_k > 0$ for $k = 1, \ldots, n$.

Exercise 4   Prove that $\mu_1, \ldots, \mu_{n-1}$ are the eigenvalues of K_{n-1} and that $\mu_n = 0$.

Exercise 5   Decide whether the quadratic form

$$Q(x, y, z) = 3x^2 - 2xy + 2yz - y^2 + 6z^2$$

is positive definite.

Another way to look at Theorem 5.1 is that in a neighborhood of a the graph of f looks enough like the graph of Q_r, so that if Q_r has a strict local minimum at 0, then f has a strict local minimum at a; if Q_r has a strict local maximum at 0, then f has a strict local maximum at a; if Q_r has neither a local minimum nor a local maximum at 0, then f has neither a local minimum nor a local maximum at a. The gap between Q_r and f shows up in the word "strict," which appears in the first two statements but not in the third.

Exercise 6   In the case of two variables, draw the graph of a Q_2 with two positive eigenvalues, of a Q_2 with two negative eigenvalues, and of a Q_2 with one of each. Observe the minimum, maximum, and saddle point that result.

PART III

13 } Integration

1 INTRODUCTION

In Chapter 4 the Riemann integral of a function f on an interval I was defined as follows: Partition the interval I into small intervals I_k. In each I_k choose a point ξ_k. Define the integral to be the limit of the sums

$$\sum_{k=1}^{n} f(\xi_k) l(I_k),$$

where $l(I_k)$ is the length of I_k, and the limit is taken as the maximum length goes to 0.

The same idea carries over to higher dimensions. If, for example, f is defined on a rectangle in the plane, partition the rectangle into small rectangles R_k. In each R_k choose a point ξ_k. Define the integral to be the limit of the sums

$$\sum_{k=1}^{n} f(\xi_k) a(R_k), \tag{1}$$

where $a(R_k)$ is the area of R_k, and the limit is taken as the maximum diameter of R_k goes to 0.

There is no real additional difficulty as long as the function f is continuous—but the resulting theory is quite unsatisfactory. Integration over rectangles is not nearly enough. To find the volume of a ball, for example, involves integration over a circle—which is much more complicated because a circle cannot be partitioned into rectangles.

One possibility is to choose some large rectangle R, which contains the circle C, and to partition R into rectangles R_k. In this case there are two different sums that are equally reasonable in formula (1). The first involves the rectangles contained in C, the second those that meet C. The problem that haunts the whole theory is to show that these two equally reasonable sums lead to the same result.

A second possibility is to extend the function to be integrated, which is defined initially on the circle, by putting it equal to 0 outside. The difficulty is ultimately the same, for the extended function is not continuous but, in general, discontinuous at each boundary point of the circle. In both cases the nature of the boundary of the circle must be analyzed very carefully.

Another way to look at the Riemann integral is this. The sum (1) can be regarded as the integral of a function that is constant [with the value $f(\xi_k)$] on each of the rectangles R_k. The problem, then, is to approximate the given function f by functions that are constant on rectangles. If f is continuous and is defined on a rectangle, then the approximation can be done quite well and quite easily; but if either condition is violated, it cannot.

A different idea, which looks similar at first but is radically better, is to approximate the given function f by functions that are constant on more general sets than intervals or rectangles. This can be accomplished as follows:

Suppose that f is nonnegative and bounded, and let $\epsilon > 0$ be given. Define

$$E_j = \{x : j\epsilon \leq f(x) < (j + 1)\epsilon\}, \tag{2}$$

and let f_ϵ be the function that takes the value $j\epsilon$ on the set E_j. The set X on which f is defined is thus partitioned into a finite number of sets E_j on each of which f_ϵ is constant; moreover,

$$f(x) - \epsilon < f_\epsilon(x) \leq f(x) \qquad \text{for every } x. \tag{3}$$

This shows that $f_\epsilon \to f$ uniformly as $\epsilon \to 0$. If we let f_k be the function f_ϵ with $\epsilon = 2^{-k}$, we not only have that $f_k \to f$ uniformly but also that the sequence $\{f_k\}$ is increasing.

With such good approximations available, the problem of defining the integral of f is reduced to that of defining the integral of f_ϵ, which in turn is obviously equivalent to the problem of defining the "length" or "area" of the sets E_j. Let us look at the two-dimensional case where the language is similar to the language we shall use in n dimensions.

Before the pioneering work of the French mathematician Henri Lebesgue (about 1900), the area of a set E in the plane was defined to be the number

$$\alpha(E) = \inf \Sigma a(R_k), \tag{4}$$

where $\{R_k\}$ is a finite sequence of rectangles covering E and the inf is taken over all such finite sequences. This is a very reasonable and intuitive definition— but it leads right back to the Riemann integral with all the problems that we have just mentioned.

Lebesgue made the apparently innocuous modification of allowing infinite sequences of rectangles as well as finite sequences, and this transformed the whole subject. Let us look at the effect of this modification in one simple example.

One of the basic properties of area is that if S and T are disjoint sets, then it should be true that the area of $S \cup T$ should be the sum of the area of S and the area of T.

Exercise 1   Show that $\alpha(S \cup T) \leq \alpha(S) + \alpha(T)$ and that $\mu(S \cup T) \leq \mu(S) + \mu(T)$, where μ is Lebesgue's modification of α.

Let R be a rectangle and let S be a sequence that is dense in R, for instance, the set of all points in R with rational coordinates. Let $T = R - S$. Any finite sequence of rectangles that covers either S or T must cover all of R. Consequently, $\alpha(S) = \alpha(T) = \alpha(R)$, and we do not at all have the desired additivity formula.

On the other hand, if $S = \{x_k\}$, and $\epsilon > 0$ is given, then let R_k be the rectangle with center x_k and area $\epsilon/2^k$. Then the sequence $\{R_k\}$ covers S, and

$$\mu(S) \leq \sum \frac{\epsilon}{2^k} = \epsilon.$$

Thus, $\mu(S) = 0$. Exercise 1 gives $\mu(T) \leq \mu(R) \leq \mu(T) + 0$, so $\mu(T) = \mu(R)$, and we do have the required additivity, $\mu(R) = \mu(S) + \mu(T)$.

Even with the Lebesgue definition of area we shall not have the additivity $\mu(S \cup T) = \mu(S) + \mu(T)$ for every pair of disjoint sets, but we shall have it for any pair that can come up in practice. The approach of Lebesgue is perhaps

Henri Lebesgue

somewhat more technical and harder to grasp in the beginning, but in the end it turns out to be simpler and much more powerful.

2   LEBESGUE MEASURE

A closed rectangle in $\mathbf{R}^n$ is a set of the form

$$Q = \{x : a_i \le x_i \le b_i, i = 1, \ldots, n\},$$

where of course $a_i \le b_i$. The corresponding open rectangles are defined similarly. The center of the rectangle is the point $a + b/2$, the side lengths are the numbers $b_i - a_i$, and the volume $v(Q)$ is the product of the side lengths. The term rectangle by itself refers to either a closed rectangle or an open rectangle.

DEFINITION 2.1   *The Lebesgue outer measure of a set $A \subset \mathbf{R}^n$ is the number*

$$\mu(A) = |A| = \inf \Sigma v(Q_k), \tag{1}$$

where $\{Q_k\}$ is a sequence of rectangles covering A and the inf *is taken over all such sequences.*

Exercise 1   The definition gives the same result if the rectangles are required to be either open or closed. [*Hint:* To get open ones, replace Q_k by a $\tilde{Q}_k$ with the same center and just slightly larger side lengths so that $v(\tilde{Q}_k) \le v(Q_k) + \epsilon/2^k$.]

Exercise 2   For any set $A \subset \mathbf{R}^n$ and any point $a \in \mathbf{R}^n$, we have $|A + a| = |A|$.

The first step in showing that the Lebesgue measure is like a volume is to show that $|Q| = v(Q)$ for any rectangle Q. It is plain from the definition that $|Q| \le v(Q)$, for there is always the covering with $Q_1 = Q$ and $v(Q_k) < \epsilon/2^k$ for $k > 1$. In getting the reverse inequality it is convenient to use special coverings. For each $\delta > 0$, let $\mathcal{L}_\delta$ be the lattice of points in $\mathbf{R}^n$ with coordinates of the form $m\delta$, m an integer, and let $\mathcal{V}_\delta$ be the set of closed cubes of side length δ with vertices in $\mathcal{L}_\delta$. The point of using these cubes is that they fit together nicely instead of at random (Figure 1). This is expressed by the following lemma.

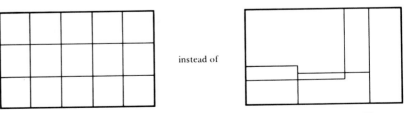

instead of

Figure 1

LEMMA 2.2

If Q is a closed rectangle with vertices in $\mathfrak{L}_\delta$, then Q is a finite union of cubes in $\mathcal{V}_\delta$ and the volume is the sum of the volumes.

Exercise 3

Try to write out a proof of this obvious lemma. Do first dimensions 1, 2, and 3, and then perhaps try induction on the dimension and on the length of the shortest side.

LEMMA 2.3

If Q is any rectangle, let Q^δ be the union of the cubes in $\mathcal{V}_\delta$ that meet $\bar{Q}$. Then Q^δ is a rectangle, and for any given $\epsilon > 0$ we have $v(Q^\delta) \leq v(Q) + \epsilon$ if δ is small enough.

Proof

If $\bar{Q} = \{x : a_j \leq x_j \leq b_j\}$, let c_j be the largest number of the form $m\delta$ that is $\leq a_j$, and let d_j be the smallest one that is $\geq b_j$. Then $Q^\delta = \{x : c_j \leq x_j \leq d_j\}$. The statement on the volumes comes from the fact that $d_j - c_j \leq b_j - a_j + 2\delta$. (Note that the δ here depends on Q as well as on ϵ.)

THEOREM 2.4

If Q is a rectangle, then $|Q| = v(Q)$.

Proof

Let Q be a closed rectangle, and let $\epsilon > 0$ be given. According to the definition and Exercise 1, there is a sequence $\{Q_k\}$ of open rectangles covering Q with

$$|Q| \geq \Sigma v(Q_k) - \epsilon. \tag{2}$$

Since Q is closed (hence compact) and the Q_k are open, a finite number of the Q_k cover Q. We can throw the rest away and suppose that the sum in (2) is finite, say with N terms. By Lemma 2.3 we can choose δ small enough so that $v(Q_k^\delta) \leq v(Q_k) + \epsilon/N$; hence

$$|Q| \geq \Sigma v(Q_k^\delta) - 2\epsilon. \tag{3}$$

It is plain that Q^δ is contained in the union of the Q_k^δ. Consequently, it is equal to the union of some of the cubes in $\mathcal{V}_\delta$ that make up the Q_k^δ. So Lemma 2.2 gives

$$v(Q^\delta) \leq \Sigma v(Q_k^\delta);$$

hence $v(Q) \leq v(Q^\delta) \leq |Q| + 2\epsilon$, which proves the theorem because ϵ is arbitrary and we know already that $|Q| \leq v(Q)$.

Exercise 4

We have proved the theorem for a closed rectangle. What about an open one?

Exercise 5

For fixed $\delta > 0$, let $\mu_\delta(A)$ be the number given by Definition 2.1 when the rectangles Q_k are required to be closed (or open) cubes of diameter $< \delta$. Show that $\mu_\delta(A) = |A|$. [*Hint:* It is plain that $|A| \leq \mu_\delta(A)$. To go the other way, suppose that $|A| < \infty$, let $\epsilon > 0$ be given, and choose a covering $\{Q_k\}$ so that

$$|A| \geq \Sigma v(Q_k) - \epsilon.$$

For each k use Lemma 2.3 to find a rectangle $\tilde{Q}_k$ that is a finite union of cubes of diameter $< \delta$ and satisfies $v(\tilde{Q}_k) \leq v(Q_k) + \epsilon/2^k$. Arrange all these little cubes in a single sequence by counting off first the ones that make up $\tilde{Q}_1$, then the ones that make up $\tilde{Q}_2$, and so on, and use Lemma 2.2.]

Exercise 6   If A and B are any two sets, then

$$|A \cup B| \leq |A| + |B|,$$

and if A and B are at a positive distance apart, then

$$|A \cup B| = |A| + |B|.$$

[*Hint:* The first part is easy from the definition.   The second part follows from the first and from Exercise 4 if you take δ smaller than the distance between A and B.   Recall that the distance is defined by $d(A, B) = \inf \{d(x, y) : x \in A$ and $y \in B\}$.]

Exercise 7   For any set A, $|A| = \inf |G|$, where the inf is taken over all open sets $G \supset A$. (*Hint:* Use the definition and Exercise 1.)

THEOREM 2.5

The Lebesgue outer measure satisfies the following conditions:
(a) $|\varnothing| = 0$ *(where $\varnothing$ is the empty set).*
(b) *If $A \subset B$, then $|A| \leq |B|$.*
(c) *If $A = \cup_{j=1}^{\infty} A_j$, then $|A| \leq \Sigma|A_j|$.*

Proof

Parts (a) and (b) are perfectly obvious, but part (c) requires some proof. It can be assumed that each $|A_j|$ is finite, for otherwise the inequality is automatic.   Let $\epsilon > 0$ be given, and for each j choose a sequence $\{Q_k^j\}$ covering A_j and satisfying

$$\sum_k v(Q_k^j) \leq |A_j| + \frac{\epsilon}{2^j};$$

therefore,

$$\sum_{j,k} v(Q_k^j) \leq \sum_j \left(|A_j| + \frac{\epsilon}{2^j}\right) = \sum_j |A_j| + \epsilon. \qquad (4)$$

When k and j both vary, $\{Q_k^j\}$ is a sequence of rectangles covering A, so $|A|$ is at most the left side of (4), which proves the theorem because ϵ is arbitrary.

Remark   In the proof above we have used the fact that the double sequence $\{Q_k^j\}$ can be arranged in an ordinary sequence.   This can be done by thinking of $\{Q_k^j\}$ as an infinite matrix and counting off the terms as they are met along the following path.

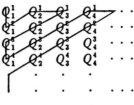

We have also used the fact that the sum of the corresponding series with terms $|Q_k^j|$ is equal to the sum

$$\sum_j \left(\sum_k |Q_k^j| \right).$$

Prove that this is so for any arrangement of the Q_k^j in an ordinary sequence by showing that in both cases the sum is the least upper bound of all finite sums of the $|Q_k^j|$.   The point here is that the numbers $|Q_k^j|$ are nonnegative.

Exercise 8   A set X is said to be countable if its points can be counted off, that is, arranged in a finite or infinite sequence.   Show that any subset of a countable set is countable and use the argument of the Remark to show that if X and Y are countable, then $X \times Y$ is countable.   Show that a countable union of countable sets is countable.

Exercise 9   Show that the rational numbers are countable, and more generally that the points in $\mathbf{R}^n$ with rational coordinates are countable.

Exercise 10   A countable union of sets of measure 0 is again a set of measure 0.   (In particular, every countable set has measure 0, so no rectangle in $\mathbf{R}^n$ is countable.)

We shall close the section with some theorems to show that certain kinds of sets must have measure 0.   Some of these have immediate interest and others will be useful later on.   When dealing with various spaces $\mathbf{R}^n$, we shall write $|\ |_n$ to display the dimension.   The following result is often useful.

THEOREM 2.6   *Let $A \subset \mathbf{R}^m$ be any set and $\varphi : A \to \mathbf{R}^n$ be any function.   If each point $a \in A$ has a neighborhood G such that $|\varphi(A \cap G)|_n = 0$, then $|\varphi(A)|_n = 0$.*

Proof   The hypothesis implies that for each point $a \in A$ there is an open ball B with rational center and radius such that $a \in B$ and $|\varphi(A \cap B)|_n = 0$, for the neighborhood G must contain such a ball.   By Exercises 8 and 9 the family $\mathfrak{B}$ of such balls is countable.   Thus, A is the countable union of the sets $A \cap B$ with $B \in \mathfrak{B}$; hence $\varphi(A)$ is the countable union of the sets $\varphi(A \cap B)$, each of which has measure 0.   By Exercise 10, $\varphi(A)$ has measure 0.

Note that when φ is the identity function, the theorem says that if A has measure 0 "locally," then A has measure 0.

**THEOREM
2.7**

*If $\varphi : \mathbf{R}^m \to \mathbf{R}^n$ satisfies $|\varphi(x) - \varphi(y)| \leq M^\rho |x - y|^\rho$, $\rho \geq m/n$, on a set
$A \subset \mathbf{R}^m$, then*

    (a) *If $\rho > m/n$, then $|\varphi(A)|_n = 0$.*
    (b) *If $\rho = m/n$, then $|\varphi(A)|_n \leq 2^n(M \sqrt{m})^m |A|_m$.*

Proof

Since the measures can be defined by cubes (Exercise 5), it is convenient to use the metric $\|x\| = \max|x_i|$ in which the ball with center a and radius r is just the cube $Q(a; 2r)$ with center a and side length $2r$.   Since $\|x\| \leq |x| \leq \sqrt{m}\,\|x\|$ in $\mathbf{R}^m$ and $\|x\| \leq |x| \leq \sqrt{n}\,\|x\|$ in $\mathbf{R}^n$, we have

$$\|\varphi(x) - \varphi(y)\| \leq (M\sqrt{m})^\rho \|x - y\|^\rho. \tag{5}$$

We shall show that this implies that for any cube Q with side length $\leq \epsilon \leq 1$ we have

$$|\varphi(A \cap Q)|_n \leq 2^n(M\sqrt{m})^{n\rho}\epsilon^{n\rho - m}|Q|_m. \tag{6}$$

Let $Q = Q(b; r)$ with $r \leq \epsilon$.   If Q does not meet A, there is nothing to prove because the left side of (6) is 0.   On the other hand, if a is a fixed point of $A \cap Q$ and x is an arbitrary point of $A \cap Q$, then $\|x - b\| \leq r/2$ and $\|a - b\| \leq r/2$, so $\|x - a\| \leq r$; hence $\|\varphi(x) - \varphi(a)\| \leq (M\sqrt{m})^\rho r^\rho$, which means that

$$\varphi(A \cap Q) \subset Q(\varphi(a); 2(M\sqrt{m})^\rho r^\rho).$$

It follows that

$$|\varphi(A \cap Q)|_n \leq 2^n(M\sqrt{m})^{n\rho} r^{n\rho} \leq 2^n(M\sqrt{m})^{n\rho}\epsilon^{n\rho - m} r^m,$$

which is exactly (6).

    Now, if $\{Q_k\}$ is any covering of A by cubes of side length $\leq \epsilon$, then (6) gives

$$|\varphi(A)|_n \leq \Sigma|\varphi(A \cap Q_k)|_n \leq 2^n(M\sqrt{m})^{n\rho}\epsilon^{n\rho - m}\,\Sigma|Q_k|_m.$$

According to Exercise 5, the measure of A is the inf of the sums on the right, so we have

$$|\varphi(A)|_n \leq 2^n (M \sqrt{m})^{n\rho} \epsilon^{n\rho-m} |A|_m.$$

If $\rho > m/n$, this gives (a) because ϵ is arbitrary, while if $\rho = m/n$ it gives (b).

Exercise 11   What about the last statement of the proof when $|A|_m = \infty$?

The theorem has a number of interesting corollaries.

COROLLARY 2.8   *The notion of measure 0 is independent of the coordinate system in $\mathbf{R}^n$. Every plane of dimension $m < n$ has measure 0.*

Proof   If T is a linear transformation, then $|Tx| \leq M|x|$, so if $m \leq n$, then the hypothesis of the theorem is satisfied for $\varphi = T$ with $\rho = 1$. If V is a subspace of $\mathbf{R}^n$ of dimension $m < n$, then $V = T(\mathbf{R}^m)$, and part (a) of the theorem shows that $|V| = 0$. If Π is a plane of dimension $m < n$, then $\Pi = V + a$, and Exercise 2 shows that $|\Pi| = 0$. If ν is the Lebesgue measure relative to some other coordinates and T is the linear transformation that changes coordinates, then $\nu(A) = |T(A)|$, and part (b) shows that if $|A| = 0$, then $\nu(A) = 0$.

COROLLARY 2.9   *Let $\varphi: \mathbf{R}^m \to \mathbf{R}^n$ be of class C^1 at each point of the set A. If $m < n$, then $|\varphi(A)| = 0$. In particular, the measure of any smooth manifold $\mathbf{M} \subset \mathbf{R}^n$ of dimension $m < n$ is 0.*

Proof   From Theorem 5.3 of Chapter 10 we know that for each point $a \in A$ there is a ball $B(a;r)$ such that

$$|\varphi(x) - \varphi(y)| \leq (\|d\varphi(a)\| + 1)|x - y| \qquad \text{for } x, y \in B(a;r),$$

so part (a) of the theorem shows that $\left|\varphi(A \cap B(a;r))\right| = 0$. Then Theorem 2.6 shows that $|\varphi(A)| = 0$. As for the smooth manifolds, if $b \in \mathbf{M}$ and φ is a local parametric representation at b with $\varphi(a) = b$, then what has been proved shows that $\left|\varphi(B(a;r))\right| = 0$, whereas $\varphi(B(a;r))$ contains a neighborhood of b in $\mathbf{M}$. Thus, $\mathbf{M}$ has measure 0 locally; so by Theorem 2.6, $\mathbf{M}$ has measure 0.

COROLLARY 2.10   *Let $\varphi: \mathbf{R}^n \to \mathbf{R}^n$ be of class C^1 at each point of the set A. If $|A| = 0$, then $|\varphi(A)| = 0$.*

Proof   The proof is the same as the one above, except that we use part (b) of Theorem 2.7 instead of part (a).

Corollary 2.9 is a good example of how measure theory can be used to prove interesting results that have nothing to do with measure theory. We have mentioned, for example, that there are paths that fill a cube in $\mathbf{R}^n$. Corollary 2.9 shows that no path of class C^1 can fill a cube in $\mathbf{R}^n$, $n > 1$, for a cube has positive measure. It also shows that if $m < n$ there is no possibility of an inverse-function theorem for functions $\varphi : \mathbf{R}^m \to \mathbf{R}^n$ of class C^1; for every open set in $\mathbf{R}^n$ contains a cube and hence has positive measure, while $\varphi(\mathbf{R}^m)$ has measure 0. (Previously, we knew that there could not be a differentiable inverse, but this shows that the range cannot even contain any open set.)

Exercise 12 Let $f : \mathbf{R}^m \to \mathbf{R}^n$ be of class C^ρ, $\rho > m/n$, at each point of a set A. If the partial derivatives $D_i f$ all vanish on A for $1 \leq |i| \leq \rho - 1$, then $|f(A)| = 0$. [*Hint:* It is enough to show that for each point $a \in A$ there is a ball $B = B(a; r)$ such that

$$|f(x) - f(y)| \leq M|x - y|^\rho \qquad \text{for } x, y \in A \cap B. \tag{7}$$

Choose B so that the partial derivatives of order ρ exist and are bounded on B, and then write Taylor's formula at y for each of the coordinate functions. Inequality (7) will result.]

Exercise 13 Let $T : \mathbf{R}^n \to \mathbf{R}^n$ be the linear transformation given by $Te_i = \lambda_i e_i$, where e_1, . . . , e_n is the usual basis of $\mathbf{R}^n$. For any set $A \subset \mathbf{R}^n$ we have

$$|T(A)| = |\lambda_1 \cdots \lambda_n| \, |A|.$$

[*Hint:* If Q is the rectangle with center a and side lengths s_1, . . . , s_n, then $T(Q)$ is the rectangle with center Ta and side lengths $|\lambda_1| s_1$, . . . , $|\lambda_n| s_n$. Cover A by rectangles to get

$$|T(A)| \leq |\lambda_1 \cdots \lambda_n| \, |A|.$$

If some λ_i is 0 you are done, and if not you can apply the same thing to T^{-1}.]

Exercise 14 $|B(a; r)| = cr^n$, with $c = |B(0; 1)|$. (*Hint:* Exercise 13 with $Tx = rx$ and Exercise 2.)

3   OUTER MEASURES

Just as it was convenient to study continuity in the abstract setting of metric spaces, it is also convenient to study measure and integration in an abstract setting. The abstract setting is much simpler than $\mathbf{R}^n$, because it involves only three simple axioms, and it also has other interesting interpretations.

DEFINITION 3.1

An outer measure on a set X is a function μ from the subsets of X to the non-negative real numbers and $+\infty$ with the three properties

(a) $\mu(\varnothing) = 0$.
(b) *If $A \subset B$, then $\mu(A) \leq \mu(B)$.*
(c) *If $A = \bigcup_{k=1}^{\infty} A_k$, then $\mu(A) \leq \Sigma\mu(A_k)$.*

These are the properties listed in Theorem 2.5 for the Lebesgue outer measure. The number $\mu(A)$ is called the *outer measure*, or more often just the measure, of the set A. Another interesting example of an outer measure on an arbitrary set X is obtained by putting $\mu(A)$ equal to the number of points in A. This one is called the *counting measure*. When X is the positive integers, the resulting theory of integration is the theory of absolutely convergent series. Still another important example is obtained by choosing any function $\varphi : X \to \mathbf{R}^n$ and putting $\mu(A) = |\varphi(A)|$.

If we think of the outer measure as a generalization of length, area, or volume (which it is in the Lebesgue case), we would like to see results to the effect that if A and B are disjoint sets, then $\mu(A \cup B) = \mu(A) + \mu(B)$. It is plain from (c) that

$$\mu(A \cup B) \leq \mu(A) + \mu(B), \tag{1}$$

but the opposite inequality is simply false when A and B are completely arbitrary. The first big job is to pick out a wide class of sets for which it is true. The definition is more technical than intuitive. It will simply have to be accepted for the sake of what can be done with it.

DEFINITION 3.2

The set $A \subset X$ is μ-measurable if for every set $S \subset X$ we have

$$\mu(S) = \mu(S \cap A) + \mu(S - A). \tag{2}$$

Exercise 1   Every set of measure 0 is measurable.

THEOREM 3.3

If A and B are measurable, then $A \cup B$, $A \cap B$, and $X - A$ are measurable.

Proof

That $X - A$ is measurable is clear, because the definition is symmetric in A and $X - A$. Let us look at $A \cap B$. First split the arbitrary set S into the part in A and the part not in A, and use the fact that A is measurable to get (2). Now split each of the sets $S \cap A$ and $S - A$ into the part in B and the part not in B, and use the fact that B is measurable to get

$$\begin{aligned} \mu(S \cap A) &= \mu(S \cap A \cap B) + \mu((S \cap A) - B), \\ \mu(S - A) &= \mu((S - A) \cap B) + \mu((S - A) - B). \end{aligned} \tag{3}$$

Since $S - (A \cap B) = \{(S \cap A) - B\} \cup \{(S - A) \cap B\} \cup \{S - A - B\}$, we have

$$\mu(S - (A \cap B)) \leq \mu((S \cap A) - B) + \mu((S - A) \cap B) + \mu(S - A - B),$$

which, together with (2) and (3), gives

$$\mu(S) \geq \mu(S \cap A \cap B) + \mu(S - (A \cap B)).$$

The opposite inequality is always true, so $A \cap B$ is measurable. $A \cup B$ can be treated in the same way, or by noticing that

$$X - (A \cup B) = (X - A) \cap (X - B)$$

and using what has been proved for complements and intersections.

THEOREM 3.4

Let $\{A_k\}$ be a sequence of disjoint measurable sets with union A. Then for any set S,

$$\mu(S) = \Sigma \mu(S \cap A_k) + \mu(S - A). \tag{4}$$

Proof

If B_n is the union of the first n A_k's, then

$$S \cap B_n = (S \cap A_n) \cup (S \cap B_{n-1}),$$

and this is exactly the decomposition of $S \cap B_n$ into the part in A_n and the part not in A_n. Therefore,

$$\mu(S \cap B_n) = \mu(S \cap A_n) + \mu(S \cap B_{n-1}),$$

and induction gives

$$\mu(S \cap B_n) = \sum_{k=1}^{n} \mu(S \cap A_k).$$

Now, B_n is measurable, by Theorem 3.3, so

$$\mu(S) = \mu(S \cap B_n) + \mu(S - B_n) \geq \sum_{k=1}^{n} \mu(S \cap A_k) + \mu(S - A).$$

[Where does $\mu(S - B_n) \geq \mu(S - A)$ come from?] Since this holds for every n, we have

$$\mu(S) \geq \sum_{k=1}^{\infty} \mu(S \cap A_k) + \mu(S - A),$$

and, as usual, the opposite inequality is automatic.

THEOREM 3.5   *If $\{A_k\}$ is a sequence of disjoint measurable sets with union A, then A is measurable and*

$$\mu(A) = \Sigma\mu(A_k). \tag{5}$$

Proof   To get (5) take $A = S$ in formula (4). To get the measurability note that the sum in (4) is at least $\mu(S \cap A)$, so (4) gives $\mu(S) \geq \mu(S \cap A) + \mu(S - A)$.

THEOREM 3.6   *The union and intersection of a sequence of measurable sets are measurable.*

Proof   Let $B_1 = A_1$, and $B_k = A_k - \bigcup_{j=1}^{k-1} B_j$. Each B_k is measurable, by Theorem 3.3 and induction. Therefore, the union of the B_k is measurable. But the union of the A_k is the same as the union of the B_k. As for the intersection, $\cap A_k = X - \cup(X - A_k)$.

Exercise 2   If $\{A_k\}$ is an increasing sequence of measurable sets with union A, then $\mu(A_k) \to \mu(A)$. [*Hint:* Put $A_0 = \varnothing$, and write $A_n = \bigcup_{k=1}^{n} (A_k - A_{k-1})$.]

Exercise 3   If $\{A_k\}$ is a decreasing sequence of measurable sets with intersection A, then $\mu(A_k) \to \mu(A)$ provided at least one $\mu(A_k)$ is finite. Show the necessity of the proviso.

Exercise 4   Let μ be the counting measure on X. Every subset of X is μ-measurable.

In Section 4 we shall show that with the Lebesgue measure on $\mathbf{R}^n$ all the open and closed sets are measurable. Combined with the theorems of this section, this means that every set that can be reached in any kind of constructive way is Lebesgue measurable. Nonmeasurable sets simply do not come up in practice. Nonconstructive examples can be given, but we prefer to skip this and to give instead an example of another natural outer measure on $\mathbf{R}^n$ which plays an important auxiliary role in the theory of surface area and which does have simple nonmeasurable sets.

Consider the problem of defining the length of a set in the plane. One idea is to cover the set with small circles and to take the sum of the diameters of the circles (Figure 2). If we take the sum of the squares of the diameters, then we get effectively back to the Lebesgue measure, for the area of a circle is just $\pi d^2/4$. This suggests that to get the area of a set in $\mathbf{R}^3$ we might cover with small balls and take the sum of the squares of the diameters. It suggests in general that to get the m-dimensional area of a set in $\mathbf{R}^n$ we might cover with small balls and take the sum of the mth powers of the diameters. However,

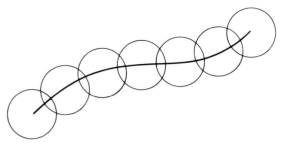

Figure 2

it is technically convenient to allow coverings by sets of any kind.   The formal definition is as follows:

DEFINITION 3.7

Let X be any metric space and let m be any positive number.   For each $\epsilon > 0$ define

$$\mu_m^\epsilon(A) = \inf \Sigma \delta(A_k)^m,$$

where $\delta(A_k)$ is the diameter of A_k, and $\{A_k\}$ is any sequence of sets of diameter $<\epsilon$ covering A.

Exercise 5   Show that μ_m^ϵ is an outer measure on X.

Exercise 6   Let Y be a subset of X, and let ν_m^ϵ be the outer measure on Y constructed above by considering Y as a metric space on its own.   Show that $\nu_m^\epsilon(A) = \mu_m^\epsilon(A)$ for every $A \subset Y$.   (This would obviously be false if we insisted on covering by balls.)

Exercise 7   On $\mathbf{R}^1$ the measure μ_1^ϵ coincides with the Lebesgue measure.   Consequently (Exercise 6), for any line segment I in $\mathbf{R}^n$, $\mu_1^\epsilon(I)$ is the length of I.

The only trouble with the measures μ_1^ϵ is that they have practically no measurable sets.   Consider, for example, two line segments I and J in $\mathbf{R}^n$, $n > 1$, with lengths between $\epsilon/2$ and ϵ and with a common midpoint.   Since $I \cup J$ has diameter $<\epsilon$, we have

$$\mu_1^\epsilon(I \cup J) < \epsilon < \mu_1^\epsilon(I) + \mu_1^\epsilon(J),$$

in spite of the fact that $\mu_1^\epsilon(I \cap J) = 0$, which shows that the segments I and J are not measurable.   The crucial point here is that I and J have length $<\epsilon$. If we fix I and J and let $\epsilon \to 0$, the trouble disappears.

DEFINITION 3.8

Let X be a metric space and let m be a positive number. The m dimensional Hausdorff measure on X is the outer measure defined by

$$\mu_m(A) = \sup_{\epsilon>0} \mu_m^\epsilon(A) = \lim_{\epsilon\to 0} \mu_m^\epsilon(A).$$

(When ϵ decreases, the admissible sequences in Definition 3.7 get fewer and the inf becomes greater. This is why the sup in Definition 3.8 is a limit.)

Exercise 8   Show that μ_m is an outer measure on X and prove the analog of Exercise 6.

Exercise 9   Show that if the sets A and B are at a positive distance apart, then $\mu_m(A \cup B) = \mu_m(A) + \mu_m(B)$. (The main theorem of Section 4 is that this is precisely the condition for every closed set to be measurable.)

Exercise 10   If $\mu_m(A) < \infty$, then $\mu_n(A) = 0$ for every $n > m$. [*Hint:* Show that $\mu_n^\epsilon(A) \leq \epsilon^{n-m}\mu_m^\epsilon(A)$.]

Exercise 11   For the measure μ_n on $\mathbf{R}^n$ there are positive constants c_1 and c_2 such that for every set A

$$c_1|A| \leq \mu_n(A) \leq c_2|A|.$$

[Later we shall show that $\mu_n(A) = c|A|$.]

The Hausdorff measures will play the fundamental role in the theory of surface area in Chapter 15. They are also useful examples to have in mind during the development of the abstract theory, for they have some bad as well as some good properties. For instance.

Exercise 12   If $m < n$, then no set of positive Lebesgue measure in $\mathbf{R}^n$ is the union of a sequence of sets of finite μ_m measure. (*Hint:* Use Exercises 10 and 11.)

Remark   The oddest example in the realm of Lebesgue nonmeasurability was discovered in the 1920s by Stefan Banach and Alfred Tarski. They showed how to cut a ball in $\mathbf{R}^3$ into a finite number of pieces and then reassemble the pieces into a larger ball.

4   MEASURABILITY IN $\mathbf{R}^n$

The purpose of the section is to show that the Lebesgue measurable sets in $\mathbf{R}^n$ are exactly the sets that can be approximated by open or closed sets. The first step is to show that the open and closed sets themselves are measurable. It is enough, of course, to consider the closed sets, for every open set is the complement of a closed set.

THEOREM
4.1

If μ is the Lebesgue measure on $\mathbf{R}^n$ (or any Hausdorff measure μ_m as in Definition 3.8), then every closed set is μ measurable.

This theorem is a consequence of an abstract theorem, whose hypothesis is taken care of by Exercise 6 of Section 2 in the case of the Lebesgue measure and by Exercise 9 of the last section in the case of the Hausdorff measures.

THEOREM
4.2

Let μ be an outer measure on a metric space X. The necessary and sufficient condition that every closed set be measurable is that $\mu(A \cup B) = \mu(A) + \mu(B)$ whenever A and B are a positive distance apart.

Proof

One half is easy. If A and B are a positive distance apart, then $\bar{A} \cap B = \varnothing$, so

$$(A \cup B) \cap \bar{A} = A \qquad \text{and} \qquad (A \cup B) - \bar{A} = B.$$

If every closed set is measurable, then in particular $\bar{A}$ is measurable, and we have $\mu(A \cup B) = \mu(A) + \mu(B)$ just by using the definition of measurability on the set $S = A \cup B$.

In proving the other half we have to show that if A is closed and S is arbitrary, then

$$\mu(S) \geq \mu(S \cap A) + \mu(S - A).$$

We can suppose that $\mu(S) < \infty$, for otherwise the inequality is obvious. The main point is to prove that if

$$G_n = \left\{ x : d(x, A) > \frac{1}{n} \right\},$$

then

$$\mu(S - A - G_n) \to 0. \tag{1}$$

Note that the G_n are increasing and have union $X - A$, so

$$S - A - G_n = \bigcup_{k=n+1}^{\infty} S \cap (G_k - G_{k-1});$$

hence

$$\mu(S - A - G_n) \leq \sum_{k=n+1}^{\infty} \mu\big(S \cap (G_k - G_{k-1})\big). \tag{2}$$

Therefore, it will suffice to show that

$$\sum_{k=1}^{\infty} \mu\big(S \cap (G_k - G_{k-1})\big) < \infty. \tag{3}$$

The picture looks as shown in Figure 3.

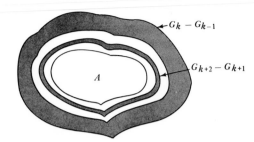

Figure 3

To prove (3) we look separately at the terms with k odd and those with k even, for the reason that the sets involved will then be at a positive distance from one another.   Doing so we get

$$\sum_{\substack{k \leq n \\ k \text{ odd}}} \mu(S \cap (G_k - G_{k-1})) = \mu \left(\bigcup_{\substack{k \leq n \\ k \text{ odd}}} S \cap (G_k - G_{k-1}) \right) \leq \mu(S).$$

The same is true for even k, so every partial sum in the series (3) is at most $2\mu(S)$, and the series converges.

Now we have established (1), and we shall use it to prove the theorem. Since $S \cap G_n = (S - A) \cap G_n$, we have

$$\mu(S \cap G_n) \leq \mu(S - A) \leq \mu(S \cap G_n) + \mu(S - A - G_n);$$

hence

$$\mu(S \cap G_n) \to \mu(S - A).$$

Since $S \cap A$ and $S \cap G_n$ are at a positive distance, this gives

$$\mu(S) \geq \mu(S \cap A) + \mu(S \cap G_n) \to \mu(S \cap A) + \mu(S - A),$$

and the theorem is proved.

From this theorem and Theorem 3.5 on sequences it follows that the union of any sequence of closed sets is measurable.   Such a set is called an F_σ (F standing for closed and σ for union).   Similarly, the intersection of any sequence of open sets is measurable.   Such a set is called a G_δ (G for open and δ for intersection).   These sets, plus and minus sets of measure 0, make up all the Lebesgue measurable sets.

THEOREM 4.3

A set $A \subset \mathbf{R}^n$ is Lebesgue measurable if and only if it has the equivalent properties:

(a) *For each $\epsilon > 0$ there is an open $G \supset A$ with $|G - A| < \epsilon$.*

(b) *A is a G_δ minus a set of measure 0.*

Proof   To see that (a) implies (b), take $G_k \supset A$ with $|G - A| < 1/k$, and let E be the intersection of the G_k. This is clearly a G_δ that contains A and satisfies $|E - A| = 0$.

We have just seen that every G_δ is measurable, and we know that every set of measure 0 is measurable, so every G_δ minus a set of measure 0 is measurable.

Now let A be measurable and write A as the union of a sequence $\{A_k\}$, each measurable and with finite measure [for example, A_k is the intersection of A with the ball $B(0; k)$]. If $\epsilon > 0$ is given, use Exercise 7 of Section 2 to find an open $G_k \supset A_k$ with $|G_k| < |A_k| + \epsilon/2^k$. Since A_k is measurable, it follows that

$$|G_k - A_k| = |G_k| - |A_k| < \frac{\epsilon}{2^k}.$$

If G is the union of the G_k, then G is open, $G \supset A$, and

$$|G - A| \leq \Sigma|G_k - A_k| < \epsilon.$$

THEOREM 4.4   *A set $A \subset \mathbf{R}^n$ is Lebesgue measurable if and only if it has the equivalent properties:*

(a) *For each $\epsilon > 0$ there is a closed $F \subset A$ with $|A - F| < \epsilon$.*
(b) *A is an F_σ plus a set of measure 0.*

Exercise 1   Prove the theorem by taking complements in Theorem 4.3.

Exercise 2   Every F_σ is a K_σ, that is, is the union of a sequence of compact sets. [*Hint:* Write the set as the union of an increasing sequence of closed sets F_j and put $K_j = F_j \cap \bar{B}(0; j)$.]

Exercise 3   What about Theorems 4.3 and 4.4 for the Hausdorff measures μ_m on $\mathbf{R}^n$?

THEOREM 4.5   *Let $\varphi : \mathbf{R}^n \to \mathbf{R}^n$ be of class C^1 at each point of a set A. If A is Lebesgue measurable, then $\varphi(A)$ is Lebesgue measurable.*

Proof   Use Theorem 4.4 and Exercise 2 to write A as the union of a set N of measure 0 and a sequence $\{K_j\}$ of compact sets. Each $\varphi(K_j)$ is measurable because it is compact, and $\varphi(N)$ is measurable because it has measure 0 (Corollary 2.10).

Remark   If φ is just continuous, it still carries compact sets into compact sets, so the validity of the above proof depends on whether it carries sets of measure 0 into sets of measure 0. It can be shown that there exist continuous functions that do not carry sets of measure 0 into sets of measure 0. When φ is such a function, there always exist measurable sets A such that $\varphi(A)$ is not measurable.

5   MEASURABLE FUNCTIONS

Throughout the section μ is an outer measure on a set X.

DEFINITION 5.1   *A function $f : X \to \mathbf{R}^n$ is measurable if $f^{-1}(G)$ is measurable for every open set $G \subset \mathbf{R}^n$.*

There is a technical advantage in allowing functions f that are defined on a subset of X rather than the whole set. In order not to multiply the notations, we shall still write $f : X \to \mathbf{R}^n$ in this case. Observe that if f is measurable, then the set D on which it is defined is necessarily a measurable set, for $D = f^{-1}(\mathbf{R}^n)$, and $\mathbf{R}^n$ is open in $\mathbf{R}^n$.

Exercise 1   A function $f : X \to \mathbf{R}^n$ is measurable if and only if $f^{-1}(F)$ is measurable for every closed set $F \subset \mathbf{R}^n$.

Exercise 2   A function $f : X \to \mathbf{R}^n$ is measurable if and only if $f^{-1}(Q)$ is measurable for every open cube $Q \subset \mathbf{R}^n$. (*Hint:* Every open set is the union of a sequence of cubes.)

Note the analogy between measurability and continuity. If X is a metric space, then $f : X \to \mathbf{R}^n$ is continuous if and only if $f^{-1}(G)$ is open for every open $G \subset \mathbf{R}^n$.

Exercise 3   Prove this statement if you have not already done so. (In this case it is assumed that f is defined on all of X, not just on a subset.)

From the above characterization of continuity and Theorem 4.2 we get the following theorem.

THEOREM 5.2   *If X is a metric space in which $\mu(A \cup B) = \mu(A) + \mu(B)$ whenever A and B are a positive distance apart, then every continuous $f : X \to \mathbf{R}^n$ is measurable. In particular, every continuous $f : \mathbf{R}^m \to \mathbf{R}^n$ is Lebesgue measurable.*

The measurable functions will turn out to be the ones suitable for integration. We shall show presently that they form a much larger class than the continuous functions (which are essentially the ones suitable for Riemann integration), but some other results come first.

THEOREM 5.3

Let $X \xrightarrow{f} \mathbf{R}^n \xrightarrow{\varphi} \mathbf{R}^m$, where f is measurable and φ is continuous. Then $\varphi \circ f$ is measurable.

Proof

If G is an open set in $\mathbf{R}^m$, then $G_1 = \varphi^{-1}(G)$ is open in $\mathbf{R}^n$; hence $(\varphi \circ f)^{-1}(G) = f^{-1}(G_1)$ is measurable.

THEOREM 5.4

$f: X \to \mathbf{R}^n$ *is measurable if and only if each coordinate function is measurable.*

Proof

Let $\varphi_k(y) = y_k$. If f is measurable, then $f_k = \varphi_k \circ f$ is measurable by Theorem 5.3. Suppose, on the other hand, that each f_k is measurable. If Q is the open cube

$$Q = \{y \in \mathbf{R}^n : a_k < y < b_k \text{ for } k = 1, \ldots, n\},$$

then $f(x)$ belongs to Q if and only if $f_k(x)$ belongs to $I_k = (a_k, b_k)$ for each k. Thus,

$$f^{-1}(Q) = \bigcap f_k^{-1}(I_k).$$

Each set on the right is measurable; therefore, so is the intersection. Exercise 2 finishes the proof.

THEOREM 5.5

Let $f, g: X \to \mathbf{R}^n$ be measurable. Then $f + g$, αf, $\langle f, g \rangle$, and $|f|$ are all measurable.

Proof

Write (f, g) for the function from X to $\mathbf{R}^{2n}$ whose first n coordinates are those of f and last n those of g. Now, $f + g = \varphi \circ (f, g)$, where $\varphi((x, y)) = x + y$. From Theorem 5.4 it follows that (f, g) is measurable, and it is plain that $\varphi: \mathbf{R}^{2n} \to \mathbf{R}^n$ is continuous, so Theorem 5.3 shows that $f + g$ is measurable.

Exercise 4   Carry out the details in the other three cases.

THEOREM 5.6

If $f: X \to \mathbf{R}^1$, then the following are equivalent.

    (a) f *is measurable.*
    (b) $\{x : f(x) < \alpha\}$ *is measurable for every real* α.
    (c) $\{x : f(x) \le \alpha\}$ *is measurable for every real* α.
    (d) $\{x : f(x) > \alpha\}$ *is measurable for every real* α.
    (e) $\{x : f(x) \ge \alpha\}$ *is measurable for every real* α.

Proof

Since the intervals

$$I_\alpha = \{y : y < \alpha\} \qquad \text{and} \qquad J_\alpha = \{y : y > \alpha\}$$

are open, and the intervals $\bar{I}_\alpha$ and $\bar{J}_\alpha$ are closed, it follows that the inverse image of each of these is measurable if f is measurable. Consequently, (a) implies each of the others. Suppose, on the other hand, that (b) (for example) holds. We have $\bar{I}_\alpha = \cap I_{\alpha+1/n}$, so $f^{-1}(\bar{I}_\alpha) = \cap f^{-1}(I_{\alpha+1/n})$ is measurable, and, therefore, so is

$$f^{-1}(\bar{J}_\alpha) = D - f^{-1}(\bar{I}_\alpha),$$

where D is the set on which f is defined. (Why is D measurable?) This shows that

$$\{x : \alpha < f(x) < \beta\} = f^{-1}(\bar{J}_\alpha) \cap f^{-1}(\bar{I}_\alpha)$$

is measurable, and then Exercise 2 shows that f is measurable. The other conditions are handled similarly.

THEOREM 5.7

If $\{f_k\}$ is a sequence of measurable real-valued functions, then the following are all measurable:

$$\sup f_k, \quad \inf f_k, \quad \limsup f_k, \quad \liminf f_k.$$

Proof

It is plain that $f(x) = \sup f_k(x)$ is $\leq \alpha$ if and only if each $f_k(x)$ is $\leq \alpha$. In other words,

$$f^{-1}(\bar{I}_\alpha) = \cap f_k^{-1}(\bar{I}_\alpha),$$

and this is measurable, since each set on the right is measurable. The inf is handled similarly. To get the lim sup, note that $g_n(x) = \sup\{f_k(x) : k \geq n\}$ is measurable by what has just been proved, and, therefore, so is $\limsup f_k = \inf g_n$.

Throughout the section we have been a little sloppy about the sets on which the various functions are defined. In Theorem 5.2, for example, the continuous function f is defined on the whole space X. In Theorem 5.3 the continuous function φ is defined on the whole space $\mathbf{R}^n$. In Theorem 5.7 the functions $\sup f_k$ and $\limsup f_k$ are defined at a point x if and only if each f_k is defined at x and the sequence $\{f_k(x)\}$ is bounded above.

Exercise 5 Look back at each of the theorems of the section to make sure that you understand the domains of the various functions.

Exercise 6 It would appear more natural to define the lim sup f_k on a different set. Tell what the different set is and reprove the theorem.

6 DEFINITION OF THE INTEGRAL

**DEFINITION
6.1**

A simple function on X is a nonnegative measurable function $\varphi: X \to \mathbf{R}^1$ which is defined everywhere and takes only a finite number of distinct values. If these are $\alpha_1, \ldots, \alpha_n$, and $E_j = \varphi^{-1}(\{\alpha_j\})$, then

$$I(\varphi) = \Sigma \alpha_j \mu(E_j). \tag{1}$$

Each E_j is measurable, since the single point $\{\alpha_j\}$ is a closed set. It may have infinite measure, of course, in which case the value of the sum in (1) is ∞—with one exception. In integration theory the product $0 \cdot \infty$ is always taken to be 0. Thus, if one of the α_j is 0, then the term $\alpha_j \mu(E_j)$ is 0, whether $\mu(E_j)$ is ∞ or not. It will turn out, of course, that $I(\varphi)$ is the integral of φ, but we shall give a general definition of the integral that applies to nonsimple functions as well and then will show that the general definition gives back $I(\varphi)$ in the case of simple functions. First it is necessary to develop a few properties of $I(\varphi)$.

**THEOREM
6.2**

Let φ be simple. If $X = \cup X_k$, where the X_k are measurable and disjoint and φ takes the constant value β_k on X_k, then

$$I(\varphi) = \Sigma \beta_k \mu(X_k). \tag{2}$$

Proof

The point is that the β_k may not be distinct. However, each β_k is equal to some α_j. If S_1 is the set of indices k for which $\beta_k = \alpha_1$, S_2 is the set of indices k for which $\beta_k = \alpha_2$, and so on, then

$$E_j = \bigcup_{k \in S_i} X_k; \qquad \text{hence } \mu(E_j) = \sum_{k \in S_i} \mu(X_k).$$

It is clear that the sum (1) is obtained from the sum (2) by grouping the terms in this way.

**THEOREM
6.3**

If φ and ψ are simple, then $I(\varphi + \psi) = I(\varphi) + I(\psi)$.

Proof

Write $X = \cup X_k$ where the X_k are measurable and disjoint and φ takes the constant value α_k on X_k, and ψ takes the constant value β_k on X_k. Then $\varphi + \psi$ takes the constant value $\alpha_k + \beta_k$ on X_k, and Theorem 6.2 gives

$$I(\varphi + \psi) = \Sigma(\alpha_k + \beta_k)\mu(X_k) = \Sigma \alpha_k \mu(X_k) + \Sigma \beta_k \mu(X_k)$$
$$= I(\varphi) + I(\psi).$$

**THEOREM
6.4**

If φ and ψ are simple and $\varphi \leq \psi$ except on a set of measure 0, then $I(\varphi) \leq I(\psi)$.

Proof   Suppose first that $\varphi \leq \psi$ everywhere. Then in the notations of the last proof we have $\alpha_k \leq \beta_k$ for each k, which obviously implies that $I(\varphi) \leq I(\psi)$. Now suppose that $\varphi \leq \psi$ except on some set N of measure 0. Let m be an upper bound for φ, and let χ be the function that is m on N and 0 everywhere else. Since $I(\chi) = m\mu(N) = 0$, and since $\varphi \leq \psi + \chi$ everywhere, we have

$$I(\varphi) \leq I(\psi + \chi) = I(\psi) + I(\chi) = I(\psi).$$

DEFINITION 6.5   *A property of points in X is said to hold almost everywhere, or a.e., if it holds except on a set of measure 0.*

In this terminology the statement of the last theorem is that if φ and ψ are simple and $\varphi \leq \psi$ a.e., then $I(\varphi) \leq I(\psi)$.

DEFINITION 6.6   *If $f : X \to \mathbf{R}^1$ is nonnegative and measurable, then*

$$\int f \, d\mu = \sup I(\varphi), \tag{3}$$

where the upper bound is taken over all simple φ that are $\leq f$ at every point where f is defined.

Exercise 1   The definition remains the same if the upper bound is taken over all simple φ that are $\leq f$ at almost every point where f is defined.

Exercise 2   If f is undefined on a set of positive measure, then $\int f \, d\mu = \infty$.

THEOREM 6.7   *If φ is simple, then $\int \varphi \, d\mu = I(\varphi)$.*

Proof   Since $\varphi \leq \varphi$, it follows that $\int \varphi \, d\mu \geq I(\varphi)$. On the other hand, if $\psi \leq \varphi$, then Theorem 6.4 gives $I(\psi) \leq I(\varphi)$. Therefore, the upper bound of $I(\psi)$, which is the integral of φ, is also $\leq I(\varphi)$.

THEOREM 6.8   *Let f and g be nonnegative and measurable and defined a.e. If $f \leq g$ a.e., then $\int f \, d\mu \leq \int g \, d\mu$.*

Proof   If a simple φ is $\leq f$ a.e., then it is also $\leq g$ a.e., so by Exercise 1 it follows that $I(\varphi) \leq \int g \, d\mu$. Since this holds for every φ, it also holds for the upper bound, which is the integral of f.

LEMMA 6.9   *Let f be nonnegative and measurable and defined a.e. There is a nondecreasing sequence of simple functions that converges to f a.e.*

Proof   It is no loss of generality to assume that f is defined everywhere, for we can simply define it to be 0 (for example) on the set where it is initially undefined. In this case we shall get a sequence of simple functions that converges everywhere to f.

If $f_k = \min(f, k)$, then the construction in Section 1 provides a simple φ_k satisfying

$$f_k(x) - \frac{1}{k} \le \varphi_k(x) \le f_k(x) \qquad \text{for all } x.$$

This is formula (3) in Section 1 with f_k for f_ϵ and $1/k$ for ϵ. Now put

$$\psi_k = \max(\varphi_1, \ldots, \varphi_k).$$

It is evident that $\{\psi_k\}$ does the job.

Exercise 3   If f happens to be bounded, then $\psi_k \to f$ uniformly.

There is sometimes occasion to integrate over a subset of X rather than over the whole space.

DEFINITION 6.10   *If f is a nonnegative measurable function and A is a measurable set, then*

$$\int_A f \, d\mu = \int \chi_A f \, d\mu,$$

where χ_A, the characteristic function of A, is the function with value 1 on A and value 0 elsewhere.

Note that if φ is a simple function taking the values $\alpha_1, \ldots, \alpha_m$ on the disjoint sets $E_1, \ldots, E_m$, then $\chi_A \varphi$ is the simple function taking the values $\alpha_1, \ldots, \alpha_m$ on the sets $A \cap E_1, \ldots, A \cap E_m$. Consequently,

$$\int_A \varphi \, d\mu = \sum_{j=1}^{m} \alpha_j \mu(A \cap E_j). \qquad (4)$$

Remark 1   It is usual in integration theory to define the "product" $0 \times ?$ to be 0. In particular, the product $\chi_A f$ is defined to be 0 at each point outside A, whether f is defined there or not.

Exercise 4   If f and A are measurable, then $\chi_A f$ is measurable when defined in this new way.

7   CONVERGENCE THEOREMS

An advantage of the Lebesgue integral over the Riemann integral is the availability of very powerful convergence theorems. The first (and the one from which the others come easily) is the following.

THEOREM 7.1

(*Monotone Convergence Theorem*) *Let $\{f_k\}$ be a nondecreasing sequence of nonnegative measurable functions with limit f.   Then*

$$\int f \, d\mu = \lim \int f_k \, d\mu. \tag{1}$$

The limit of the sequence of functions is to be taken in the strict sense: the limit f is defined at a point x if and only if each of the f_k is defined there and the limit of $\{f_k(x)\}$ exists. It may as well be assumed that each f_k is defined a.e., for otherwise both sides of (1) are automatically $+\infty$. It is not assumed, however, that f is defined a.e., so the theorem gives a powerful method for proving that limits exist a.e.: If the limit of the integrals is finite, then the integral of f is finite, and so f must be defined a.e. The slick proof given here comes from W. Rudin. It begins with a lemma, which is in fact a special case of the theorem.

LEMMA 7.2

If φ is simple and $\{A_k\}$ is an increasing sequence of measurable sets with union A, then

$$\int_A \varphi \, d\mu = \lim \int_{A_k} \varphi \, d\mu.$$

Proof

Application of formula (4) of the last section gives

$$\int_A \varphi \, d\mu = \sum_{j=1}^m \alpha_j \mu(A \cap E_j) \tag{2}$$

and

$$\int_{A_k} \varphi \, d\mu = \sum_{j=1}^m \alpha_j \mu(A_k \cap E_j). \tag{3}$$

Now, Exercise 2 of Section 3 does the job, for it shows that each term in the finite sum (3) converges to the corresponding term in (2).

Proof of the Theorem

It is plain from Theorem 6.8 that

$$\int f \, d\mu \geq \lim \int f_k \, d\mu.$$

To prove the opposite inequality it is enough to show that if φ is simple and $\leq f$ where f is defined, then

$$\int \varphi \, d\mu \leq \lim \int f_k \, d\mu. \tag{4}$$

As observed above, it can be assumed that each f_k is defined a.e., in which case there is a set A with complement of measure 0 such that each f_k is defined everywhere on A and for every point x in A the sequence $\{f_k(x)\}$ is nondecreasing. If c is a positive number less than 1, and if

$$A_k = \{x \in A : f_k(x) \geq c\varphi(x)\},$$

then $\{A_k\}$ is a nondecreasing sequence with union A. The lemma gives

$$c \int \varphi \, d\mu = \int c\varphi \, d\mu = \int_A c\varphi \, d\mu = \lim \int_{A_k} c\varphi \, d\mu$$
$$\leq \lim \int_{A_k} f_k \, d\mu \leq \lim \int_A f_k \, d\mu.$$

Since this holds for every positive c less than 1, it proves formula (4) and hence the theorem.

A second basic convergence theorem is the following.

THEOREM 7.3

(*Fatou's Lemma*) *If $\{f_k\}$ is a sequence of nonnegative measurable functions defined a.e., then*

$$\int (\liminf f_k) \, d\mu \leq \liminf \int f_k \, d\mu. \tag{5}$$

Proof

If $g_n(x) = \inf \{f_k(x) : k \geq n\}$, then $\{g_n\}$ is nondecreasing and its limit is $\liminf f_k$. The monotone convergence theorem gives

$$\int (\liminf f_k) \, d\mu = \lim \int g_n \, d\mu = \liminf \int g_n \, d\mu \leq \liminf \int f_n \, d\mu,$$

the last inequality coming from the fact that $g_n \leq f_n$.

THEOREM 7.4

If f and g are nonnegative and measurable, then

$$\int (f + g) \, d\mu = \int f \, d\mu + \int g \, d\mu.$$

Proof

It can be assumed that f and g are defined a.e., for otherwise both sides are $+\infty$. Use Lemma 6.9 to find nondecreasing sequences $\{\varphi_k\}$ and $\{\psi_k\}$ of simple functions that converge to f and g a.e. The monotone convergence theorem gives

$$\int (f + g) \, d\mu = \lim \int (\varphi_k + \psi_k) \, d\mu = \lim \int \varphi_k \, d\mu + \lim \int \psi_k \, d\mu$$
$$= \int f \, d\mu + \int g \, d\mu.$$

THEOREM 7.5

Any series of nonnegative measurable functions can be integrated term by term.

Exercise 1   Prove the theorem.

THEOREM 7.6

If f is nonnegative and measurable, and $\{E_k\}$ is a disjoint sequence of measurable sets with union E, then

$$\int_E f \, d\mu = \sum_{k=1}^{\infty} \int_{E_k} f \, d\mu.$$

Exercise 2   Prove the theorem.

There is still one more basic convergence theorem, called the *dominated convergence theorem*, that applies to functions that are not necessarily nonnegative. This is discussed in Section 8.

8   INTEGRABLE FUNCTIONS

So far we have considered the integration of nonnegative functions. Now we take up the rest.

DEFINITION 8.1   *A function $f: X \to \mathbf{R}^n$ is integrable if it is measurable and $\int |f| \, d\mu < \infty$.*

Exercise 1 $f: X \to \mathbf{R}^n$ is integrable if and only if each coordinate function is integrable.

The natural way to integrate a function with values in $\mathbf{R}^n$ is to integrate each coordinate function separately. The natural way to integrate a real-valued function is to express it as a difference of nonnegative functions. If $f: X \to \mathbf{R}^1$ is measurable, then so are the functions

$$f^+ = \max(f, 0) = \frac{|f| + f}{2} \quad \text{and} \quad f^- = \max(-f, 0) = \frac{|f| - f}{2}.$$

Both are nonnegative, and, moreover,

$$f = f^+ - f^- \quad \text{and} \quad |f| = f^+ + f^-. \tag{1}$$

DEFINITION 8.2   *If $f: X \to \mathbf{R}^1$ is integrable, then*

$$\int f \, d\mu = \int f^+ \, d\mu - \int f^- \, d\mu.$$

Exercise 2 f is integrable if and only if both f^+ and f^- are integrable.

The first step is to show that f can be split into positive and negative parts in any reasonable way at all—not just in the way of formula (1).

LEMMA 8.3   *If $f = g - h$, where g and h are nonnegative and integrable, then f is integrable and*

$$\int f \, d\mu = \int g \, d\mu - \int h \, d\mu.$$

Proof f is measurable, since both g and h are, and then it is integrable, since $|f| \le g + h$. The fact that $f^+ - f^- = f = g - h$ gives $f^+ + h = f^- + g$; then Theorem 7.3 gives

$$\int f^+ \, d\mu + \int h \, d\mu = \int f^- \, d\mu + \int g \, d\mu$$

from which the formula in the lemma follows directly.

DEFINITION 8.4

If $f: X \to \mathbf{R}^n$, the integral of f is obtained by integrating each coordinate separately. That is, if f_k is the kth coordinate of f and $I_k = \int f_k \, d\mu$, then

$$\int f \, d\mu = (I_1, \ldots, I_n).$$

THEOREM 8.5

If $f, g: X \to \mathbf{R}^n$ are integrable and α is a real number, then $f + g$ and αf are integrable, and

$$\int (f + g) \, d\mu = \int f \, d\mu + \int g \, d\mu, \qquad \int (\alpha f) \, d\mu = \alpha \int f \, d\mu.$$

Proof

It is clearly enough to treat each coordinate separately, that is, to treat the case $n = 1$. In this case we have $f = f^+ - f^-$ and $g = g^+ - g^-$, and then $f + g = (f^+ + g^+) - (f^- + g^-)$, and Lemma 8.3 does the job.

Exercise 3 Why is Lemma 8.3 necessary here rather than just the original Definition 8.2?

Exercise 4 Integrability and the value of the integral of an $f: X \to \mathbf{R}^n$ are both independent of the coordinate system used in $\mathbf{R}^n$.

The third basic convergence theorem is the following.

THEOREM 8.6

(Dominated Convergence Theorem) *For each k, let $f_k: X \to \mathbf{R}^n$ be integrable and satisfy $|f_k(x)| \leq g(x)$ a.e., where $g: X \to \mathbf{R}^1$ is some fixed integrable function. If $f_k \to f$ a.e., then f is integrable, and*

$$\int f_k \, d\mu \to \int f \, d\mu.$$

The function g is said to dominate the sequence $\{f_k\}$. For this reason the theorem is called the *dominated convergence theorem*.

Proof

It is plain that f is integrable, for it satisfies $|f| \leq g$ a.e. In proving the convergence, it is enough to treat each coordinate separately, that is, to treat the case $n = 1$. The advantage of this is to make available Fatou's lemma for use with the nonnegative functions $g + f_k$. It gives

$$\int g \, d\mu + \int f \, d\mu = \int (g + f) \, d\mu \leq \liminf \int (g + f_k) \, d\mu$$
$$= \int g \, d\mu + \liminf \int f_k \, d\mu;$$

hence

$$\int f \, d\mu \leq \liminf \int f_k \, d\mu. \tag{2}$$

If this formula is applied to the function $-f$ and the sequence $\{-f_k\}$ (which, of course, satisfy the conditions of the theorem), it gives

$$\int -f \, d\mu \leq \liminf \int -f_k \, d\mu = -\limsup \int f_k \, d\mu;$$

hence

$$\int f \, d\mu \geq \limsup \int f_k \, d\mu. \tag{3}$$

Since lim inf is always $\leq$ lim sup, it follows from (2) and (3) that the two are equal and are equal to the integral of f. This implies that the limit exists and is equal to the integral of f.

Exercise 5   Prove the fact used above—that

$$\lim \inf(-a_k) = -\lim \sup a_k.$$

THEOREM 8.7   *Let $X = [a, b]$ be a closed bounded interval with the Lebesgue measure. A function $f: X \to \mathbf{R}^n$ is Riemann integrable if and only if it is bounded and continuous a.e.; if this is the case, the Riemann and Lebesgue integrals are equal.*

Proof   Suppose that f is bounded, say $|f| \leq M$, and continuous a.e. If p is a partition, let φ_p be the function that takes the constant value $f(\xi_i)$ on the interval $[x_{i-1}, x_i)$. It is obvious that φ_p is measurable, that $|\varphi_p| \leq M$, and that

$$S(f; p) = \int \varphi_p \, d\mu. \tag{4}$$

(φ_p is almost a simple function—indeed, it is one if f is real valued and nonnegative.)   Now, let $\{p_k\}$ be any sequence of partitions with $|p_k| \to 0$. It is plain that $\varphi_{p_k}(x) \to f(x)$ for every x at which f is continuous, hence for almost every x.   The dominated convergence theorem gives

$$S(f; p_k) = \int \varphi_{p_k} \, d\mu \to \int f \, d\mu,$$

which shows that the Riemann integral exists and is equal to the Lebesgue integral.

As for the converse, it has already been seen that a Riemann integrable function must be bounded, so what remains is to show that it must be continuous a.e.   Since the coordinate functions can be treated separately, it is all right to suppose that f is real valued.   Let D be the set of discontinuities, and let

$$D_n = \left\{ x : \lim_{y \to x} \sup f(y) - \lim_{y \to x} \inf f(y) \geq \frac{1}{n} \right\}. \tag{5}$$

Exercise 6   Strictly speaking, the limits superior and inferior have not been defined in exactly this situation.   Define them and show that

$$D = \bigcup_{n=1}^{\infty} D_n. \tag{6}$$

If D has positive measure, then at least one D_n must have positive measure. (Does this assertion require that the D_n be measurable?   Are they measurable)?

Choose such an n and fix it.   We shall show that for every partition p,

$$\bar{S}(f; p) - \underline{S}(f; p) \geq \frac{\mu(D_n)}{n},$$

which will show that f is not Riemann integrable.

Let $I_1, \ldots , I_m$ be the intervals in p that meet D_n in at least one interior point (of the interval, that is).   These cover D_n except possibly for the finite number of end points of the intervals in p.   Therefore,

$$\sum_{j=1}^{m} l(I_j) \geq \mu(D_n).$$

Moreover, if M_j and m_j are the upper and lower bounds of f on I_j, then $M_j - m_j \geq 1/n$.   It follows that

$$\bar{S}(f; p) - \underline{S}(f; p) \geq \sum_{j=1}^{m} (M_j - m_j) l(I_j) \geq \frac{\mu(D_n)}{n}.$$

Example   Calculate $\int_0^1 x^\alpha \, dx$.

If $\alpha \geq 0$, then x^α is Riemann integrable, and the old formulas with primitives give the value $(1/\alpha + 1)$.   Suppose, however, that $\alpha < 0$, in which case x^α is not Riemann integrable, because it is not bounded.   Define

$$f_n(x) = x^\alpha \quad \text{if } x \geq \frac{1}{n}, \qquad f_n(x) = 0 \quad \text{if } x < \frac{1}{n}.$$

Now, f_n is Riemann integrable, and

$$\int_0^1 f_n \, dx = \int_{1/n}^1 x^\alpha \, dx = \frac{1}{\alpha + 1}\left(1 - \frac{1}{n^{\alpha+1}}\right).$$

This converges if and only if $\alpha > -1$; then the monotone convergence theorem shows that x^α is integrable on $[0, 1]$ if and only if $\alpha > -1$.   If this is the case, the value of the integral is $1/(\alpha + 1)$.

An important function, called the *gamma function*, is defined by

$$\Gamma(x) = \int_0^\infty t^{x-1} e^{-t} \, dt \qquad \text{for } x > 0.$$

Exercise 7   Show that $t^{x-1} e^{-t}$ is integrable on $[0, \infty)$ if $x > 0$.   Show that $\Gamma(1) = 1$ and $\Gamma(x + 1) = x\Gamma(x)$.

The exercise shows that if n is a positive integer, then $\Gamma(n + 1) = n!$.

Exercise 8 If f is nonnegative and improperly Riemann integrable, then it is Lebesgue integrable and the two integrals are equal. However, the function $\sin x/x$ is improperly Riemann integrable on $[0, \infty)$ and is not Lebesgue integrable.

THEOREM 8.8 *If $f: X \to \mathbf{R}^n$ is integrable, then*

$$|\textstyle\int f \, d\mu| \leq \int |f| \, d\mu,$$

and equality holds if and only if $f(x) = |f(x)|v$ a.e. for some constant vector v.

Proof The theorem is obvious if f is real valued. We shall reduce the general case to this one by taking suitable inner products. Note first that for any $w \in \mathbf{R}^n$, $w \neq 0$, there is one and only one $v \in \mathbf{R}^n$, namely $v = w/|w|$, such that $|v| = 1$ and $\langle v, w \rangle = |w|$.

To prove the theorem, take $w = \int f \, d\mu$ and $v = w/|w|$, as above. (If $w = 0$, the theorem is obvious, so we can suppose that this is not the case.) We have

$$|\textstyle\int f \, d\mu| = \langle v, \int f \, d\mu \rangle = \int \langle v, f \rangle \, d\mu \leq \int |f| \, d\mu$$

by Cauchy–Schwarz and the fact that $|v| = 1$. Now, if equality holds, then since $|\langle v, f(x) \rangle| \leq |f(x)|$, we must have $\langle v, f(x) \rangle = |f(x)|$ a.e.; then by the remark at the beginning we must have $f(x) = 0$ or else $v = f(x)/|f(x)|$. In either case $f(x) = |f(x)|v$.

COROLLARY 8.9 *Let $f: X \to \mathbf{R}^n$ be integrable and satisfy $|f(x)| \leq M$ a.e. If $|\int f \, d\mu| = M\mu(X) < \infty$, then f is constant a.e.*

Proof We have $\int |f| \, d\mu \leq M\mu(X) = |\int f \, d\mu|$. The first inequality shows that $|f(x)| = M$ a.e., and Theorem 8.8 shows that $f(x) = |f(x)|v$ a.e.

9 PRODUCT MEASURES

In dimension one the practical way to calculate an integral is to find a primitive. In higher dimensions it is to reduce the integral to a succession of one-dimensional ones. The theorem that tells how to do this is not only important in calculations but plays a fundamental theoretical role as well. It is simple to state and to use, but the proof is long and technical. Unfortunately, we shall omit it.

If X and Y are sets, then $X \times Y$ is the set of pairs (x, y) with $x \in X$ and $y \in Y$. If $X = \mathbf{R}^m$ and $Y = \mathbf{R}^n$, the pair (x, y) can be considered as a point in $\mathbf{R}^{m+n}$—the point whose first m coordinates are the coordinates of x and whose last n are the coordinates of y. Thus, $\mathbf{R}^m \times \mathbf{R}^n = \mathbf{R}^{m+n}$. The problem is this: Given measures μ and ν on X and Y, construct a natural measure $\mu \times \nu$ on

$X \times Y$. It can be solved if the measures μ and ν are σ-finite in the following sense:

DEFINITION 9.1 *An outer measure μ on a set X is σ-finite if X is the union of a sequence of measurable sets of finite measure.*

The definition of the measure $\mu \times \nu$ is similar to that of the Lebesgue measure.

DEFINITION 9.2 *Let μ and ν be σ-finite outer measures on X and Y. For any set $E \subset X \times Y$, define*

$$(\mu \times \nu)(E) = \inf \Sigma \mu(A_k)\nu(B_k),$$

where the inf *is taken over all sequences $\{A_k\}$ and $\{B_k\}$ such that $E \subset \bigcup A_k \times B_k$.*

The first basic theorem is the following:

THEOREM 9.3 *If μ and ν are σ-finite outer measures on X and Y, then $\mu \times \nu$ is a σ-finite outer measure on $X \times Y$. If A is a μ-measurable set in X, and B is a ν-measurable set in Y, then $A \times B$ is a $(\mu \times \nu)$-measurable set, and*

$$(\mu \times \nu)(A \times B) = \mu(A)\nu(B). \tag{1}$$

If μ and ν are the Lebesgue measures on $\mathbf{R}^m$ and $\mathbf{R}^n$, then $\mu \times \nu$ is the Lebesgue measure on $\mathbf{R}^{m+n}$.

For the second basic theorem we need a little notation. If $f: X \times Y \to Z$, and x is a point of X, then $f_x: Y \to Z$ is the function defined by

$$f_x(y) = f(x, y).$$

f_x is called the *section of f through x*.

THEOREM 9.4 *(**Fubini's Theorem**)   Let μ and ν be σ-finite, and let $f: X \times Y \to \mathbf{R}^1$ be nonnegative and $(\mu \times \nu)$-measurable. Then*
(a) *For almost every $x \in X$, the section f_x is ν-measurable.*
(b) *The function $F(x) = \int f_x \, d\nu$ is μ-measurable, and*

$$\int f \, d(\mu \times \nu) = \int F \, d\mu.$$

(c) *The same statements hold with X and Y interchanged.*

Exercise 1 Take X to be the positive integers and μ to be the counting measure. Show that the notion of a measurable function from $X \times Y$ to $\mathbf{R}^1$ is equivalent to the notion of a sequence of measurable functions from Y to $\mathbf{R}^1$. Use Fubini's theorem to deduce that a series of nonnegative measurable functions can be integrated term by term. Use this to deduce the monotone convergence theorem (in the σ-finite case). This very special case gives some illustration of the power of the theorem.

Ordinarily, parts (b) and (c) of Fubini's theorem are written in the following way:

$$\int f \, d(\mu \times \nu) = \int \{\int f(x, y) \, d\nu(y)\} \, d\mu(x)$$
$$= \int \{\int f(x, y) \, d\mu(x)\} \, d\nu(y). \tag{2}$$

In fact, $d\mu(x)$ and $d\nu(y)$ are often written simply dx and dy, in which case the formula becomes

$$\int f \, d(\mu \times \nu) = \int \{\int f(x, y) \, dy\} \, dx = \int \{\int f(x, y) \, dy\} \, dx. \tag{3}$$

Of course, the theorem applies also to measurable functions that are not necessarily nonnegative.

Exercise 2   Prove the following theorem by splitting each coordinate function of f into its positive and negative parts.

THEOREM
9.5

Let μ and ν be σ-finite, and let $f: X \times Y \to \mathbf{R}^n$ be $(\mu \times \nu)$-measurable. If one of the three integrals below is finite, then formula (2) holds:

$$\int |f| \, d(\mu \times \nu), \qquad \int \{\int |f(x, y)| \, d\nu(y)\} \, d\mu(x),$$
$$\int \{\int |f(x, y)| \, d\mu(x)\} \, d\nu(y).$$

One frequent application of Fubini's theorem is to the case when f is the characteristic function of a set E in $X \times Y$. In this case the section f_x is the characteristic function of the set

$$E_x = \{y \in Y : (x, y) \in E\},$$

which is called the section of E through x. Fubini's theorem shows that if E is $(\mu \times \nu)$-measurable, then E_x is ν-measurable for almost every x, and

$$(\mu \times \nu)(E) = \int \nu(E_x) \, d\mu(x). \tag{4}$$

In particular, E has $(\mu \times \nu)$-measure 0 if and only if almost all sections have ν-measure 0. This is one of the most common ways to show that a set in $\mathbf{R}^n$ has measure 0.

Quite often the key problem in the applications of Fubini's theorem is to prove that the set or function involved is $(\mu \times \nu)$-measurable. To prove that a set $E \subset X \times Y$ has measure 0, for example, it is not enough to prove that almost all sections have measure 0. It must be proved also that the set is $(\mu \times \nu)$-measurable. This is something of an anomaly, because all the sets and functions that actually do turn up in practice are measurable!

Exercise 3 Let $X = \mathbf{R}^n$ and μ be the Lebesgue measure, and let $Y = \mathbf{R}^1$ and ν be the Lebesgue measure. A function $f:X \to \mathbf{R}^1$ is μ-measurable if and only if its graph is $(\mu \times \nu)$-measurable. Notice that the sections of the graph are just single points and, consequently, have ν-measure 0. Hence, the function is μ-measurable if and only if the graph has $(\mu \times \nu)$-measure 0.

As a first example of the use of Fubini's theorem, let us show that the volume of a cone in $\mathbf{R}^n$ is $1/n$ times the area of the base times the height. The first task is to put straight what the result says.

Let B' be a measurable set in $\mathbf{R}^{n-1}$ and let $h > 0$. The set B' is moved up to the plane $x_n = h$ as follows (see Figure 4):

$$B = \{x \in \mathbf{R}^n : x' \in B' \text{ and } x_n = h\},$$

where, as usual, x' denotes the first n coordinates of x. Then the cone with vertex 0, base B', and height h is the set

$$C = \{\alpha x : x \in B \text{ and } 0 \leq \alpha \leq 1\}.$$

The result is that

$$\mu_n(C) = \frac{1}{n} h\mu_{n-1}(B'), \tag{5}$$

where, of course, μ_n and μ_{n-1} are the Lebesgue measures in $\mathbf{R}^n$ and $\mathbf{R}^{n-1}$. Note that the cone is the union of the line segments joining points of B to 0.

The section C_t through a point $t \in \mathbf{R}^1$ is obtained as follows: The point (y', t) belongs to C if and only if $(y', t) = \alpha(x', h)$, where $x' \in B'$ and $0 \leq \alpha \leq 1$,

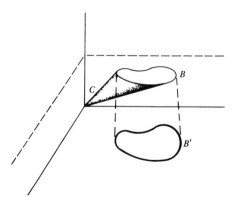

Figure 4

and, therefore, if and only if

$$y' = \frac{t}{h} x', \qquad x' \in B' \quad \text{and} \quad 0 \leq t \leq h. \tag{6}$$

Consequently, $C_t = T(B')$, where $T: \mathbf{R}^{n-1} \to \mathbf{R}^{n-1}$ is the linear transformation given by (6). It follows from Exercise 13 of Section 2 that

$$\mu_{n-1}(C_t) = \left(\frac{t}{h}\right)^{n-1} \mu_{n-1}(B'), \qquad 0 \leq t \leq h,$$

and hence from Fubini that

$$\mu_n(C) = \int_0^h \mu_{n-1}(C_t)\, dt = \frac{1}{n} h \mu_{n-1}(B').$$

Strictly speaking, the calculations are not justified until it is shown that the cone C is measurable. If the base B' is closed, then the cone is closed. If B' is open, then $C - (B' \cup \{0\})$ is open, so there is no problem in these cases. The general case can be handled by these two statements and formula (5) itself.

Exercise 4   Do this.

Next let us calculate the volume (i.e., measure) of the unit ball $B^n = B(0; 1)$ in $\mathbf{R}^n$. If we write $\mathbf{R}^n = \mathbf{R}^{n-1} \times \mathbf{R}^1$, then the section $B^n_{x'}$ through a point $x' \in \mathbf{R}^{n-1}$ is the interval

$$-\sqrt{1 - |x'|^2} \leq x_n \leq \sqrt{1 - |x'|^2}.$$

Therefore,

$$\mu_n(B^n) = \int_{B^{n-1}} \int_{-\sqrt{1-|x'|^2}}^{\sqrt{1-|x'|^2}} dx_n\, dx' = \int_{B^{n-1}} 2\sqrt{1 - |x'|^2}\, dx',$$

which suggests that for the purpose of induction it will be advantageous to start with the integral

$$I(m, n) = \int_{B^n} (1 - |x|^2)^{m/2}\, dx = \int_{B^{n-1}} \int_{-\sqrt{1-|x'|^2}}^{\sqrt{1-|x'|^2}} (1 - |x'|^2 - x_n^2)^{m/2}\, dx_n\, dx'.$$

If we write $a = \sqrt{1 - |x'|^2}$, and $t = x_n$, the inner integral is

$$\int_{-a}^{a} (a^2 - t^2)^{m/2}\, dt = a^{m+1} \int_{-\pi/2}^{\pi/2} \cos^{m+1} \theta\, d\theta.$$

Therefore,

$$I(m, n) = I(m + 1, n - 1) \int_{-\pi/2}^{\pi/2} \cos^{m+1} \theta\, d\theta.$$

Using this formula to evaluate $I(m + 1, n - 1)$, then $I(m + 2, n - 2)$, and so on, we get

$$I(m, n) = \int_{-\pi/2}^{\pi/2} \cos^{m+1} \theta \, d\theta \cdots \int_{-\pi/2}^{\pi/2} \cos^{m+n} \theta \, d\theta;$$

then, taking $m = 0$, we get

$$|B^n|_n = \int_{-\pi/2}^{\pi/2} \cos \theta \, d\theta \cdots \int_{-\pi/2}^{\pi/2} \cos^n \theta \, d\theta. \tag{7}$$

The integrals of the powers of the cosine are evaluated in Section 4 of Chapter 4. (But they are evaluated in a better way below.)

THEOREM 9.6

If $f:\mathbf{R}^2 \to \mathbf{R}^1$ *is nonnegative and measurable, then*

$$\int_{\mathbf{R}^2} f(x, y) \, dx \, dy = \int_0^{2\pi} \int_0^\infty f(r \cos \theta, r \sin \theta) r \, dr \, d\theta.$$

The theorem says that if we put $x = r \cos \theta$ and $y = r \sin \theta$ in a double integral, then we should put $dx \, dy = r \, d\theta \, dr = r \, dr \, d\theta$. The theorem is useful when the function f has some radial symmetry and in various other ways. The pair (r, θ) is called the *polar coordinates* of the point (x, y).

Proof

First we shall integrate over the right half-plane, and we shall use Fubini to integrate first with respect to y, then with respect to x. In the first integral x is constant, and we make the change of variable $y = x \tan \theta$. This gives

$$\int_0^\infty \int_{-\infty}^\infty f(x, y) \, dy \, dx = \int_0^\infty \int_{-\pi/2}^{\pi/2} f(x, x \tan \theta) x \sec^2 \theta \, d\theta \, dx.$$

Now use Fubini to integrate first with respect to x and then with respect to θ. In the first integral, where θ is constant, make the change of variable $x = r \cos \theta$ to get the theorem.

Exercise 5   What about the measurability required to use Fubini the second time?

Now we shall use Theorem 9.6 to get a nice formula relating the gamma function to integrals of powers of sines and cosines. Recall that

$$\Gamma(x) = \int_0^\infty t^{x-1} e^{-t} \, dt. \tag{8}$$

We have already established the basic formula

$$\Gamma(x + 1) = x\Gamma(x), \quad \Gamma(1) = 1. \tag{9}$$

If we put $t = s^2$ in formula (8), we get

$$\Gamma(x) = 2 \int_0^\infty s^{2x-1} e^{-s^2} \, ds. \tag{10}$$

Now calculate the product $\Gamma(x)\Gamma(y)$ by using first Fubini and then polar coordinates. The result is

$$
\begin{aligned}
\Gamma(x)\Gamma(y) &= 2\int_0^\infty s^{2x-1}e^{-s^2}\,ds\ 2\int_0^\infty t^{2y-1}e^{-t^2}\,dt \\
&= 4\int_0^\infty \int_0^\infty s^{2x-1}t^{2y-1}e^{-(s^2+t^2)}\,ds\,dt \\
&= 4\int_0^{\pi/2}\int_0^\infty r^{2x-2y-1}e^{-r^2}\cos^{2x-1}\theta\ \sin^{2y-1}\theta\,dr\,d\theta \\
&= 2\Gamma(x+y)\int_0^{\pi/2}\cos^{2x-1}\theta\ \sin^{2y-1}\theta\,d\theta.
\end{aligned}
$$

Thus, we have the following nice formula:

$$
\frac{\Gamma(x)\Gamma(y)}{\Gamma(x+y)} = 2\int_0^{\pi/2}\cos^{2x-1}\theta\ \sin^{2y-1}\theta\,d\theta. \tag{11}
$$

In particular, if we take $2x - 1 = m$ and $2y - 1 = 0$, we get

$$
\int_{-\pi/2}^{\pi/2}\cos^m\theta\,d\theta = \frac{\Gamma\left(\dfrac{m+1}{2}\right)\Gamma\left(\dfrac{1}{2}\right)}{\Gamma\left(\dfrac{m+2}{2}\right)}. \tag{12}
$$

Notice that formula (11) also shows ($2x - 1 = 0$, $2y - 1 = 0$) that

$$
\Gamma(\tfrac{1}{2}) = \sqrt{\pi}. \tag{13}
$$

Exercise 6   Use formulas (7), (12), and (13) to show that

$$
|B^n|_n = \frac{\pi^{n/2}}{\Gamma\left(\dfrac{n+2}{2}\right)}. \tag{14}
$$

Write this out using formula (9).

Exercise 7   There are some important examples of Fubini-like theorems where the measures μ and ν are not σ-finite, but in general this hypothesis cannot be avoided. Make an example in which $X = \mathbf{R}^1$ with Lebesgue measure and $Y = \mathbf{R}^1$ with counting measure, and f is the characteristic function of a suitable set such as is shown in Figure 5.

Exercise 8   The Hausdorff measure μ_m is not σ-finite on $\mathbf{R}^n$ if $m < n$. (*Hint:* Exercise 12 of Section 3.)

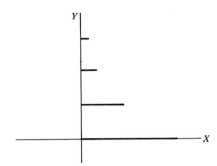

Figure 5

10   FUNCTIONS DEFINED BY INTEGRALS

We shall consider two kinds of functions—the indefinite integrals

$$F(s) = \int_a^s f(t)\, dt$$

and the functions of the form

$$F(t) = \int f(x, t)\, d\mu(x).$$

The results here are by no means the best possible ones, but they are good enough to have interesting applications.

THEOREM   *Let $f : \mathbf{R}^1 \to \mathbf{R}^n$ be integrable on each compact subinterval of an open interval I,*
10.1   *and let*

$$F(s) = \int_a^s f(t)\, dt.$$

Then F is continuous at each point of I and satisfies $F'(s) = f(s)$ at each point s where f is continuous.

Proof   Let χ_s be the characteristic function of the interval $[a, s]$.   If $s_k \to b$, then $\chi_{s_k}(t) \to \chi_b(t)$ at every point t, except perhaps $t = b$, so the dominated convergence theorem gives

$$F(s_k) = \int f\chi_{s_k}\, dt \to \int f\chi_b\, dt = F(b).$$

(Strictly speaking, this presupposes that $b > a$.   What if $b \leq a$?)

Exercise 1   Prove the statement on differentiability by the same proof that was used for the Riemann integral.

Remark   The function f in Theorem 10.1 may not be continuous at any point, in which case the statement on differentiability becomes vacuous.   It is one of the fundamental theorems in integration theory that $F'(s) = f(s)$ a.e., whether f is continuous at any point or not.   This is proved in Chapter 14.

Exercise 2   Let $f, g : \mathbf{R}^1 \to \mathbf{R}^n$ be integrable on each compact subinterval of an open interval I.   If

$$\int_a^b f \, dt = \int_a^b g \, dt$$

for every two points a and b of I, then $f = g$ a.e. on I.   [*Hint:* The proof is tricky.   The difference $h = f - g$ has integral 0 over every compact subinterval, and the proof depends on the fact that every open subset of I is the union of a disjoint sequence of subintervals (which are not compact).]

Now consider functions of the form

$$F(t) = \int f(x, t) \, d\mu(x),$$

where μ is a measure on a set X, and $f : X \times \mathbf{R}^1 \to \mathbf{R}^n$.

THEOREM 10.2   *Let I be an interval, and let $f : X \times I \to \mathbf{R}^n$ satisfy the following conditions:*
   (a) For each $t \in I$, the section $f(x, t)$ is measurable and satisfies $|f(x, t)| \le g(x)$ a.e., where $g : X \to \mathbf{R}^1$ is a fixed integrable function.
   (b) For almost every x, the section $f(x, t)$ is continuous on I.   Then $F(t) = \int f(x, t) \, d\mu(x)$ is continuous on I.

Proof   By condition (a) the function F is defined everywhere on I.   Let N be a set of measure 0 in X such that if $x \notin N$, then the section $f_x(t) = f(x, t)$ is continuous on I.   If $x \notin N$, and $t_k \to a$, then $f(x, t_k) \to f(x, a)$.   Therefore, $f(x, t_k) \to f(x, a)$ a.e. on X, and the dominated convergence theorem gives

$$F(t_k) = \int f(x, t_k) \, d\mu(x) \to \int f(x, a) \, d\mu(x) = F(a).$$

Exercise 3   The gamma function

$$\Gamma(t) = \int_0^\infty x^{t-1} e^{-x} \, dx$$

is continuous on $0 < t < \infty$.   (*Hint:* In applying Theorem 10.2 you will want to take I to be an arbitrary compact interval $[a, b]$ with $a > 0$.)

Exercise 4   Discuss

$$F(t) = \int_0^\infty \frac{e^{-tx} \sin x}{x} \, dx.$$

The next question is the differentiation of

$$F(t) = \int f(x, t) \, d\mu(x).$$

It is to be hoped that

$$F'(t) = \int \frac{\partial f(x, t)}{\partial t} \, d\mu(x),$$

that is, that the derivative is obtained simply by differentiating under the integral. If this is true (which it is under suitable conditions), then for the gamma function it gives

$$F'(t) = \int_0^\infty x^{t-1} \log x e^{-x} \, dx, \tag{1}$$

and for the function F in Exercise 4 it gives

$$F'(t) = - \int_0^\infty e^{-tx} \sin x \, dx. \tag{2}$$

The theorem is as follows.

THEOREM 10.3   *Let μ be a σ-finite measure on X, let I be an open interval in $\mathbf{R}^1$, and let $f: X \times I \to \mathbf{R}^n$. Assume*

(a) *For almost every $t \in I$, the section $f(x, t)$ is measurable on X, and for some t it is integrable.*

(b) *For almost every $x \in X$, the section $f(x, t)$ is C^1 on I.*

(c) *There is an integrable $g: X \to \mathbf{R}^1$ such that*

$$\left| \frac{\partial f(x, t)}{\partial t} \right| \leq g(x) \qquad \textit{for all } t \in I \textit{ and almost all } x \in X.$$

Then the function

$$F(t) = \int f(x, t) \, d\mu(x)$$

is of class C^1 on I and satisfies

$$F'(t) = \int \frac{\partial f(x, t)}{\partial t} \, d\mu(x).$$

Exercise 5   Verify formulas (1) and (2) by checking the conditions in this theorem.

The first step in the proof of the theorem is a lemma to show that the combination of measurability in one variable and continuity in the other is stronger than it looks.

LEMMA 10.4

Suppose that $g: X \times I \to \mathbf{R}^n$ is continuous on I for almost every $x \in X$ and is measurable on X for almost every $t \in I$. Then g is measurable on X for every $t \in I$, and g is measurable on $X \times I$.

Proof

Let N be a set of measure 0 in X such that if $x \notin N$, then $g(x, t)$ is continuous on I. Let $t_0 \in I$ be given, and choose a sequence $t_k \to t_0$ such that $g(x, t_k)$ is measurable. If $x \notin N$, then $g(x, t_k) \to g(x, t_0)$, so $g(x, t_0)$ is measurable.

To prove that g is measurable on $X \times I$, divide I into n equal intervals I_k^n, choose a point t_k^n in I_k^n, and set

$$g_n(x, t) = g(x, t_k^n) \qquad \text{if } t \in I_k^n.$$

(To avoid overlap we can take the I_k^n closed on the left and open on the right.) It is plain that each g_n is measurable on $X \times I$ and that $g_n(x, t) \to g(x, t)$ if $x \notin N$. Therefore, g is measurable on $X \times I$.

Proof of the Theorem

First we show that $\partial f / \partial t$ is measurable on $X \times I$. Let N be a subset of X of measure 0 such that if $x \notin N$, then $f(x, t)$ is C^1 on I. Let $h_k \to 0$ and set

$$g_k(x, t) = \frac{f(x, t + h_k) - f(x, t)}{h_k}.$$

It is easy to see that g_k is measurable on $X \times I$ and that $g_k(x, t) \to \partial f(x, t)/\partial t$ if $x \notin N$. Thus, $\partial f / \partial t$ is measurable on $X \times I$.

According to the lemma and condition (c) in the theorem, $\partial f / \partial t$ satisfies the hypotheses of Theorem 10.2. Hence, the function

$$G(t) = \int \frac{\partial f(x, t)}{\partial t} \, d\mu(x) \tag{3}$$

is continuous on I. For any points $a, s \in I$, Fubini's theorem gives

$$\int_a^s G(t) \, dt = \int \left\{ \int_a^s \frac{\partial f}{\partial t} \, dt \right\} d\mu(x). \tag{4}$$

On the other hand, condition (b) in the theorem shows that for almost every x we have

$$f(x, s) - f(x, a) = \int_a^s \frac{\partial f(x, t)}{\partial t} \, dt. \tag{5}$$

It follows that $f(x, s)$ is integrable over X for every s, for we can choose a so that $f(x, a)$ is integrable, while the integral on the right certainly is by

Fubini and condition (c).   Substitution of (5) into (4) gives

$$\int_a^s G(t) \, dt = \int [f(x, s) - f(x, a)] \, d\mu(x) = F(s) - F(a). \qquad (6)$$

Since G is continuous, Theorem 10.1 shows that $F'(s) = G(s)$.

Theorem 10.3 is of theoretical importance, but it also leads to interesting calculations.

Example   Calculate

$$F(t) = \int_0^\infty \frac{e^{-tx} \sin x}{x} \, dx.$$

According to the theorem, we have

$$F'(t) = -\int_0^\infty e^{-tx} \sin x \, dx, \qquad t > 0.$$

Integration by parts twice (differentiating e^{-tx} and integrating the other factor) shows that $F'(t) = -t^2 F'(t) - 1$ and, consequently, that

$$F'(t) = \frac{-1}{1 + t^2}, \qquad t > 0.$$

Therefore, since F is C^1,

$$F(b) - F(s) = -\int_s^b \frac{dt}{1 + t^2} = \arctan s - \arctan b, \quad s, b > 0. \qquad (7)$$

Exercise 6   Show that $\lim_{b \to \infty} F(b) = 0$.   (Dominated convergence theorem!)

Letting $b \to \infty$ in formula (7), we get

$$F(s) = \frac{\pi}{2} - \arctan s, \qquad \text{for } s > 0.$$

In other words,

$$\int_0^\infty \frac{e^{-sx} \sin x}{x} \, dx = \frac{\pi}{2} - \arctan s \qquad \text{for } s > 0. \qquad (8)$$

If we could put $s = 0$ in this formula, we would obtain the formula

$$\int_0^\infty \frac{\sin x}{x} \, dx = \frac{\pi}{2}. \qquad (9)$$

But we cannot.   Nevertheless, formula (9) is correct, and the way to establish it is to start from the beginning with the function

$$F_r(t) = \int_0^r \frac{e^{-tx} \sin x}{x} \, dt.$$

The same calculations lead to

$$F_r'(t) = -\frac{1 - e^{-tr} \cos r - te^{-tr} \sin r}{1 + t^2}.$$

This time 0 does not have to be excluded, and we get

$$F_r(b) - F_r(0) = -\int_0^b \frac{1 - e^{-tr} \cos r - te^{-tr} \sin r}{1 + t^2} \, dt.$$

Letting $b \to \infty$ (dominated convergence theorem again!), we get

$$F_r(0) = \int_0^\infty \frac{1 - e^{-tr} \cos r - te^{-tr} \sin r}{1 + t^2} \, dt.$$

By the definition of the improper integral and once more the dominated convergence theorem, we have

$$\int_0^\infty \frac{\sin x}{x} \, dx = \lim_{r \to \infty} F_r(0) = \int_0^\infty \frac{dt}{1 + t^2} = \frac{\pi}{2}.$$

This is a good example, and a typical one, of how difficult theorems from the general theory are needed to do apparently simple explicit calculations. It is also a typical example of how even the powerful general theorems do not usually fill the bill exactly but have to be twisted around to fit the particular problem.

Exercise 7   What does Theorem 10.3 say when X is the positive integers and μ is the counting measure?

11   CONVOLUTION

The convolution of two functions $f, g : \mathbf{R}^n \to \mathbf{R}^1$ is the function $f * g$ defined by

$$f * g(x) = \int f(x - y)g(y) \, dy \tag{1}$$

at any point x where the integrand $f(x - y)g(y)$ is integrable.   The integral is of course the Lebesgue integral.   The convolution has many important properties, some of which will be proved in this section.   Before starting, it is necessary to obtain a couple of very simple formulas for changes of variable.

THEOREM 11.1

If $f: \mathbf{R}^n \to \mathbf{R}^1$ is Lebesgue integrable, then

    (a) *For every point $a \in R^n$, $f(x + a)$ is Lebesgue integrable, and $\int f(x + a)\, dx = \int f(x)\, dx$.*

    (b) *The function $f(-x)$ is Lebesgue integrable, and $\int f(-x)\, dx = \int f(x)\, dx$.*

    (c) *For each $\rho > 0$, the function $f(x/\rho)$ is Lebesgue integrable, and $\int f(x/\rho)\, dx = \rho^n \int f(x)\, dx$.*

Proof

The procedure in proving such formulas is always the same. The first step is to prove the formula when f is the characteristic function of a measurable set. Then the rest follows automatically: If f is a simple function, then it is a linear combination of characteristic functions and the formula follows by linearity. If f is nonnegative and measurable, then it is an increasing limit of simple functions and the formula follows from the monotone convergence theorem. Finally, if f is integrable, it is a difference of nonnegative integrable functions. Therefore, what we have to do is to prove the formula in each case when f is the characteristic function of a measurable set.

    Case a. If f is the characteristic function of the set E, then $f(x + a)$ is the characteristic function of the set $E - a$, and it is obvious that the measure of $E - a$ is equal to the measure of E.

    Case b. $f(-x)$ is the characteristic function of the set $-E$, and it is obvious that the measure of $-E$ is equal to the measure of E.

    Case c. Let $Tx = \rho x$. Then $f(x/\rho)$ is the characteristic function of $T(E)$, and according to Exercise 13 of Section 2, the measure of $T(E)$ is ρ^n times the measure of E.

Some of the main properties of the convolution are as follows.

THEOREM 11.2

Let f, g, $h: \mathbf{R}^n \to \mathbf{R}^1$ be integrable. Then

    (a) *$f * g$ is defined almost everywhere and is integrable.*

    (b) *$f * g = g * f$ a.e.*

    (c) *$(f * g) * h = f * (g * h)$ a.e.*

Proof

Everything depends on Fubini's theorem, so before starting we have to show that $f(x - y)g(y)$ is measurable on $\mathbf{R}^n \times \mathbf{R}^n$. It is clear that $g(y)$ is, so all we have to look at is $f(x - y)$.

    What we have to show is that if G is an open set in $\mathbf{R}^1$ and $F(x, y) = f(x - y)$, then $F^{-1}(G)$ is a measurable set in $\mathbf{R}^{2n}$. Let $U: \mathbf{R}^{2n} \to \mathbf{R}^{2n}$ be the linear transformation defined by

$$U(x, y) = (x - y, x + y).$$

Then $F(x, y) \in G$ if and only if $x - y \in f^{-1}(G)$, and this is true if and only if $U(x, y) \in f^{-1}(G) \times \mathbf{R}^n$. In other words,

$$F^{-1}(G) = U^{-1}(f^{-1}(G) \times \mathbf{R}^n).$$

Now, $f^{-1}(G)$ is measurable in $\mathbf{R}^n$, because f is measurable. Hence, $f^{-1}(G) \times \mathbf{R}^n$ is measurable in R^{2n} by Theorem 9.3. Finally, $F^{-1}(G)$ is measurable by Theorem 4.5 (with $\varphi = U^{-1}$).

To prove part (a), use Fubini's theorem and Theorem 11.1 as follows:

$$\int |f| * |g|(x)\, dx = \int \{ \int |f(x - y)|\, |g(y)|\, dy \}\, dx$$
$$= \int |g(y)| \{ \int |f(x - y)|\, dy \}\, dx = \int |f(x)|\, dx \int |g(x)|\, dx. \quad (2)$$

Since f and g are integrable, the right side is finite, which shows that $f(x - y)g(y)$ is integrable for almost all x, that is, that $f * g$ is defined a.e. By Fubini's theorem, $f * g$ is measurable, and since $|f * g(x)| \leq |f| * |g|(x)$ and the latter is integrable by (2), it follows that $f * g$ is integrable.

Remark If we define

$$\|f\| = \int |f|\, dx, \quad (3)$$

then formula (2) gives

$$\|f * g\| \leq \|f\|\, \|g\|. \quad (4)$$

The number $\|f\|$ plays the role of an absolute value on the space of integrable functions.

Exercise 1 For which f is $\|f\| = 0$?

To prove part (b) of the theorem we use parts (a) and (b) of Theorem 11.1 to write

$$f * g(x) = \int f(x - y)g(y)\, dy = \int f(z)g(x - z)\, dz = g * f(x).$$

To prove part (c) write

$$f * g(x - z) = g * f(x - z) = \int g(x - z - y)f(y)\, dy;$$

therefore,

$$(f * g) * h(x) = \int f * g(x - z)h(z)\, dz = \int\int g(x - z - y)f(y)h(z)\, dy\, dz$$
$$= \int g * h(x - y)f(y)\, dy = (g * h) * f(x) = f * (g * h)(x).$$

Exercise 2 Justify this calculation by Fubini.

THEOREM 11.3 *Let f be integrable on each compact set and let g be of class C^1 and vanish outside a compact set. Then $f * g$ is of class C^1 and $D_j(f * g) = f * D_j g$.*

Proof   It is simply a question of differentiating

$$f * g(x) = \int f(y)g(x - y)\, dy$$

under the integral sign, which is immediately possible by Theorem 10.3.

DEFINITION
11.4

A function $f : \mathbf{R}^n \to \mathbf{R}^1$ is of class C_0^m if it is of class C^m and vanishes outside a compact set.

THEOREM
11.5

If f is integrable on each compact set and g is of class C_0^m, then $f * g$ is of class C^m and

$$D^k(f * g) = f * D^k g$$

for any derivative D^k of order $k \leq m$.

Exercise 3   Prove the theorem.   (It is an immediate consequence of the previous one, of course.)

12   APPROXIMATION THEOREMS

Some very nice approximation theorems are possible by convolution.   The idea is that if g looks as shown in Figure 6, that is, is very small outside a neighborhood of 0 but peaks sharply at 0 so that its integral is 1, then $f * g$ is close to f in various senses, depending on what kind of function f is.   The process of approximating the function f in this way is called *mollifying* or *regularizing* the function.

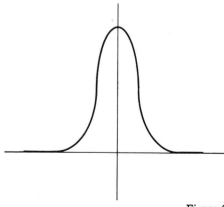

Figure 6

To make the process systematic, we shall choose a fixed nonnegative function e in C_0^∞ with integral 1, and set

$$e_\rho(x) = \rho^{-n}e\left(\frac{x}{\rho}\right), \qquad \rho > 0.$$

According to Theorem 11.1(c), the integral of e_ρ is 1. Moreover, if e vanishes outside the ball $B(0; r)$, then e_ρ vanishes outside the ball $B(0; \rho r)$. Therefore, as $\rho \to 0$, the functions e_ρ look more or less like Figure 6, with sharper and taller peaks. As a matter of convenience we shall suppose that e vanishes outside $B(0; 1)$, so e_ρ vanishes outside $B(0; \rho)$. This is really immaterial, but it does simplify occasional formulas.

THEOREM 12.1

Let $\rho \to 0$. Then
  (a) *If f is uniformly continuous, then $f * e_\rho \to f$ uniformly.*
  (b) *If f is continuous, then $f * e_\rho \to f$ uniformly on each compact set.*

Proof

Since the integral of e_ρ is 1, we have

$$f(x) - f * e_\rho(x) = \int [f(x) - f(y)]e_\rho(x - y)\, dy. \tag{1}$$

Given $\epsilon > 0$, choose $\delta > 0$ so that if $|x - y| < \delta$, then $|f(x) - f(y)| < \epsilon$. If $\rho < \delta$, then $e_\rho(x - y) = 0$ unless $|x - y| \leq \rho < \delta$; so (1) gives

$$|f(x) - f * e_\rho(x)| \leq \int \epsilon e_\rho(x - y)\, dy = \epsilon \int e_\rho(x)\, dx = \epsilon,$$

which proves part (a).

We shall prove (b) by showing that $f * e_\rho \to f$ uniformly on each ball $\bar{B}(0; r)$. Choose φ continuous and equal to 1 on $\bar{B}(0; r + 1)$ and equal to 0 outside $\bar{B}(0; r + 2)$, and set $g = \varphi f$. Part (a) applies to g, which is certainly uniformly continuous. Moreover, $g = f$ on $\bar{B}(0; r + 1)$. Therefore, the result follows from the following lemma, which says that $g * e_\rho = f * e_\rho$ on $\bar{B}(0; r)$ if $\rho < 1$.

LEMMA 12.2

Let E be any set and let E_δ be the set of points at distance $\leq \delta$ from E. If $g = f$ on E_δ and $\rho < \delta$, then $g * e_\rho = f * e_\rho$ on E.

Proof

If $h = g - f$, then $h(y)e_\rho(x - y)$ is identically 0 when $x \in E$. Indeed, $e_\rho(x - y) = 0$ unless $|x - y| \leq \rho < \delta$—and if this is the case, then $y \in E_\delta$, so $h(y) = 0$.

Exercise 1

If E is measurable and $\rho < \delta$, then $\chi = \chi_{E_\delta} * e_\rho$ is 1 on E, 0 outside $E_{2\delta}$, and between 0 and 1 everywhere.

Exercise 2 If F is compact and G is open and $G \supset F$, then there is a C^∞ function χ that is 1 on F, 0 outside G, and between 0 and 1 everywhere.

Exercise 3 State and prove an analog of Theorem 12.1 with continuity replaced by class C^k.

THEOREM 12.3 *If f is integrable, then $\|f - f * e_\rho\| \to 0$, where $\|f\| = \int |f|\, dx$.*

Proof According to formula (4) of Section 11, we have

$$\|f * e_\rho - g * e_\rho\| \le \|f - g\| \, \|e_\rho\| = \|f - g\|. \tag{2}$$

Therefore, in order to prove the theorem for a given f, it is enough to show that for each $\epsilon > 0$ there is a g such that the theorem holds for g and such that $\|f - g\| < \epsilon$. In particular, it is enough to prove the theorem when f is an integrable simple function. And then by linearity it is enough to prove the theorem when f is the characteristic function of a measurable set E of finite measure. Given $\epsilon > 0$, we can choose a compact $F \subset E$ so that $\|\chi_E - \chi_F\| = \mu(E - F) < \epsilon$; so, in fact, it is enough to prove the theorem when f is the characteristic function of a compact set F. And this we proceed to do.

Given $\epsilon > 0$, choose an open $G \supset F$ so that $\mu(G - F) < \epsilon$, and then choose $\delta > 0$ so that $F_{2\delta} \subset G$. According to Exercise 1,

$$\|\chi_{F_\delta} * e_\rho - \chi_F\| \le \mu(G - F) < \epsilon.$$

According to formula (2),

$$\|\chi_{F_\delta} * e_\rho - \chi_F * e_\rho\| \le \|\chi_{F_\delta} - \chi_F\| \le \mu(G - F) < \epsilon.$$

The two together give

$$\|\chi_F * e_\rho - \chi_F\| < 2\epsilon \qquad \text{if } \rho < \delta,$$

which completes the proof.

Theorems 12.1 and 12.3 and Exercise 2 indicate the value of approximation by convolution. It gives the best possible approximation within the class of functions considered. That is, if f is uniformly continuous, then the approximation is uniform. If f is uniformly continuous along with all derivatives of orders $\le m$, then the approximation is uniform along with all derivatives of orders $\le m$. If f is integrable, the approximation is in the natural sense of Theorem 12.3. This convolution approximation has the same character with respect to almost all of the important classes of functions (of which there are many that we shall not have the time to introduce).

Exercise 4   Go back to Section 8 of Chapter 6 and redo the proof of the Weierstrass approximation theorem in the light of the present ideas on approximation by convolution. (But do it in $\mathbf{R}^n$, of course. The result is known in $\mathbf{R}^n$ by virtue of the Stone–Weierstrass theorem.) What can you say when f is of class C^r instead of just continuous?

13   MULTIPLE SERIES

There are many ways to sum a double series

$$\sum_{j,k=0}^{\infty} a_{jk}.$$

One is to sum first on j, holding k fixed, and then to sum on k. Another is to sum first on k, holding j fixed, and then to sum on j. A third is to sum over all j and k with $j + k \le r$, and then let $r \to \infty$. A fourth is to sum over all j and k with $j \le r$ and $k \le r$, and then let $r \to \infty$.

Exercise 1   Give an example where the four methods lead to different results.

The example shows that there is no reasonable theory of convergence of such series—without some additional restriction. There *is* a very reasonable theory of *absolute convergence*. In fact, it is already included in our theory of integration.

Let $\mathbf{N}$ denote the set of nonnegative integers, and let $\mathbf{N}^p$ be the set of p-tuples of nonnegative integers. If $k = (k_1, \ldots, k_p) \in \mathbf{N}^p$, let $|k| = k_1 + \cdots + k_p$. The theory of absolutely convergent series of dimension p is exactly the theory of integration on $\mathbf{N}^p$ with respect to the counting measure ν. Note that if μ is the counting measure on $\mathbf{N}$, then $\nu = \mu \times \mu \times \cdots \times \mu$ (p factors), so Fubini will be applicable. Recall that all sets and functions are measurable with respect to a counting measure.

DEFINITION 13.1   *A multiple sequence of dimension p in $\mathbf{R}^n$ is a function $a : \mathbf{N}^p \to \mathbf{R}^n$.*

As in the case of ordinary sequences we sometimes write a_k or $a_{k_1 \cdots k_p}$ for $a(k) = a(k_1, \ldots, k_p)$, and $\{a_k\}$ or $\{a_{k_1 \cdots k_p}\}$ for the sequence itself, i.e., for a.

DEFINITION 13.2   *The series associated with a multiple sequence $a : \mathbf{N}^p \to \mathbf{R}^n$ is absolutely convergent if the function a is ν integrable, ν being the counting measure on $\mathbf{N}^p$. If this is the case, we call the integral the sum of the series and write*

$$\Sigma a_k = \Sigma a_{k_1 \cdots k_p} = \int a \, d\nu.$$

Let us see what Fubini has to say about the four ways to sum a double series suggested at the beginning. (The fact that $p = 2$ simplifies the notation, but the arguments are general.) In the first and second methods what we have are the repeated integrals

$$\int\{\int a(j,k)\,d\mu(j)\}\,d\mu(k) \quad \text{and} \quad \int\{\int a(j,k)\,d\mu(k)\}\,d\mu(j),$$

μ being the counting measure on **N**. Since $\nu = \mu \times \mu$, Fubini says that if a is ν integrable, then these two integrals are equal to the sum of the series, that is, to $\int a\,d\nu$. It also says, and this is very important, that if either of the above integrals is finite when a is replaced by $|a|$, then a must be ν integrable.

The other two methods can be described like this: There is an increasing sequence $\{E_r\}$ of subsets of $\mathbf{N}^2$ with union $\mathbf{N}^2$, and the sum envisioned is the number

$$\lim_{r\to\infty} \int_{E_r} a\,d\nu.$$

In the third method E_r consists of the (j,k) with $j + k \le r$, and in the fourth it consists of the (j,k) with $j \le r$ and $k \le r$.

Exercise 2 If $E_r \nearrow \mathbf{N}^p$, then a is ν-integrable if and only if $\lim_{r\to\infty}\int_{E_r}|a|\,d\nu < \infty$; if this is the case, then

$$\int a\,d\nu = \lim_{r\to\infty}\int_{E_r} a\,d\nu$$

(*Hint:* Use Fatou's lemma and the dominated convergence theorem.)

The outcome of this discussion is that all four methods (and indeed any others you can think of!) lead to the same result if a is ν integrable, and that it can be decided whether a is ν-integrable by the methods themselves simply by replacing a by $|a|$.

One important example of a multiple series is a multiple power series, which is a series of the form

$$f(x) = \Sigma a_{k_1\cdots k_n} x_1^{k_1} \cdots x_n^{k_n}. \tag{1}$$

If we write

$$x^k = x_1^{k_1} \cdots x_n^{k_n} \quad \text{for } x \in \mathbf{R}^n \quad \text{and} \quad k \in \mathbf{N}^n,$$

the series becomes

$$f(x) = \Sigma a_k x^k \tag{2}$$

so that it looks just like an ordinary power series. The results also look like the results for ordinary power series. To state them it is helpful to introduce some notation. If $r, \rho \in R^n$, we write $r > \rho$ to mean that $r_j > \rho_j$ for each j. If $r > 0$, we write D_r for the rectangle

$$D_r = \{x \in R^n : |x_j| < r_j \text{ for each } j\}.$$

THEOREM 13.3

If the series (2) converges absolutely at some point $x = \xi$ with $|\xi_j| = r_j \neq 0$ for each j, then it converges absolutely and uniformly on D_r, the function f is C^∞ on D_r, and the series can be differentiated term by term; i.e.,

$$\frac{\partial f}{\partial x_j}(x) = \sum k_j a_k x^{k-e_j}$$

with absolute convergence on D_r. (As usual, e_j is the vector with jth coordinate 1 and all the others 0.)

Proof

It is obvious that the series converges absolutely and uniformly on D_r, for if $x \in D_r$, then $|a_k x^k| \leq |a_k r^k|$, and the function on the right is integrable over $\mathbf{N}^n$ by hypothesis. To prove the assertion on differentiability, fix any $\rho < r$ and put $h_j = \rho_j/r_j$. If $x \in D_\rho$, then

$$|a_k x^k| \leq |a_k \rho^k| = |a_k r^k| h^k \leq M h^k,$$

the last inequality coming from the fact that each term of the convergent series $\sum |a_k r^k|$ is at most the sum. Now, the series $\sum k_j h^{k-e_j}$ converges. Indeed, the terms are nonnegative, so Fubini says that it can be summed first with respect to k_1, then with respect to k_2, and so on. The resulting series are all geometric, except for the sum with respect to k_j, which is the derivative of a geometric series. If $j = 1$, for example, the result is that

$$\sum k_1 h^{k-e_1} = \frac{1}{(1-h_1)^2(1-h_2) \cdots (1-h_n)}.$$

Now Theorem 10.3 does the job if we take $X = \mathbf{N}^n$, $I = (-\rho_j, \rho_j)$, and $g(k) = M k_j h^{k-e_j}$. This can be done with each of the variables, so we are back in the initial position, but with regard to the derivatives of f and the differentiated series (on any smaller rectangle). Therefore, f is of class C^∞ on D_r and the series can be differentiated term by term at will with no loss of absolute convergence on D_r.

Exercise 3

If f is given by (2) with absolute convergence on D_r, $r > 0$, then the series is the Taylor series of f; that is,

$$a_k = \frac{D_1^{k_1} \cdots D_n^{k_n} f(0)}{k_1! \cdots k_n!}.$$

14   REGULAR VALUES AND SARD'S THEOREM

Let $f:\mathbf{R}^n \to \mathbf{R}^m$ be of class C^1 on an open set G. A point $y \in f(G)$ is called a *regular value* of f if every point of the set

$$\mathbf{M}_y = \{x \in G : f(x) = y\}$$

is a regular point of f.   If y is a regular value, then of course $\mathbf{M}_y$ is a smooth manifold of dimension $n - m$.

Since it is required that every point of $\mathbf{M}_y$ be a regular point of f, it might be expected that the regular values are pretty sparse, but **A.** Sard found a remarkable theorem.

THEOREM
14.1

(Sard's Theorem)   *If f is of sufficiently high class C^r, then almost every value of f is a regular value.*

A point $x \in G$ is called a *critical point* of f if it is not a regular point, and a point $y \in f(G)$ is called a *critical value* of f if it is the value of f at some critical point (i.e., if the set $\mathbf{M}_y$ contains at least one critical point).   In terms of critical points and critical values, Sard's theorem is as follows:

THEOREM
14.2

(Sard's Theorem)   *If f is of sufficiently high class C^r, then the set of critical values has measure 0.   Equivalently, if K is the set of critical points, then $f(K)$ has measure 0.*

Note that the theorem is already known (and not very interesting) when $n < m$, for Corollary 2.9 shows that the whole set $f(G)$ has measure 0.   Therefore, we shall suppose from now on that $n \geq m$.   In this case the set K of critical points is the set where the differential df has rank $< m$; if we write K_j for the set of points where df has rank j, then what we shall have to prove is that $f(K_j)$ has measure 0 for $j < m$.   The first step is to handle $f(K_0)$.

LEMMA
14.3

If f is of class C^r with $r > n(n + 1)/2m$, then $f(K_0)$ has measure 0.

Proof

The proof goes by induction on the dimension n.   (m is fixed.)   The induction is started by the remark above which shows that the lemma holds for $n < m$.

Let A_k be the set of points where the partial derivatives $D_i f$ vanish for $1 \leq |i| \leq k$, and let ρ be the first integer $> n/m$.   (Note that $n/m \leq n(n + 1)/2m$, so $\rho \leq r$.)   Since $K_0 = A_1$, we have

$$\left| f(K_0) \right| = \left| f(A_1) \right| \leq \sum_{k=1}^{\rho-2} \left| f(A_k - A_{k+1}) \right| + \left| f(A_{\rho-1}) \right|,$$

so it is enough to prove that each $f(A_k - A_{k+1})$ has measure 0 and that $f(A_{\rho-1})$ has measure 0.   The latter is Exercise 12 of Section 2, so we can concentrate on a fixed $f(A_k - A_{k+1})$.   By Theorem 2.6, it is enough to show that each point $a \in A_k - A_{k+1}$ has a neighborhood G_0 such that $\left| f((A_k - A_{k+1}) \cap G_0) \right| = 0$.   Therefore, let a be a fixed point of $A_k - A_{k+1}$.

By the definitions of A_k and A_{k+1} there exist indices i_0 with $|i_0| = k$ and j_0 with $1 \leq j_0 \leq m$ such that $\nabla D_{i_0} f_{j_0}(a) \neq 0$. Therefore, the set

$$\mathbf{N} = \{x \in G : D_{i_0} f_{j_0}(x) = 0\}$$

is a smooth manifold of dimension $n - 1$ in $\mathbf{R}^n$ in a neighborhood of the point a, and it is of class C^s with

$$s = r - k \geq r - (\rho - 2) > \frac{n(n+1)}{2m} - \frac{n}{m} = \frac{n(n-1)}{2m}.$$

Let φ be a local parametric representation of $\mathbf{N}$ at a which is of class C^s on an open $G' \subset \mathbf{R}^{n-1}$ and set $g = f \circ \varphi$. Let K'_0 be the set of points in G' where dg has rank 0 (that is, $dg = 0$). Since $dg(t) = df(\varphi(t)) \, d\varphi(t)$, it follows that if $\varphi(t) \in K_0$, then $t \in K'_0$, that is, that $\varphi(K'_0) \supset K_0 \cap \varphi(G')$, and hence that

$$g(K'_0) \supset f(K_0 \cap \varphi(G')) \supset f(A_k \cap \varphi(G')).$$

Now, $\varphi(G')$ contains a neighborhood of a in $\mathbf{N}$, so there is a neighborhood G_0 of a in $\mathbf{R}^n$ such that $\varphi(G') \supset \mathbf{N} \cap G_0 \supset A_k \cap G_0$. Consequently,

$$g(K'_0) \supset f(A_k \cap G_0).$$

The induction hypothesis gives that $|g(K'_0)| = 0$, and the lemma is proved.

Now we turn to the other $f(K_j)$.

LEMMA 14.4

If f is of class C^r with $r > n(n+1)/2$, then $f(K_j)$ has measure 0 for $j < m$.

Proof

We fix j and use bars to denote the first j coordinates and primes to denote the rest (both in $\mathbf{R}^n$ and in $\mathbf{R}^m$). First we prove the lemma when f has the special form

$$f(x) = (\bar{x}, h(x)), \qquad \text{where } h : \mathbf{R}^n \to \mathbf{R}^{m-j}, \tag{1}$$

and then we show how to reduce the general case to this one.

If $\bar{x} \in \mathbf{R}^j$, then $h_{\bar{x}}$ is the section of h defined by

$$h_{\bar{x}}(x') = h(\bar{x}, x').$$

If A is a set, then $A_{\bar{x}}$ is the section of A defined by

$$A_{\bar{x}} = \{x' : (\bar{x}, x') \in A\}.$$

When f has the form (1) we have

$$f(A)_{\bar{x}} = h_{\bar{x}}(A_{\bar{x}}). \tag{2}$$

We apply this to $A = K_j$. Since the Jacobi matrix of f is

$$\begin{pmatrix} I & 0 \\ \dfrac{\partial h}{\partial \bar{x}} & \dfrac{\partial h}{\partial x'} \end{pmatrix},$$

it follows that $A = \{x : \partial h/\partial x' = 0\}$, and hence that $A_{\bar{x}} = \{x' : dh_{\bar{x}}(x') = 0\}$. From formula (2) and Lemma 14.3 we get that

$$\left| f(A)_{\bar{x}} \right|_j = 0;$$

from Fubini we get that $\left| f(A) \right|_m = 0$. [The differentiability needed to use the first lemma is class C^r with $r > (n - j)(n - j + 1)/2(m - j)$, and this is obvious for $r > n(n + 1)/2$.]

Now we proceed to the general case. As usual it is enough to show that each point $a \in K_j$ has a neighborhood in K_j whose image has measure 0, so let a be a fixed point in K_j. The Jacobi matrix of f has j linearly independent rows. By relabeling the coordinates in $\mathbf{R}^m$ we can assume that the first j are linearly independent. The Jacobi matrix of $\bar{f}$ has j linearly independent columns. By relabeling the coordinates in $\mathbf{R}^n$ we can assume that the first j are linearly independent. With this choice of coordinates the Jacobi matrix $\partial \bar{f}/\partial \bar{x}$ is nonsingular at the point a, so by the implicit-function theorem we can solve the equation $\bar{f}(\bar{x}, x') = \bar{y}$ for $\bar{x}$ in terms of $(\bar{y}, x')$. The solution is a function $\varphi : \mathbf{R}^n \to \mathbf{R}^j$ that is of class C^r on a neighborhood of $(\bar{f}(a), a')$ and maps any such neighborhood on a neighborhood of $\bar{a}$. We define

$$\psi(\bar{y}, x') = (\varphi(\bar{y}, x'), x') \qquad \text{and} \qquad g = f \circ \psi.$$

It is plain from the construction that g does have the form (1); so if we can show that $g(\tilde{K}_j) \supset f(K_j \cap G_0)$, $\tilde{K}_j$ being the set of points where dg has rank j and G_0 being a neighborhood of a, then we shall be finished.

The function ψ is regular at the point $(\bar{f}(a), a')$. Indeed, its Jacobi matrix is

$$\begin{pmatrix} \dfrac{\partial \varphi}{\partial \bar{y}} & \dfrac{\partial \varphi}{\partial x'} \\ 0 & I \end{pmatrix},$$

which is nonsingular because $\partial \varphi/\partial \bar{y}$ is nonsingular. (Why is $\partial \varphi/\partial \bar{y}$ nonsingular?) Since $dg(z) = df(\psi(z))\, d\psi(z)$, it follows that the rank of dg at z is equal to the rank of df at $\psi(z)$, hence that $\psi(\tilde{K}_j) \supset K_j \cap G_0$, and finally that $g(\tilde{K}_j) \supset f(K_j \cap G_0)$, G_0 being some neighborhood of a.

Remark Lemma 14.4 establishes Sard's theorem when f is of class C^r with $r > n(n + 1)/2$. The theorem is actually true if $r \geq n - m + 1$ (and of course $r \geq 1$ if $n < m$).

Arthur Sard

With this minimum differentiability the outline of the proof remains the same; but there is one very sticky point, and the theorem is usually called "Hard Sard."

Let $\mathbf{N}$ be a smooth manifold of dimension n and class C^r with $r > n(n + 1)/2$, and let $f: \mathbf{N} \to \mathbf{R}^m$ be a function of class C^r. We have seen in Section 4 of Chapter 11 that for any point $a \in \mathbf{N}$ $df(a)$ is a linear transformation from the tangent space $T_a(\mathbf{N})$ into $\mathbf{R}^m$. The point a is called a *regular point* if $df(a)$ has the maximum possible rank (i.e., the smaller of n and m) and a *critical point* otherwise. Regular values and critical values are defined as before. In this context Sard's theorem remains perfectly correct. Indeed, it is an immediate consequence of the original version and the following exercise.

Exercise 1   The point $a \in \mathbf{N}$ is a regular point of f if and only if it is a regular point of $f_\varphi = f \circ \varphi$, where φ is a local parametric representation of $\mathbf{N}$ at a.

Exercise 2   Write out the proof of Sard's theorem for functions on manifolds. Why assume that $\mathbf{N} \in C^r$ with $r > n(n + 1)/2$?

Example   Let $\mathbf{N}$ be the torus obtained by revolving the circle $x^2 + (y - 2)^2 = 1$ around the x axis and let $f(x, y, z) = z$. The gradient of f is $(0, 0, 1)$, so the critical points are those where this vector is orthogonal to the tangent plane (i.e., the points where the tangent plane is horizontal). These are the four points $(0, 0, \pm 3)$ and $(0, 0, \pm 1)$. The critical values are the numbers ± 3 and ± 1. (If these statements are not apparent, it will be helpful to look back at Section 4 of Chapter 11.)

It is interesting to look at the smooth manifolds

$$\mathbf{M}_\alpha = \{(x, y, z) : f(x, y, z) = \alpha\}.$$

When α is one of the critical values, $\mathbf{M}_\alpha$ is not a smooth manifold of dimension $2 - 1 = 1$. When α is a regular value, $\mathbf{M}_\alpha$ is, of course, a smooth manifold of dimension 1. The interesting thing is that the character of $\mathbf{M}_\alpha$ remains the same as long as α does not pass through a critical value, and the character changes when α does pass through a critical value (Figure 7). When $\alpha = \pm 3$, $\mathbf{M}_\alpha$ is a single point—obviously not a smooth one-dimensional manifold. When $\alpha = \pm 1$, $\mathbf{M}_\alpha$ is essentially a pair of tangent circles. (Use the results of Section 3 of Chapter 11 to show that this is not a smooth one-dimensional manifold.) When α is between 1 and 3 or between -1 and -3, $\mathbf{M}_\alpha$ is essentially a circle; when α is between -1 and 1, $\mathbf{M}_\alpha$ is essentially a pair of disjoint circles.

This phenomenon—that the character of $\mathbf{M}_\alpha$ changes only when α passes through a critical value—is a general one, not an accidental feature of the example. It provides one of the powerful methods for studying compact manifolds.

Exercise 3   Let $\mathbf{N}$ be a smooth manifold of dimension n, and let $f : \mathbf{N} \to \mathbf{R}^m$ be of class C^1. If $y \in \mathbf{R}^m$ is a regular value of f, then the set

$$\mathbf{M}_y = \{x \in \mathbf{N} : f(x) = y\}$$

is a smooth manifold of dimension $n - m$. State and prove the result also when things are of class C^r.

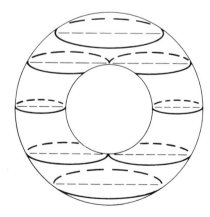

Figure 7

Exercise 4   Try to formulate a Sard's theorem when $f: \mathbf{N} \to \mathbf{M}$, where $\mathbf{N}$ and $\mathbf{M}$ are smooth manifolds of dimensions n and m of suitable class C^r. The point here is that $df(a)$ maps the tangent space $T_a(\mathbf{N})$ into the tangent space $T_b(\mathbf{M})$, $b = f(a)$, which, of course, has to be proved. Now define regular value and define the notion of a set of m-dimensional measure 0 on $\mathbf{M}$, and state and prove Sard's theorem. (You can define the notion of measure 0 either by using local parametric representations or by using the Hausdorff measure μ_m. The best thing is to do it both ways and to show that the two notions are equivalent. This involves going back to Section 2 and redoing the results for Hausdorff measures.)

Sard's theorem and Exercise 7, Section 2 of Chapter 12, give an immediate proof of the functional-dependence theorem, which asserts that functions f_1, $\ldots$, f_m are functionally dependent on a compact set K if their gradients are linearly dependent at each point of K. The precise statement is as follows.

Exercise 5   Let $f_1, \ldots, f_m$ be real-valued functions of sufficiently high class C^r on a neighborhood of a compact set $K \subset \mathbf{R}^n$. If the gradients are linearly dependent at each point of K, then there is a function $F: \mathbf{R}^m \to \mathbf{R}^1$ such that
  (a) F is of class C^∞ on $\mathbf{R}^m$ and does not vanish identically on any open set.
  (b) $F(f_1(x), \ldots, f_m(x)) = 0$ for each $x \in K$.

Exercise 6   State and prove the functional-dependence theorem when $\mathbf{R}^n$ is replaced by a smooth manifold of class C^r.

Here is an interesting example.

Exercise 7   The functions $f_1(x, y) = x$, $f_2(x, y) = xy$, $f_3(x, y) = xye^y$ are functionally dependent in the above sense. Nevertheless, even though these functions are analytic, there is no analytic relation between them on any neighborhood of 0. (A function is analytic if it has a power-series expansion in a neighborhood of each point.)

14 } Differentiation

1   REGULAR BOREL MEASURES

The object of this chapter is to study some important measures on $\mathbf{R}^n$ and their relation to the Lebesgue measure.

DEFINITION 1.1

A regular Borel measure on an open set $\Omega \subset \mathbf{R}^n$ is an outer measure ν on Ω such that every open set is measurable, every point has a neighborhood of finite measure, and every set is contained in a G_δ of the same measure.

Exercise 1   For any $A \subset \Omega$, $\nu(A) = \inf \nu(G)$ over the open $G \supset A$. A is ν measurable if and only if A differs from a G_δ by a set of measure 0 and, also, if and only if A differs from a K_σ by a set of measure 0.

The Lebesgue measure is the basic regular Borel measure.   Unqualified terms such as measurable, integrable, and so on, will always refer to it.   As usual, the Lebesgue measure of a set A will be written $|A|$, or $|A|_n$ if it is necessary to display the dimension, and the Lebesgue integral will be written $\int f \, dx$.

Example 1 (Regular Borel Measures on $\mathbf{R}^1$)   A regular Borel measure ν on $\mathbf{R}^1$ determines a function α as follows:

$$\alpha(x) = \nu\big((0, x]\big) \quad \text{if } x \geq 0, \qquad \alpha(x) = -\nu\big((x, 0]\big) \quad \text{if } x < 0.$$

Exercise 2   The function α is nondecreasing, continuous on the right, and takes the value 0 at 0.   When is α continuous?

Conversely, let α be any nondecreasing function and define the "length" of an open interval $I = (a, b)$ by

$$l_\alpha(I) = \alpha(b - 0) - \alpha(a + 0),$$

where $\alpha(b - 0)$ and $\alpha(a + 0)$ denote the left- and right-hand limits of α at b and a. Then define a measure ν_α, as in the Lebesgue case, by

$$\nu_\alpha(A) = \inf \Sigma l_\alpha(I_k),$$

where $\{I_k\}$ is a sequence of open intervals covering A. It is not necessary to use open intervals, but some care must be taken to get things straight at points where α is not continuous. For instance, if I is the half-open interval $(a, b]$, put $l_\alpha(I) = \alpha(b + 0) - \alpha(a + 0)$, then cover by half-open intervals, and so on.

Exercise 3   Show that ν_α is a regular Borel measure and that $\nu_\alpha(I) = l_\alpha(I)$ when I is an open or half-open interval. What is it when I is half-open on the other end, or closed?

Exercise 4   Discuss the correspondence between increasing functions and regular Borel measures furnished by Exercises 2 and 3.

As far as $\mathbf{R}^1$ is concerned, this chapter is essentially the theory of increasing functions.

Example 2
(Indefinite Integrals)   The measure ν of Theorem 1.2 is called the *indefinite integral* of f. As mentioned above, the unqualified terms measurable, integrable, and so on, refer to the Lebesgue measure.

THEOREM
1.2   *If f is nonnegative, measurable, and integrable over every compact set in Ω then there is a unique regular Borel measure ν on Ω such that for every measurable set $E \subset \Omega$*

$$\nu(E) = \int_E f \, dx. \tag{1}$$

Exercise 5   Prove the theorem. [*Hint:* When A is not measurable, define $\nu(A) = \inf \nu(E)$ over all measurable $E \supset A$. Use Theorem 7.6 of Chapter 13.]

Exercise 6   If a given regular Borel measure ν is the indefinite integral of f and also the indefinite integral of g, then $f = g$ a.e. on Ω.

Exercise 7   If the regular Borel measure ν is an indefinite integral, then every Lebesgue measurable set is ν measurable. [*Hint:* If $|E| = 0$, then $\nu(E) = 0$.]

When ν is an indefinite integral, every integral with respect to ν can be expressed as a Lebesgue integral as follows:

THEOREM
1.3   *Let the regular Borel measure ν be the indefinite integral of f. If g is nonnegative and ν measurable, then gf is Lebesgue measurable, and*

$$\int g \, d\nu = \int gf \, dx. \tag{2}$$

Proof   To simplify a little, we shall suppose that g is defined a.e. with respect to ν and shall leave the case where it is not to the reader. This does not mean, however, that g is defined a.e. with respect to the Lebesgue measure. We shall have to use the convention that the product gf is 0 wherever f is 0, whether g is defined there or not.

First suppose that $g = \chi_N$, where $\nu(N) = 0$. By the definition of a regular Borel measure, there is a G_δ set $E \supset N$ with $\nu(E) = 0$. By formula (1), $\chi_E f = 0$ a.e.; therefore, $\chi_N f = 0$ a.e., so $\chi_N f$ is measurable.

Next, suppose that $g = \chi_A$, where A is ν measurable. Then there is a G_δ set $E \supset A$ such that $\nu(E - A) = 0$; hence

$$\chi_A f = \chi_E f - \chi_{E-A} f$$

is measurable. Integration gives formula (2).

If φ is simple with respect to ν, then φ is a linear combination of characteristic functions of ν measurable sets, so φf is measurable. Clearly (2) holds, since it holds for each of the characteristic functions.

Now let $\{\varphi_k\}$ be an increasing sequence of simple functions with respect to ν, which converge to g a.e. with respect to ν. Then the sequence $\{\varphi_k f\}$ is an increasing sequence, and by what has been proved it converges to gf a.e., so gf is measurable. The monotone convergence theorem gives formula (2).

Remark   This is the pattern for the proofs of many formulas. First come characteristic functions of sets of measure 0, then characteristic functions of measurable sets, then simple functions, and finally increasing sequences.

The basic problems are the differentiability of regular Borel measures and the relationship between the derivative and the indefinite integral.

DEFINITION 1.4   *The regular Borel measure ν is differentiable at the point x if the limit*

$$D\nu(x) = \lim_{|B|\to 0} \frac{\nu(B)}{|B|}$$

exists, where B denotes an open ball containing x. $D\nu(x)$ is called the derivative of ν at x.

Exercise 8   The definition is the same if the balls are required to be closed instead of open. (*Hint:* The point here is that the open and closed balls have the same Lebesgue measure, though not necessarily the same ν measure.)

Exercise 9   If α is a nondecreasing function on $\mathbf{R}^1$, and ν_α is the corresponding measure, then $D\nu_\alpha(x) = \alpha'(x)$.

The basic theorem is as follows.

THEOREM 1.5   *Every regular Borel measure ν is differentiable almost everywhere. It is an indefinite integral if and only if $\nu(N) = 0$ whenever $|N| = 0$, and in this case it is the integral of its derivative.*

This theorem, which is quite hard, is established in Sections 2 and 3. A measure with the property that $\nu(N) = 0$ whenever $|N| = 0$ is called *absolutely continuous*. Let us look at a couple of examples of measures that are not absolutely continuous. As a very simple example there is the counting type measure where $\nu(A)$ is the number of points in A with integer coordinates. In this case $D\nu(x) = 0$ if x does not have integer coordinates, and $D\nu(x)$ is undefined (or is $+\infty$, if you prefer) if x does have integer coordinates.

A more interesting example is the Cantor function (named after Georg Cantor), which is a function α defined on the interval $J_0 = [0, 1]$ by repeating the following construction: If J is a closed interval such that α is already defined at the end points but not in the interior, then on the closed middle third of J define α to have the constant value equal to the average of the values at the end points. Start with $J_0 = [0, 1]$ and with $\alpha(0) = 0$ and $\alpha(1) = 1$. Let I_1 be the open middle third and let J_1 be the rest. Define α on $\bar{I}_1$ by the prescription above, which means $\alpha = \frac{1}{2}$. Now, J_1 is the union of two closed intervals. Let I_2 be the union of the two open middle thirds, and on each of these closed middle thirds define α by the prescription above. (On the left-hand one it will be $\frac{1}{4}$ and on the right-hand one it will be $\frac{3}{4}$.) Let $J_2 = J_1 - I_2$. In general J_k is the union of 2^k closed intervals, each of length $1/3^k$, I_{k+1} is the union of the 2^k open middle thirds, and then $J_{k+1} = J_k - I_{k+1}$. If α is already defined on I_k, then it is defined on each interval of I_{k+1} by the prescription above, that is, as the average of its values at the two end points of the corresponding interval in J_k. The first few stages look as shown in Figure 1. By this procedure α is

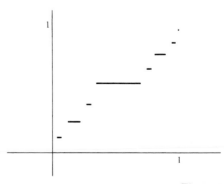

Figure 1

defined on the open set

$$G = \cup I_k$$

(actually on the union of the $\bar{I}_k$) and is undefined on a subset of the compact set

$$C = J_0 - G = \cap J_k.$$

Since J_k is the union of 2^k intervals each of length $1/3^k$, its measure is $2^k/3^k$, so the measure of C is 0.

Exercise 10   Show that the function α has a unique continuous extension to the whole interval $[0, 1]$.

Exercise 11   Show that $\alpha'(x) = 0$ for each $x \in G$; hence $\alpha'(x) = 0$ a.e. in spite of the fact that α is continuous and nondecreasing, and actually increases from 0 to 1. In particular, α is not the integral of its derivative.

If ν_α is the corresponding measure, then clearly $\nu_\alpha(I) = 0$ for each of the intervals I that make up I_k. Hence $\nu_\alpha(I_k) = 0$; therefore, $\nu_\alpha(G) = 0$. On the other hand, $\nu_\alpha([0, 1]) = 1$ because $\alpha(0) = 0$ and $\alpha(1) = 1$, so $\nu_\alpha(C) = 1$. Since $|C| = 0$, this shows that ν_α is not absolutely continuous. In fact, this is the extreme opposite of absolute continuity. The interval $[0, 1]$ splits into the

Henri Poincaré

"In the old days people invented new functions with some practical aim in mind. Nowadays they do it just to poke holes in the reasonings of their fathers— and they won't ever get more from these functions than that." In spite of such opinions the work of Cantor was fundamental.

Georg Cantor

two disjoint parts G and C such that $\nu_\alpha(C) = 1$, $|C| = 0$, while $\nu_\alpha(G) = 0$, $|G| = 1$. It is possible to make similar examples where the function α is not just nondecreasing but strictly increasing.

It was toward the end of the last century that such bizarre functions began appearing. They were viewed with alarm by many of the leading mathematicians of the day, one of the greatest of whom was Henri Poincaré.

Example 3 (Change of Variable) For the purpose of calculus the most important regular Borel measure is the one that arises from a change of variable $y = \varphi(x)$ in an integral $\int f(y)\,dy$. Let $\varphi : \mathbf{R}^n \to \mathbf{R}^n$ be continuous and one to one on an open set Ω and define

$$\nu(A) = |\varphi(A)| \qquad \text{for } A \subset \Omega. \tag{3}$$

Exercise 12 If $\varphi(A)$ is Lebesgue measurable, then A is ν measurable. (Just use the definition of measurability.) This implies that every compact set in Ω is ν measurable and has finite measure.

Exercise 13 If E is a G_δ that contains $\varphi(A)$ and has the same Lebesgue measure, then $\varphi^{-1}(E)$ is a G_δ that contains A and has the same ν measure.

The two exercises together show that ν is a regular Borel measure on Ω. Now we shall establish the formula

$$\int_{\varphi(\Omega)} f(y)\,dy = \int_\Omega f(\varphi(x))\,d\nu \tag{4}$$

whenever f is Lebesgue measurable on $\varphi(\Omega)$. If f is the characteristic function of a Lebesgue measurable set $B \subset \varphi(\Omega)$, then $f \circ \varphi$ is the characteristic function of $A = \varphi^{-1}(B)$, so formula (4) is just the definition of ν.   Knowing the formula in this case, we obtain it immediately for simple f by linearity, for nonnegative measurable f by the monotone convergence theorem, and for arbitrary measurable f by linearity again.

The basic theorem to be proved in this situation is the following.

THEOREM 1.6   *If φ is differentiable at a point x, then*

$$D\nu(x) = |\det d\varphi(x)|.$$

The proof is given in Section 4 for the case where φ is of class C^1 at x and in Section 4 of Chapter 16 for the general case.

Suppose, finally, that φ is of class C^1 on Ω.   In this case Corollary 2.10 of Chapter 13 shows that ν is absolutely continuous.   (The Cantor function shows that it is not absolutely continuous in general!)   Consequently, formula (4), along with Theorems 1.3, 1.5, and 1.6, gives the change-of-variable formula

$$\int_{\varphi(\Omega)} f(y)\, dy = \int_{\Omega} f(\varphi(x)) |\det d\varphi(x)|\, dx. \tag{5}$$

As an illustration, consider the change from rectangular to polar coordinates, in which the function φ is given by $x = r \cos\theta$, $y = r \sin\theta$, $0 < r < \infty$ and $0 < \theta < 2\pi$.   In this case $\varphi(\Omega)$ is the whole plane except for the nonnegative x axis, which has measure 0, and $\det d\varphi(r, \theta) = r$; so formula (5) gives

$$\iint_{R^2} f(x, y)\, dx\, dy = \int_0^\infty \int_0^{2\pi} f(r \cos\theta, r \sin\theta) r\, d\theta\, dr, \tag{6}$$

which is exactly the formula we have found already for changing from rectangular to polar coordinates.

In the one-dimensional case the function φ must be either strictly increasing or strictly decreasing (since it is one to one), and $\det d\varphi(x) = \varphi'(x)$.   How is the absolute value in formula (5) to be reconciled with the formula that we have known all along for changing variable in the one-dimensional case?

Exercise 14   In the definition of a regular Borel measure we have used the condition

A. *Every point has a neighborhood of finite measure.*

Later we shall want to consider regular Borel measures on arbitrary metric spaces, not just open sets in $\mathbf{R}^n$.   In our present situation the condition A is equivalent to the condition

B. *Every compact set has finite measure,*

but in general it is not.

2 DIFFERENTIABILITY THEOREMS

In proving the differentiability of regular Borel measures it is convenient to introduce upper and lower derivatives which always exist and provide something to work with.

DEFINITION 2.1

Let ν be a regular Borel measure on Ω. For each $\epsilon > 0$ let

$$D^\epsilon\nu(x) = \sup \frac{\nu(B \cap \Omega)}{|B|}, \qquad D_\epsilon\nu(x) = \inf \frac{\nu(B \cap \Omega)}{|B|},$$

where the sup and inf are taken over all open balls that contain x and have radius $< \epsilon$. Let

$$\bar{D}\nu(x) = \lim_{\epsilon \to 0} D^\epsilon\nu(x), \qquad \underline{D}\nu(x) = \lim_{\epsilon \to 0} D_\epsilon\nu(x).$$

If $\underline{D}\nu(x) = \bar{D}\nu(x) < \infty$, then ν is differentiable at x, and its derivative $D\nu(x)$ is the common value.

Exercise 1   This definition of $D\nu(x)$ is consistent with the one in Section 1.

Remark 1   Closed balls, cubes, or various other sets can be used in place of the open balls, but the latter suit our present purposes best. The proofs are not very different in these other approaches.

Remark 2   It is clear that if $\epsilon < \eta$, then $D^\epsilon\nu(x) \le D^\eta\nu(x)$ and $D_\epsilon\nu(x) \ge D_\eta\nu(x)$. Therefore,

$$\bar{D}\nu(x) = \inf_{\epsilon > 0} D^\epsilon\nu(x) \qquad \text{and} \qquad \underline{D}\nu(x) = \sup_{\epsilon > 0} D_\epsilon\nu(x).$$

This has two important consequences. One is that the upper and lower derivatives $\bar{D}\nu$ and $\underline{D}\nu$ do exist at every point, at least if $+\infty$ is allowed as a value. The second is that it is not necessary to consider all $\epsilon > 0$. Any sequence which approaches 0 will do.

THEOREM 2.2

$\bar{D}\nu$ and $\underline{D}\nu$ are measurable functions not only with respect to Lebesgue measure, but with respect to any regular Borel measure.

Proof

By the second part of Remark 2 it is enough to show that $D^\epsilon\nu$ and $D_\epsilon\nu$ are measurable. If $D^\epsilon\nu(a) > \alpha$, then a belongs to an open ball B of radius $< \epsilon$ such that $\nu(B \cap \Omega)/|B| > \alpha$. Since the ball is open, any point x sufficiently close to a belongs to the same ball and, therefore, also satisfies $D^\epsilon\nu(x) > \alpha$. In other words, the set $\{x : D^\epsilon\nu(x) > \alpha\}$ is an open set for each real α; hence, its complement $\{x : D^{\epsilon}\nu(x) \leq \alpha\}$ is a closed set for each real α.   It follows from Theorem 5.6 of Chapter 13 that $D^{\epsilon}\nu$ is measurable with respect to any regular Borel measure μ, for each closed set is μ measurable. $D_{\epsilon}\nu$ is handled similarly.

Remark 3   What we have actually shown is that $D^{\epsilon}\nu$ is lower semicontinuous on Ω and that $D_{\epsilon}\nu$ is upper semicontinuous on Ω.   There is one point that might not have been apparent in the proof.   For the validity of the statement that $\{x : D^{\epsilon}\nu(x) > \alpha\}$ is open, it is necessary to interpret $D^{\epsilon}\nu(x)$ as $+\infty$ at the points where the definition makes it natural to do so.   On the other hand, the value $+\infty$ was not considered in the discussion of measurable functions.   This is why the final deduction from Theorem 5.6 is based on the fact that the complement $\{x : D^{\epsilon}\nu(x) \leq \alpha\}$ is closed.   Later it will be important to know that $\bar{D}\nu$ and $\underline{D}\nu$ are ν measurable as well as Lebesgue measurable.

The study of regular Borel measures rests mainly on an important technical result called the Vitali covering theorem.

DEFINITION 2.3   *A family $\mathfrak{F}$ of balls covers a set E in the sense of Vitali if for every $x \in E$ and every $\epsilon > 0$ there is some ball in $\mathfrak{F}$ that contains x and has radius $<\epsilon$.*

THEOREM 2.4   (***Vitali's Covering Theorem***)   *Let $\mathfrak{F}$ be a family of balls that covers a set E in the sense of Vitali.   Then $\mathfrak{F}$ contains a disjoint sequence of balls that covers almost all of E.*

The theorem does look technical, but the power of it will be apparent shortly.   The true Vitali theorem is somewhat more general, but this version will suffice.   It does not matter whether the balls are open or closed.   In the proof we shall treat a family of closed balls; but if $\mathfrak{F}$ is a family of open balls, then the corresponding family $\bar{\mathfrak{F}}$ of closed balls also covers in the sense of Vitali.   Therefore, it contains a disjoint sequence $\{\bar{B}_k\}$ that covers almost all of E.   Then also the disjoint sequence $\{B_k\}$ covers almost all of E, because $|\bar{B} - B| = 0$ for any ball B.

Exercise 2   Prove the last statement.

Proof   To begin with we shall suppose that E is bounded.   Let G be a bounded open set containing E, and throw out all the balls in $\mathfrak{F}$ that are not contained in G.   The new family, which we will still call $\mathfrak{F}$, continues to cover E in the sense of Vitali.

The disjoint sequence is defined as follows: Let M_1 be the upper bound of the radii of all balls in $\mathfrak{F}$, and choose a ball $\bar{B}_1 = \bar{B}(a_1; r_1)$ in $\mathfrak{F}$

with $r_1 \geq M_1/2$. Assuming that $\bar{B}_1, \ldots, \bar{B}_k$ and $M_1, \ldots, M_k$ are already defined, let M_{k+1} be the upper bound of the radii of all the balls in $\mathfrak{F}$ that are disjoint from $\bar{B}_1, \ldots, \bar{B}_k$, and then choose such a ball $\bar{B}_{k+1} = \bar{B}(a_{k+1}; r_{k+1})$ with $r_{k+1} \geq M_{k+1}/2$. (If M_{k+1} turns out to be 0, then $r_{k+1} = 0$ and $\bar{B}_{k+1}$ is defined to be empty.)   It is clear that the sequence $\{M_k\}$ is decreasing; furthermore, it converges to 0, for $M_k \leq 2r_k$, and the sequence $\{r_k\}$ converges to 0 because of the fact that $|B_k| = cr_k^n$, while

$$\sum_{k=1}^{\infty} |\bar{B}_k| \leq |G| < \infty. \tag{1}$$

Suppose that a point $x \in E$ is not in any of the balls $\bar{B}_1, \ldots, \bar{B}_{k_0}$. Then it is at a positive distance δ from their union (which is compact). Choose a ball $\bar{B}(a; r)$ in $\mathfrak{F}$ that contains x and has radius $r < \delta/2$. Then $\bar{B}(a; r)$ is disjoint from $\bar{B}_1, \ldots, \bar{B}_{k_0}$ because every point of $\bar{B}(a; r)$ is at distance $< \delta$ from x.   Consequently,

$$M_{k_0+1} \geq r. \tag{2}$$

Since $\{M_k\}$ decreases to 0, there is an index j such that

$$M_j \geq r > M_{j+1} \qquad \text{and} \qquad j > k_0. \tag{3}$$

Since $r > M_{j+1}$, it follows that $\bar{B}(a; r)$ meets one of the sets $\bar{B}_1, \ldots, \bar{B}_j$, say $\bar{B}_m$, and necessarily $m > k_0$. Hence, $|x - a_m| \leq 2r + r_m$, while $r \leq M_j \leq M_m \leq 2r_m$. Thus, $|x - a_m| \leq 5r_m$, and we have established the following formula:

$$E - \bigcup_{k=1}^{k_0} \bar{B}_k \subset \bigcup_{k>k_0} \bar{B}(a_k, 5r_k), \tag{4}$$

and hence the inequality

$$\left| E - \bigcup_{k=1}^{k_0} \bar{B}_k \right| \leq \sum_{k=k_0+1}^{\infty} |\bar{B}(a_k; 5r_k)| = 5^n \sum_{k=k_0+1}^{\infty} |\bar{B}_k|. \tag{5}$$

If k_0 is large enough, the number on the right is as small as we please by virtue of (1), so $|E - \bigcup_{k=1}^{\infty} \bar{B}_k| = 0$, as required.

This takes care of the case when E is bounded. Now let E be arbitrary.   Let

$$G_n = \{x : n - 1 < |x| < n\},$$

let $E_n = E \cap G_n$, and let $\mathfrak{F}_n$ consist of the balls in $\mathfrak{F}$ that are contained in G_n. It is plain that $\mathfrak{F}_n$ covers E_n in the sense of Vitali, so there is a disjoint sequence $\{B_{nk}\}$ in $\mathfrak{F}_n$ that covers almost all of E_n. When n and k both vary, the B_{nk} form a disjoint sequence in $\mathfrak{F}$ that covers almost all of $\bigcup E_n$, and this is almost all of E.

Exercise 3   Prove the last statement.

Now, let us see the use of the Vitali theorem.

LEMMA 2.5   *If $\bar{D}\nu(\mathrm{x}) \geq \alpha$ a.e. on a set E, then*

$$\nu(E) \geq \alpha|E|.$$

Proof   It can be assumed that $\bar{D}\nu(x) \geq \alpha$ everywhere on E; for if the lemma is true in this case and if

$$E' = \{x : x \in E \text{ and } \bar{D}\nu(x) \geq \alpha\},$$

then we have

$$\nu(E) \geq \nu(E') \geq \alpha|E'| = \alpha|E|.$$

Take any $\beta < \alpha$ and any open $G \supset E$, and let $\mathfrak{F}$ be the family of all open balls that are contained in G and satisfy

$$\frac{\nu(B)}{|B|} > \beta. \tag{6}$$

If x is any point of E, and ϵ is any positive number, then by the definition of $\bar{D}\nu(x)$ there is a ball of radius $<\epsilon$ that contains x and satisfies (6). This means that the family $\mathfrak{F}$ covers E in the sense of Vitali. If $\{B_k\}$ is a disjoint sequence in $\mathfrak{F}$ that covers almost all of E, then

$$\nu(G) \geq \Sigma\nu(B_k) \geq \beta\Sigma|B_k| \geq \beta|E|.$$

Since this is true for every open $G \supset E$, it follows that $\nu(E) \geq \beta|E|$; and since this is true for every $\beta < \alpha$, it follows that $\nu\ (E) \geq \alpha\ |E|$.

Exercise 4   Use the lemma to show that $\bar{D}\nu(x) < \infty$ a.e.

We need a corresponding lemma for the lower derivative but have to be satisfied with something a little weaker.

LEMMA 2.6   *If $\underline{D}\nu(x) \leq \alpha$ a.e. on a set E, then there is a set N of measure 0 such that*

$$\nu(E - N) \leq \alpha|E|.$$

Example   Take ν to be a counting measure on $\mathbf{R}^1$, say $\nu(A) = $ number of integers in A. It is evident that if x is not an integer, then $\bar{D}\nu(x) = \underline{D}\nu(x) = 0$. The set N in the lemma is the set of integers. What happens with the Cantor measure?

Proof   It can be assumed that $\underline{D}\nu(x) \leq \alpha$ everywhere on E, for this just involves throwing out another set of measure 0. It can also be assumed that E is measurable, for if E' is a measurable set containing E and with the same measure and if $F = \{x : \underline{D}\nu(x) \leq \alpha\}$, then F is also a measurable set containing E. If we apply the lemma to the measurable set $E' \cap F$, we get

$$\nu(E - N) \leq \nu((E' \cap F) - N) \leq \alpha|E' \cap F| = \alpha|E|.$$

Finally, it can be assumed that E is bounded, for if the lemma holds for each $E \cap B(0; n)$, then it also holds for E.

Exercise 5   Prove this last statement.

Let $\epsilon > 0$ be given and choose an open $G \supset E$ with $|G| < |E| + \epsilon$. Take any $\beta > \alpha$ and let $\mathfrak{F}$ be the family of open balls contained in G and satisfying $\nu(B)/|B| < \beta$. As before, $\mathfrak{F}$ covers E in the sense of Vitali. Let $\{B_k\}$ be a disjoint sequence that covers almost all of E. If F_β is the union of the B_k and $N_\beta = E - F_\beta$, then

$$\nu(E - N_\beta) = \nu(E \cap F_\beta) \leq \nu(F_\beta)$$
$$= \Sigma\nu(B_k) \leq \beta\Sigma|B_k| \leq \beta|G| \leq \beta(|E| + \epsilon).$$

Now, let N' be the union of the N_β for $\beta = \alpha + 1/m$, where m is a positive integer. Each N_β has measure 0, so N' does also, and $\nu(E - N') \leq \nu(E - N_\beta) \leq \beta(|E| + \epsilon)$. Since this holds for each β, it follows that $\nu(E - N') \leq \alpha(|E| + \epsilon)$.

The set N' depends on the initial choice of ϵ. Form N'_k for $\epsilon = 1/k$, and let N be the union of the N'_k. Then N also has measure 0, and $\nu(E - |N) \leq \nu(E - N'_k) \leq \alpha(|E| + 1/k)$ for every k, which proves the lemma.

THEOREM 2.7   *Every regular Borel measure is differentiable a.e.*

Proof   It is plain that $\underline{D}\nu(x) \leq \bar{D}\nu(x)$, and it has been seen already in Exercise 4 that $\bar{D}\nu$ is finite a.e., so what has to be proved is that the set

$$E = \{x : \underline{D}\nu(x) < \bar{D}\nu(x)\}$$

has measure 0. Fix any two numbers α and β, $\alpha < \beta$, and let

$$F = \{x : \underline{D}\nu(x) < \alpha < \beta < \bar{D}\nu(x)\} \cap B(0; m).$$

By Lemma 2.6 there is a set N of measure 0 such that $\nu(F - N) \leq \alpha|F|$, while, by Lemma 2.6, $\nu(F - N) \geq \beta|F - N| = \beta|F|$. Since $\alpha < \beta$, it follows that $|F| = 0$. Now, E is the union of the sets F with rational α and β, and integral m, and these can be arranged in a sequence. Hence $|E| = 0$.

Exercise 6   Expand the idea of the last proof to prove the following theorem.

THEOREM 2.8   *If $\{v_k\}$ is an increasing sequence of regular Borel measures with limit v, then v is a regular Borel measure and $Dv_k \to Dv$ a.e.*

[The hypothesis means that for every set E, the sequence $\{v_k(E)\}$ is increasing and has limit $v(E) < \infty$.]

3   INTEGRATION OF DERIVATIVES

The main theorem on the integration of derivatives is the following:

THEOREM 3.1   *If v is a regular Borel measure on Ω, then there is a G_δ set N of measure 0 such that for every measurable set $E \subset \Omega$,*

$$v(E) = \int_E Dv \, dx + v(E \cap N). \tag{1}$$

We begin with a lemma.

LEMMA 3.2   *For every measurable set E, we have*

$$v(E) \geq \int_E Dv \, dx. \tag{2}$$

Proof   It can be assumed that E is open; for if the inequality holds in this case, then for every open $G \supset E$ we have

$$v(G) \geq \int_G Dv \, dx \geq \int_E Dv \, dx,$$

and taking the lower bound over such G gives (2). In the second place, it can be assumed that E is bounded; for if (2) holds for each $E \cap B(0; n)$, then it holds for E itself. As a consequence of these two remarks, it can be assumed that E is bounded and both Lebesgue measurable and v measurable.

Let $\epsilon > 0$ be given and set

$$E_j = \{x : x \in E \text{ and } j\epsilon \leq \bar{D}v(x) < (j+1)\epsilon\}. \tag{3}$$

Note that E_j is both Lebesgue measurable and v measurable. Indeed, E is measurable in both senses, and so is $\bar{D}v$, by virtue of Theorem 2.2. According to Lemma 2.5, we have $v(E_j) \geq j\epsilon|E_j|$, and obviously we have $\int_{E_j} \bar{D}v \, dx \leq (j+1)\epsilon|E_j|$; combining the two we get

$$v(E_j) \geq \int_{E_j} \bar{D}v \, dx - \epsilon|E_j|.$$

Summing over j (and this is where E_j needs to be ν measurable) gives

$$\nu(E) \geq \int_E \bar{D}\nu \, dx - \epsilon|E| = \int_E D\nu \, dx - \epsilon|E|,$$

which in turn gives (2), since ϵ is arbitrary.

Exercise 1 The lemma is also true if E is ν measurable. (The main point is that if E is ν measurable, then, although it is not necessarily Lebesgue measurable, $\chi_E \, D\nu$ is Lebesgue measurable. This fact follows from Lemma 3.2.)

Exercise 2 What is the reason for using $\bar{D}\nu$ instead of $D\nu$ in formula (3)?

Proof of the Theorem To begin with, let us take E to be bounded and find a corresponding set N that depends on E. Let $\epsilon > 0$ be given, and define E_j as in the proof of the lemma, formula (3). According to Lemma 2.6, there is a set N_j of measure 0 such that $\nu(E_j - N_j) \leq (j + 1)\epsilon|E_j|$, while obviously $\int_{E_j} D\nu \, dx \geq j\epsilon|E_j|$; combining the two we get

$$\nu(E_j - N_j) \leq \int_{E_j} D\nu \, dx + \epsilon|E_j|.$$

Taking N_ϵ to be the union of the N_j and summing on j, we get

$$\nu(E - N_\epsilon) \leq \int_E D\nu \, dx + \epsilon|E|,$$

and, taking N to be the union of the N_ϵ for $\epsilon = 1/k$, we get

$$\nu(E - N) \leq \int_E D\nu \, dx. \tag{4}$$

Now we shall show that the same set N works for any subset F of E that is both Lebesgue measurable and ν measurable. Formula (4) and Lemma 3.2 give

$$\int_E D\nu \, dx \geq \nu(E - N) = \nu(F - N) + \nu(E - F - N)$$
$$\geq \int_{F-N} D\nu \, dx + \int_{E-F-N} D\nu \, dx = \int_E D\nu \, dx.$$

Consequently, equality holds throughout, so that

$$\nu(F - N) = \int_{F-N} D\nu \, dx = \int_F D\nu \, dx. \tag{5}$$

Now, let E_k be the ring

$$E_k = \{x : k \leq |x| < k + 1\},$$

which is clearly both Lebesgue measurable and ν measurable, let N_k be the set that has just been found for $E = E_k \cap \Omega$, and let N be a G_δ of measure 0 that contains the union of the N_k.   If F is both Lebesgue measurable and ν measurable, then so is $F \cap E_k$, and by (5) we have

$$\nu(F - N) \le \sum \nu((F \cap E_k) - N) \le \sum \int_{F \cap E_k} D\nu \, dx = \int_F D\nu \, dx.$$

Finally, if E is any Lebesgue measurable set, let F be a G_δ that contains it and satisfies $|F - E| = 0$.   Then

$$\nu(E - N) \le \nu(F - N) \le \int_F D\nu \, dx = \int_E D\nu \, dx.$$

On account of Lemma 3.2, equality must hold throughout; therefore, as N is ν measurable,

$$\nu(E) = \nu(E - N) + \nu(E \cap N) = \int_E D\nu \, dx + \nu(E \cap N),$$

which proves the theorem.

Exercise 3   Formula (1) holds when E is ν measurable as well as when E is Lebesgue measurable.   (As in Exercise 1, the point is that $\chi_E D\nu$ is Lebesgue measurable, even though χ_E itself is not.)

Theorem 3.1 gives an immediate solution to the problem of which regular Borel measures are indefinite integrals.   They are those which are absolutely continuous in the following sense.

DEFINITION 3.3   *A regular Borel measure ν is absolutely continuous if $\nu(E) = 0$ whenever $|E| = 0$.*

THEOREM 3.4   *A regular Borel measure ν on Ω is an indefinite integral if and only if it is absolutely continuous.   If this is the case, then for every ν-measurable set $E \subset \Omega$ we have*

$$\nu(E) = \int_E D\nu \, dx,$$

and for every nonnegative ν-measurable function f we have

$$\int f \, d\nu = \int f \, D\nu \, dx.$$

It is plain on the face of it that if ν is an indefinite integral, then it is absolutely continuous; and it is plain from Theorem 3.1 that if ν is absolutely continuous, then it is the indefinite integral of $D\nu$, for the term $\nu(E \cap N)$ drops out.   The last assertion comes from Theorem 1.3.   Note that Theorem 1.5 now is proved completely.

COROLLARY 3.5   *If v is the indefinite integral of f, then $Dv = f$ a.e. on Ω.*

Exercise 4   Prove the corollary.   (See Exercise 5 of Section 1.)

Exercise 5   If v is absolutely continuous, then every Lebesgue measurable set or function is v measurable.

Exercise 6   A regular Borel measure v is absolutely continuous if and only if it has the following property: For every set F, with $v(F) < \infty$, and every $\epsilon > 0$, there is a $\delta > 0$ such that if $E \subset F$ and $|E| < \delta$, then $v(E) < \epsilon$.

Exercise 7   Interpret the property in Exercise 6 for increasing functions on $\mathbf{R}^1$.

Exercise 8   For any set $A \subset \mathbf{R}^n$, set $v_A(E) = |A \cap E|$. Show that $Dv_A(x) = 1$ a.e. on A and that if A is measurable, then also $Dv_A(x) = 0$ a.e. on $\mathbf{R}^n - A$. [*Hint:* If A is measurable, then v_A is the indefinite integral of the characteristic function of A. If A is not measurable, find a measurable $B \supset A$ such that $v_B(E) = v_A(E)$ for every measurable E. Why doesn't the argument show that $Dv_A(x) = 0$ a.e. on $\mathbf{R}^n - A$?] If $Dv_A(x) = 1$, then x is called a *point of density* of the set A.

Exercise 9   If $f : \mathbf{R}^n \to \mathbf{R}^m$ is locally integrable, then for almost every point x

$$\lim_{|B| \to 0} \frac{1}{|B|} \int_B |f(x) - f(y)| \, dy = 0, \tag{6}$$

the limit being taken over the balls that contain x. [*Hint:* Suppose that the limit superior in (6) is $> \alpha > 0$ on a set E of positive measure. Choose a small ϵ and let $E_j = \{x \in E : j\epsilon \le f(x) < (j + 1)\epsilon\}$. Some E_j must have positive measure, and hence by Exercise 8 a point of density. Show that the limit superior cannot be $> \alpha$ at a point of density of E_j at which corollary 3.5 holds for $|f|$.] The points where (6) holds are called the *Lebesgue points* of the function f.

The Lebesgue points of a function are important in many theorems, such as the following.

Exercise 10   Let $e : \mathbf{R}^n \to \mathbf{R}^1$ be bounded and measurable and vanish for $|x| \ge 1$. Let $\int e(x) \, dx = 1$ and set $e_\rho(x) = \rho^{-n} e(x)$. Show that if f is locally integrable, then $f * e_\rho(x) \to f(x)$ at each Lebesgue point of f. (This sort of thing is useful in connection with the process of regularization described in Chapter 13.)

Remark   The theorems of these first three sections can be carried through in an abstract setting where the space $\mathbf{R}^n$ is replaced by a metric space X, and the Lebesgue measure is replaced by some basic regular Borel measure μ which satisfies the following conditions:

A. *There is a constant c such that $\mu(B(a; 2r)) \leq c\mu(B(a; r))$ for every ball $B(a; r)$.*
B. *If $\{B_k\}$ is a sequence of balls with $\mu(B_k) \to 0$, then $\delta(B_k) \to 0$.*
C. *If $\{B_k\}$ is a sequence of balls with $\delta(B_k) \to \infty$, then $\mu(B_k) \to \infty$.*

The results are particularly interesting when X is a smooth n-dimensional manifold and μ is the Hausdorff area measure, which is the subject of the next chapter. They also show that the metric in $\mathbf{R}^n$ can be changed so that balls become cubes, for example, while the theorems remain the same.

Exercise 11 Go back through the chapter in the light of this remark. (Almost everything will remain unchanged.)

4 CHANGE OF VARIABLE

In this section we shall carry out the details of the change of variable formula

$$\int_{\varphi(\Omega)} f(y)\, dy = \int_{\Omega} f(\varphi(x)) |\det d\varphi(x)|\, dx \tag{1}$$

that was discussed somewhat briefly in Section 1. The ideas in the proof will appear again in the next chapter in a more elaborate form, and the theorem itself will be improved. To begin with, let us improve Theorem 2.7.

THEOREM 4.1

If $\varphi : \mathbf{R}^n \to \mathbf{R}^n$ satisfies $|\varphi(x) - \varphi(y)| \leq M|x - y|$ on a set A, then

$$|\varphi(A)| \leq M^n |A|.$$

Proof Clearly we can assume that $|A| < \infty$. Let $\epsilon > 0$ be given and choose an open $G \supset A$ such that $|G| < |A| + \epsilon$. Let $\mathcal{F}$ be the family of balls that are contained in G and have center in A. This family covers A in the sense of Vitali, so there is a disjoint sequence $\{B_k\}$ that covers almost all of A. By Theorem 2.7, φ takes sets of measure 0 into sets of measure 0, so we have

$$|\varphi(A)| \leq \Sigma |\varphi(A \cap B_k)|.$$

Now, if $B = B(a; r)$ is a ball with center in A, then $\varphi(A \cap B) \subset B(\varphi(a); Mr)$; therefore, $|\varphi(A \cap B)| \leq M^n |B|$, so

$$|\varphi(A)| \leq M^n \Sigma |B_k| \leq M^n |G| < M^n(|A| + \epsilon).$$

THEOREM 4.2

If $T : \mathbf{R}^n \to \mathbf{R}^n$ is a linear transformation, then for any set A

$$|T(A)| = |\det T|\, |A|.$$

Proof   If U is an orthogonal transformation, then the previous theorem applies with $M = 1$ to give $|U(A)| \leq |A|$.   It also applies to U^{-1} to give $|U(A)| \geq |A|$. This proves the theorem when U is orthogonal, for the determinant of an orthogonal transformation is ± 1.   It also proves the interesting fact that the Lebesgue measure in $\mathbf{R}^n$ is independent of the (orthonormal) coordinate system, for a coordinate change is effected by an orthogonal transformation.

Let H be self-adjoint, and choose the coordinate axes along the eigenvectors of H (which can be done without changing the measures by the previous part of the proof).   Then Exercise 13, Section 2 of Chapter 13, gives

$$|H(A)| = |\lambda_1 \cdots \lambda_n| \, |A| = |\det H| \, |A|.$$

If T is nonsingular, then $T = UH$, so

$$|T(A)| = |\det U| \, |H(A)| = |\det U| \, |\det H| \, |A| = |\det T| \, |A|.$$

If T is singular, then $\det T = 0$ and $|T(A)| = 0$, because the range of T is contained in a subspace of dimension $< n$.

THEOREM 4.3

If $\varphi : \mathbf{R}^n \to \mathbf{R}^n$ is of class C^1 at a point a, then

$$\lim_{r \to 0} \frac{|\varphi(B(a; r))|}{|B(a; r)|} = |\det d\varphi(a)|.$$

Proof   We begin with the case where $d\varphi(a) = I$.   In this case Theorem 7.2 of Chapter 10 shows that

$$\big|B(\varphi(a); (1 + \epsilon)r)\big| \geq \big|\varphi(B(a; r))\big| \geq \big|B(\varphi(a); (1 - \epsilon)r)\big|$$

if r is small enough.   Dividing by $|B(a; r)|$ we get

$$(1 + \epsilon)^n \geq \frac{\big|\varphi(B(a; r))\big|}{|B(a; r)|} \geq (1 - \epsilon)^n.$$

Letting $r \to 0$ we see that the limits superior and inferior are both between $(1 + \epsilon)^n$ and $(1 - \epsilon)^n$.   Since $\epsilon > 0$ is arbitrary, this implies that the limit exists and is equal to 1.

Next suppose that $T = d\varphi(a)$ is nonsingular, and put $\psi = T^{-1} \circ \varphi$ so that $\varphi = T \circ \psi$ and $d\psi(a) = I$.   By Theorem 4.2 and what has just been proved, we have

$$\lim_{r \to 0} \frac{\big|\varphi(B(a; r))\big|}{|B(a; r)|} = |\det T| \lim_{r \to 0} \frac{\big|\psi(B(a; r))\big|}{|B(a; r)|} = |\det T|.$$

Finally, suppose that $T = d\varphi(a)$ is singular.   From the definition of the differential we have

$$|\varphi(x) - \varphi(a) - T(x - a)| \leq \epsilon|x - a| \text{ if } |x - a| < r$$

with r small enough.   If we choose the coordinates in $\mathbf{R}^n$ so that the range of T is contained in $\mathbf{R}^{n-1}$ and use primes to denote the first $n-1$ coordinates, then we get

$$|\varphi'(x) - \varphi'(a)| \leq (M + \epsilon)|x - a|,$$

and

$$|\varphi_n(x) - \varphi_n(a)| \leq \epsilon|x - a|$$

with $M = \|T\|$; hence

$$\varphi(B(a; r)) \subset B'(\varphi'(a); (M + \epsilon)r) \times [\varphi_n(a) - \epsilon r, \varphi_n(a) + \epsilon r].$$

Fubini's theorem gives

$$\left|\varphi(B(a; r))\right| \leq c'(M + \epsilon)^{n-1}r^{n-1}2\epsilon r \leq 2c'(M + 1)^{n-1}r^n\epsilon$$

with $c' = B'(0; 1)$; consequently,

$$\lim_{r \to 0} \frac{\left|\varphi(B(a; r))\right|}{|B(a; r)|} = 0,$$

which proves the theorem.

Now we are ready to change variables.   If $\varphi : \mathbf{R}^n \to \mathbf{R}^n$ is continuous and one to one on the open set Ω, the discussion in Section 1 shows that

$$\nu(A) = |\varphi(A)|$$

is a regular Borel measure on Ω such that

$$\int_{\varphi(\Omega)} f(y)\, dy = \int_\Omega f(\varphi(x))\, d\nu$$

for every Lebesgue measurable function f on $\varphi(\Omega)$.   If ν is absolutely continuous (and by Corollary 2.10 of Chapter 13 this is the case if φ is of class C^1), then Theorem 3.4 gives

$$\int_{\varphi(\Omega)} f(y)\, dy = \int_\Omega f(\varphi(x))\, D\nu(x)\, dx.$$

Finally, if φ is of class C^1 at almost every point, Theorem 4.3 shows that $D\nu(x) = |\det d\varphi(x)|$ almost everywhere.

As far as the absolute continuity of ν is concerned, it is a little unsatisfactory to achieve this by assuming that φ is of class C^1 at all points.   For instance, $\varphi(x) = \sqrt[3]{x}$ is of class C^1 at each point $x \neq 0$, but not of class C^1 at 0.   The corresponding ν is certainly absolutely continuous, and φ makes a perfectly good change of variable.   In order to state the final theorem on change of variable we shall just define away this problem—but then we shall make some remarks on the definition afterward.

DEFINITION
4.4

A function $\varphi : \mathbf{R}^n \to \mathbf{R}^n$ is absolutely continuous on a set A if φ is continuous and $\varphi(N)$ has measure 0 whenever $N \subset A$ has measure 0.

The final theorem reads:

THEOREM
4.5

If $\varphi : \mathbf{R}^n \to \mathbf{R}^n$ is one to one and absolutely continuous on the open set Ω and is of class C^1 at almost every point of Ω, then

$$\int_{\varphi(\Omega)} f(y)\ dy = \int_{\Omega} f(\varphi(x))|\det d\varphi(x)|\ dx$$

for every Lebesgue measurable function f.

The usual way to establish absolute continuity is to decompose the set Ω into two parts such that φ is of class C^1 on one part and takes the other part into a set of measure 0.

Exercise 1 If φ is continuous on $A \cup B$ and absolutely continuous on A and B separately, then φ is absolutely continuous on $A \cup B$.

Exercise 2 If φ is continuous on A and $|\varphi(A)| = 0$, then φ is absolutely continuous on A.

Exercise 3 Let φ be continuous on A and suppose that there is a subset N such that $|\varphi(N)| = 0$ and φ is of class C^1 on $A - N$. Then φ is absolutely continuous on A.

The result of the last exercise is the one that usually is used to prove absolute continuity. In the example $\varphi(x) = \sqrt[3]{x}$, for instance, it applies with $N = \{0\}$. It is important to realize that $\varphi(N)$ must have measure 0. Whether N has measure 0 is irrelevant. If α is the Cantor function described in Section 1, then α is of class C^1 on the set G (which means almost everywhere), but it is not absolutely continuous. To get a similar example where the function φ is one to one, just take $\varphi(x) = \alpha(x) + x$.

Remark There is no standard definition of absolute continuity except in the one-dimensional case where the function φ is defined on an interval. Unfortunately, the standard definition in this case is a little bit different from the one above. It requires in addition that the curve $y = \varphi(x)$ have finite length on every compact subinterval. For example, the function $\varphi(x) = x \sin 1/x$ is absolutely continuous in our sense but not in the classical sense. The term originated in the study of increasing functions and differences of increasing functions. In this setting the finiteness of the length is automatic, but it is not a natural condition to impose in general. In Chapter 15 we shall study absolutely continuous functions from $\mathbf{R}^m$ to $\mathbf{R}^n$ for any $m \leq n$.

5   DIFFERENTIABILITY OF LIPSCHITZ FUNCTIONS

H. Rademacher found a fundamental theorem on the differentiability of Lipschitz functions.

THEOREM 5.1

(*H. Rademacher*) *If* $f: \mathbf{R}^n \to \mathbf{R}^m$ *is locally Lipschitzian on an open set* Ω, *then* f *is differentiable almost everywhere on* Ω.

Actually he proved a bit more, but this will suffice. Before taking up the proof we gather a few preliminaries.

Since a function is differentiable if and only if each coordinate function is, we can assume that $m = 1$. Since differentiability is local, we can assume that f is Lipschitz, not just locally Lipschitz, and that Ω is a cube. Thus, f is real-valued, Ω is a cube, and

$$|f(x) - f(y)| \leq M|x - y|. \tag{1}$$

Exercise 1   Prove the theorem when $n = 1$ (in which case it is a much earlier theorem of Lebesgue). Hint: if f is nondecreasing, this follows from Exercise 3, Section 1, and Theorem 2.7. In general, f is a difference of nondecreasing Lipschitz functions.

Exercise 2   For each θ, $D_\theta f$ exists almost everywhere and is measurable. In particular,

$$\frac{f(y', y_n + h) - f(y', y_n)}{h} \to D_n f(y) \qquad \text{as } h \to 0 \tag{2}$$

almost everywhere on Ω.

This will be combined with the following.

Exercise 3   (Egoroff's theorem) Let μ be a finite measure on a set X. Let g_h be defined a.e. and measurable, and let $g_h \to g$ a.e. For $\lambda > 0$ there is a measurable set K, with $\mu(X - K) < \lambda$, on which the convergence is uniform. If X is a subset of $\mathbf{R}^n$, K can be taken compact.

Exercise 4   The set A of points $a = (a', a_n)$ such that the section f_{a_n} is differentiable at a' is measurable. (Recall that $f_{a_n}(x') = f(x', a_n)$.)

Proof of the Theorem

Assume, for purposes of induction, that the theorem holds in dimension $n - 1$. (The case $n = 1$ is covered by Exercise 1.) Let $\lambda > 0$ be given, and use (2) and Egoroff's theorem to find a compact K, $|\Omega - K| < \lambda$, on which the convergence in (2) is uniform. Then $D_n f$ is continuous, hence uniformly continuous, on K. We will show that f is differentiable almost everywhere on K. Then we can take a sequence $\lambda_m \to 0$ and conclude that f is differentiable almost everywhere on the union of the corresponding K_m, hence almost everywhere on Ω.

Let D be the set of points of density of K, and let A be the set described in Exercise 4. According to Exercise 8, Section 3, $K - D$ has measure 0. According to the induction hypothesis, each section of $\Omega - A$ has $(n-1)$-dimensional measure 0, so by Exercise 4 and Fubini, $\Omega - A$ has measure 0. Therefore, $K - (A \cap D)$ has measure 0, and we will show that f is differentiable at each point a of $A \cap D$. Henceforth, a is fixed in $A \cap D$.

LEMMA 5.2

For each ϵ there is a δ' such that if $y \in K$ and $|y - a| < \delta'$, then

$$|f(y', y_n) - f(y', a_n) - D_n f(a)(y_n - a_n)| \leq \epsilon |y - a|. \tag{4}$$

Proof of the Lemma

Set $h = a_n - y_n$. By the uniform convergence in (2), there is a δ' such that (4) holds if $D_n f(a)$ is replaced by $D_n f(y)$. By the uniform continuity of $D_n f$ on K, this replacement makes no difference.

LEMMA 5.3

For each ϵ there is a δ' such that if $y \in K$ and $|y - a| < \delta'$, then

$$|f(y) - f(a) - \langle \nabla f(a), y - a \rangle| < \epsilon |y - a|.$$

Proof of the Lemma

Write the left side in the form

$$f(y', y_n) - f(y', a_n) - D_n f(a)(y_n - a_n)$$
$$+ f(y', a_n) - f(a', a_n) - \langle \nabla f_{a_n}(a'), y' - a' \rangle.$$

The first term is covered by Lemma 5.2, the second by the fact that the section f_{a_n} is differentiable at a'.

LEMMA 5.4

For each ϵ there is a δ' such that if $|x - a| < \delta'$, then

$$B(x, \epsilon |x - a|) \cap K \neq \emptyset.$$

Proof of the Lemma

Let $r = (1 + \epsilon)|x - a|$. If the intersection is empty, then

$$|B(a, r)| \geq |B(a, r) \cap K| + |B(x, \epsilon |x - a|)|$$

so that

$$1 \geq \frac{|B(a, r) \cap K|}{|B(a, r)|} + \left(\frac{\epsilon}{1 + \epsilon} \right)^n$$

which is impossible for small r, since a is a point of density of K.

Back to Proof of the Theorem

Let $\epsilon > 0$ be given, let δ' be chosen in accordance with Lemmas 5.3 and 5.4, and let $\delta < \delta'/(1 + \epsilon)$. If $|x - a| < \delta$, use Lemma 5.4 to choose $y \in K$ with $|y - x| \leq \epsilon |x - a|$, so that

$$|y - a| \leq |y - x| + |x - a| \leq (1 + \epsilon)|x - a| < \delta'.$$

Then we have

$$|f(x) - f(a) - \langle \nabla f(a), x - a \rangle|$$

$$\leq |f(x) - f(y)| + |\langle \nabla f(a), x - y \rangle| + |f(y) - f(a) - \langle \nabla f(a), y - a \rangle|$$

$$\leq 2M|x - y| + \epsilon|y - a|$$

$$\leq 2M\epsilon|x - a| + \epsilon(1 + \epsilon)|x - a|$$

$$\leq (2M + 1 + \epsilon)\epsilon|x - a|,$$

which shows that f is differentiable at a.

15 | Surface Area

1   AREA MEASURES

In this chapter we shall develop formulas for the m-dimensional area of sets in $\mathbf{R}^n$ and for the area of m-dimensional parametric surfaces $\varphi : \mathbf{R}^m \to \mathbf{R}^n$. It is assumed of course that $m \leq n$. When $m = n$ the theory is just the theory of Lebesgue measure, and the formula for the area of the parametric surface φ is just the change of variable formula of the last section (with $f = 1$):

$$|\varphi(\Omega)| = \int_\Omega |\det d\varphi(x)| \, dx.$$

(In this case, of course, we do not ordinarily think of area but rather of volume, and we think of φ as a solid rather than a surface. When $m < n$ it is more natural to think of surfaces and areas.)

When $m < n$ the right way to define the m-dimensional area of a set is to use the m-dimensional Hausdorff measure μ_m that was introduced in Section 3 of Chapter 13. The reason for this is the following theorem.

THEOREM 1.1

For each m there is a constant γ_m such that

$$\mu_m(A) = \gamma_m |A|_m \qquad if \ A \subset \mathbf{R}^m.$$

Exercise 1   Go back to Section 3 of Chapter 13 to review the definition and elementary properties of the Hausdorff measures. Then show that μ_m is an absolutely continuous regular Borel measure on $\mathbf{R}^m$. (It is not true that μ_m is a regular Borel measure on $\mathbf{R}^n$ if $m < n$, for compact sets do not have finite measure.)

Proof of the Theorem   Since μ_m is an absolutely continuous regular Borel measure, Theorem 3.4 of the last chapter shows that it is the integral of its derivative. But the derivative is plainly constant, for $\mu_m(E + a) = \mu_m(E)$ for any set E and any point a. Thus, γ_m is just the constant $D\mu_m$.

DEFINITION 1.2

In any metric space we shall write

$$|A|_m = \alpha_m(A) = \frac{1}{\gamma_m} \mu_m(A)$$

and shall call α_m the m-dimensional area measure on the metric space.

In order to see that this definition is acceptable, suppose first that $A \subset \mathbf{R}^n$ is a subset of an m-dimensional subspace V of $\mathbf{R}^n$. Any choice of an orthonormal basis of V turns V into $\mathbf{R}^m$ and determines, therefore, a Lebesgue measure on V which gives the natural m-dimensional area of A. Theorem 1.1 shows that this is just $\alpha_m(A)$.

Exercise 2

How does Exercise 8, Section 3 of Chapter 13, come in here?

It is plain that the m-dimensional area of a set A ought to be the same as that of the translate $A - a$, and it is plain that α_m has this property too. Therefore, $\alpha_m(A)$ is the right thing if A is contained in any m-dimensional plane. Finally, if A is contained in a finite union of distinct m-dimensional planes Π_j, then the m-dimensional area of A ought to be the sum of areas of the $A \cap \Pi_j$.

Exercise 3

Show that if A is contained in the union of the m-dimensional planes Π_j, then

$$\alpha_m(A) = \Sigma \alpha_m(A \cap \Pi_j).$$

(Even for a sequence, as a matter of fact.) (*Hint:* You will have to show that each Π_j is α_m measurable and that the intersections $\Pi_j \cap \Pi_k$ have measure 0.)

This shows that $\alpha_m(A)$ does coincide with the intuitive notion of the m-dimensional area whenever A is contained in a finite union of planes of dimension m—which is about as far as intuition carries.

Exercise 4

The value of the constant γ_m is not important, but it is an interesting exercise to calculate it.

THEOREM 1.3

If $\varphi : X \to Y$ satisfies $d(\varphi(x), \varphi(y)) \leq M d(x, y)$, then for any set $A \subset X$ we have

$$|\varphi(A)|_m \leq M^m |A|_m.$$

Exercise 5

Prove the theorem.

COROLLARY 1.4

Any compact subset of a smooth m-dimensional manifold has finite m-dimensional area.

Exercise 6   Prove the corollary, and also the following one.

COROLLARY 1.5   *Any smooth manifold of dimension $< m$ has m-dimensional area 0.*

Now we shall develop some formulas for the area of polyhedra.

THEOREM 1.6   *If $T : \mathbf{R}^m \to \mathbf{R}^n$ is a linear transformation, then*
$$|T(A)|_m = \sqrt{\det T^*T}\,|A|_m.$$

Proof   If T is not one to one, then the left side is 0 because the range of T is a subspace of dimension $< m$; the right side is 0 because T^*T is not one to one either. If T is one to one, then we can write $T = UH$, where $H = \sqrt{T^*T}$ is self-adjoint from $\mathbf{R}^m$ to $\mathbf{R}^m$, and $U = TH^{-1}$ is orthogonal from $\mathbf{R}^m$ to $\mathbf{R}^n$.

Now, if $U : \mathbf{R}^m \to \mathbf{R}^n$ is orthogonal, then the sets E and $U(E)$ have precisely the same diameter, so it is clear that
$$|U(A)|_m = |A|_m.$$

Therefore, by Theorem 4.2 of Chapter 14, we have
$$|T(A)|_m = |H(A)|_m = \det H|A|_m;$$
and since $H^2 = T^*T$, we have $(\det H)^2 = \det T^*T$.

DEFINITION 1.7   *Let $v_0, \ldots, v_m$ be affinely independent points in $\mathbf{R}^n$. The m-dimensional simplex with vertices $v_0, \ldots, v_m$ is the set*
$$\sigma = \left\{ x : x = \sum_{i=0}^m t_i v_i \text{ with } 0 \le t_i \le 1 \text{ and } \sum_{i=0}^m t_i = 1 \right\}.$$

Recall (Section 2 of Chapter 9) that $v_0, \ldots, v_m$ are affinely independent if and only if $v_1 - v_0, \ldots, v_m - v_0$ are linearly independent. When $m = 1$ the simplex is simply the line segment with vertices v_0 and v_1. When $m = 2$ it is the triangle with vertices $v_0, v_1,$ and v_2. When $m = 3$ it is the tetrahedron with vertices $v_0, v_1, v_2,$ and v_3. And so on.

Consider first the unit m-dimensional simplex σ_m in $\mathbf{R}^m$, that is, the one with vertices $0, e_1, \ldots, e_m$. It is easy to see that the section through the point $x_n = t$ is given by
$$(\sigma_m)_t = (1 - t)\sigma_{m-1}.$$
Therefore, Fubini's theorem gives
$$|\sigma_m|_m = \int_0^1 |(\sigma_m)_t|_{m-1}\, dt = |\sigma_{m-1}| \int_0^1 (1 - t)^{m-1}\, dt = \frac{|\sigma_{m-1}|_{m-1}}{m}, \qquad (1)$$

and then induction gives

$$|\sigma_m|_m = \frac{1}{m!}.$$

(2)

Exercise 7   In formula (1) we used the fact that $|(1 - t)\sigma_{m-1}|_{m-1} = (1 - t)^{m-1}|\sigma_{m-1}|_{m-1}$. This follows from Theorem 1.6 by considering the linear transformation T: $\mathbf{R}^{m-1} \to \mathbf{R}^{m-1}$ given by $Tx = \rho x$. However, the same idea has been used before back in Exercise 13, Section 2 of Chapter 13.

Now consider an arbitrary σ with vertices $v_0, \ldots, v_m$. If $T:\mathbf{R}^m \to \mathbf{R}^n$ is the linear transformation with $Te_i = v_i - v_0$, then $T(\sigma_m) = \sigma - v_0$; therefore,

$$|\sigma|_m = |\sigma - v_0|_m = \sqrt{\det T^*T}\,|\sigma_m|_m = \frac{1}{m!}\sqrt{\det T^*T}.$$

The matrix of T has the coordinates of $Te_j = v_j - v_0$ down the jth column. Therefore, the matrix of T^* has the coordinates of $v_i - v_0$ along the ith row. Consequently, the matrix of T^*T has $\langle v_i - v_0, v_j - v_0 \rangle$ in the ith row, jth column, and we get the following result:

THEOREM 1.8   *If σ is the m-dimensional simplex in $\mathbf{R}^n$ with vertices $v_0, \ldots, v_m$, then*

$$|\sigma|_m = \frac{1}{m!}\sqrt{\det\{\langle v_i - v_0, v_j - v_0 \rangle\}}.$$

Exercise 8   Find the area of the triangle in $\mathbf{R}^3$ with vertices $(1, 0, 0)$, $(0, 1, 0)$, and $(0, 0, 1)$.

When $m = n$ we can use T directly instead of T^*T, for $\det T^*T = |\det T|^2$. If we write $(w_1, \ldots, w_m)$ for the matrix that has the coordinates of w_1 down the first column, those of w_2 down the second column, and so on, then we get the following:

THEOREM 1.9   *If σ is the m-dimensional simplex in $\mathbf{R}^m$ with vertices $v_0, \ldots, v_m$, then*

$$|\sigma|_m = \frac{1}{m!}|\det(v_1 - v_0, \ldots, v_m - v_0)|.$$

Still in the same situation (that is, $m = n$), there is another interesting formula that is obtained in the following way: Let τ be the $(m + 1)$-dimensional simplex in $\mathbf{R}^{m+1}$ with vertices $0, (1, v_0), \ldots, (1, v_m)$, where $(1, v)$ is the vector with first coordinate 1 and the rest equal to those of v. What we are doing here is moving σ up to the plane $x_0 = 1$ and then joining it to 0. For the section through the point $x_0 = t$, we have

$$\tau_t = t\sigma;$$

therefore,

$$|\tau|_{m+1} = \int_0^1 t^m |\sigma|_m \, dt = \frac{|\sigma|_m}{m+1}.$$

Using Theorem 1.9 to calculate $|\tau|_{m+1}$ we get

THEOREM 1.10 *If σ is the m-dimensional simplex in $\mathbf{R}^m$ with vertices $v_0, \ldots, v_m$, then*

$$|\sigma|_m = \frac{1}{m!} \left| \det \begin{pmatrix} 1 \\ v_0 \end{pmatrix}, \ldots, \begin{pmatrix} 1 \\ v_m \end{pmatrix} \right|.$$

(The notation means that the first column of the matrix is composed of a 1, and then the coordinates of v_0, and so on.)

The theorems on simplexes lead immediately to the possibility of calculating the area of any polyhedron—just because of the definition of the term polyhedron.

DEFINITION 1.11 *A face of the simplex with vertices $v_0, \ldots, v_m$ is a simplex whose vertices are among these.*

DEFINITION 1.12 *An m-dimensional polyhedron is a finite union of simplexes of dimension m such that the intersection of any two is either empty or is a face of both.*

If σ and τ are simplexes in a polyhedron of dimension m, then $\sigma \cap \tau$ is contained in a plane of dimension $m-1$, so $|\sigma \cap \tau|_m = 0$. Therefore, the m-dimensional area of the polyhedron is the sum of the m-dimensional areas of the simplexes of which it is a union.

Remark 1 A given polyhedron may be cut up into simplexes in many different ways, but the argument above shows that the sum of the areas always remains the same—the area of the polyhedron itself.

Remark 2 In the definition of a polyhedron it is not really necessary to insist that the intersection of any two simplexes be either empty or a face of both. The reason is that the union of any two simplexes can be cut into smaller simplexes that do satisfy this condition. We shall not go into this, however.

Remark 3 Often a polyhedron of dimension m is defined to be a union of simplexes of dimension $\leq m$, with at least one of dimension $= m$. Since a simplex of dimension $< m$ has m-dimensional area 0, the ones of dimension $< m$ do not enter the picture. (That is, in area problems they do not. In other kinds of problems they do.)

2   PARAMETRIC SURFACES—INTRODUCTORY REMARKS

An m-dimensional parametric surface in $\mathbf{R}^n$ is a continuous function $\varphi : \mathbf{R}^m \to \mathbf{R}^n$ defined on some set $E \subset \mathbf{R}^m$. The surface is not the set $\varphi(E)$, but if φ is one to one, then it is natural to define the m-dimensional area of φ to be the m-dimensional area of the set $\varphi(E)$. If φ is not one to one, this will not do. For instance, the path to the dentist's office and home again does not have the same length as the path one way, in spite of the fact that both have the same range. To get the length of the round trip each point of the range must be counted twice.

**DEFINITION
2.1**

Let $\varphi : \mathbf{R}^m \to \mathbf{R}^n$ be continuous on the set $E \subset \mathbf{R}^m$. For each point $y \in \mathbf{R}^n$, let $N(E; y)$ be the number (possibly $+\infty$) of points $x \in E$ with $\varphi(x) = y$. The area of the surface φ is the number

$$\text{area } \varphi = \int N(E; y) \, d\alpha_m. \tag{1}$$

Consider more closely the case where φ is one to one. In this case $N(E; y)$ is 1 if $y \in \varphi(E)$ and is 0 otherwise; so $N(E; y)$ is the characteristic function of the set $\varphi(E)$, and formula (1) gives

$$\text{area } \varphi = |\varphi(E)|_m \qquad \text{if } \varphi \text{ is one to one.} \tag{2}$$

Furthermore, if

$$\nu(A) = |\varphi(A)|_m,$$

then under some fairly mild additional conditions ν is an absolutely continuous regular Borel measure, and the results of Chapter 14 give

$$\text{area } \varphi = |\varphi(E)|_m = \int_E D\nu \, dx.$$

Thus, the problem is reduced to that of calculating the derivative

$$D\nu(a) = \lim_{r \to 0} \frac{|\varphi(B(a; r))|_m}{|B(a; r)|_m}. \tag{3}$$

The limit on the right, which is important whether φ is one to one or not, is called the Jacobian of φ at x and is written $J_\varphi(x)$. The fundamental theorem is that if φ is of class C^1 at a point a, then the Jacobian of φ at the point a is the same as the Jacobian of its differential. Thus, by Theorem 1.6,

$$J_\varphi(a) = J_{d\varphi(a)} = \sqrt{\det d\varphi(a)^* \, d\varphi(a)}.$$

(This shows the relation between the Jacobian and the Jacobi matrix and explains the name.)

Combining the various formulas above and supposing first that φ is one to one, we get

$$\text{area } \varphi = \int N(E; y) \, d\alpha_m = \int_E J_\varphi(x) \, dx. \tag{4}$$

It is reasonable to suspect, and it turns out to be true under the mild additional conditions on φ, that the integral on the right performs just the same counting process as does $N(E; y)$. The formula is correct whether φ is one to one or not and is the basic formula for surface area.

Remark 1   There is a theory of area that applies to functions φ that are simply continuous. Here we shall have to assume more than that, but the results will still have quite good generality. The continuous theory is extremely subtle and difficult. Formulas like (4) do not make sense in general, and even when they do they are not true. Take, for instance, a continuous increasing function $\varphi : \mathbf{R}^1 \to \mathbf{R}^1$ on $[0, 1]$ with $\varphi(0) = 0$, $\varphi(1) = 1$, and $\varphi'(x) = 0$ a.e., for example, the Cantor function. It is clear that the length of φ should be at least the length of the segment $[0, 1]$, that is, at least 1. But the integral of $J_\varphi(x) = \varphi'(x)$ is 0, since $\varphi'(x) = 0$ a.e.

Exercise 1   In this last discussion we are looking at φ as a path on the line itself. It is a little more striking to look at the graph of φ as a path in the plane. Do this.

Remark 2   In the early days it was thought that the theory of surface area was very much like the theory of length, and, in particular, that the area should be defined as the limit of the areas of approximating polyhedra. The approximating polyhedra are obtained as follows: Suppose that the parametric surface φ is defined on a polyhedron—to be definite, let us say on a rectangle Q. Cut Q into small simplexes such that the intersection of any two is empty or is a common face. A small piece of Q might look as shown in Figure 1.

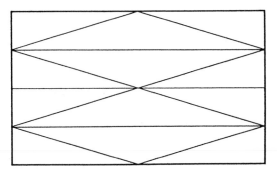

Figure 1

Let σ be one of the simplexes in this "triangulation" of Q, and let it have vertices $a_0, \ldots, a_m$. Each point $x \in \sigma$ can be written uniquely in the form

$$x = \sum_{j=0}^{m} t_j a_j \quad \text{where} \quad t_j \geq 0 \quad \text{and} \quad \sum_{j=0}^{m} t_j = 1.$$

Define

$$\Phi(x) = \sum t_j \varphi(a_j).$$

Then $\Phi(\sigma)$ is the simplex with vertices $\varphi(a_0), \ldots, \varphi(a_m)$, and it is not hard to see that Φ is a well-defined continuous function on Q. Φ is an inscribed parametric polyhedron that approximates the initial φ if the simplexes are small enough.

It was thought initially that the area of Φ should approximate the area of φ, but consider the simple example of a cylinder given by the function

$$\varphi(s, t) = (\cos s, \sin s, t), \qquad 0 \leq s \leq 2\pi, \quad 0 \leq t \leq 1.$$

This just wraps the rectangle Q, $0 \leq s \leq 2\pi$, $0 \leq t \leq 1$, around into a cylinder whose base is the circle of radius 1 and whose height is 1. If we cut Q into simplexes ($=$ triangles) where the base is parallel to the s axis and is long relative to the height, then the corresponding simplexes $\Phi(\sigma)$ are nearly perpendicular to the cylinder. It suddenly appears very unreasonable that the area of Φ is an approximation to the area of the cylinder. The remarkable fact is the following:

Exercise 2   For every number $\alpha \geq 2\pi$ ($=$ the area of the cylinder) and every number $\epsilon > 0$, there is a triangulation of the rectangle into simplexes of diameter $< \epsilon$ such that for the corresponding inscribed polyhedron Φ we have

$$|\text{area } \Phi - \alpha| < \epsilon.$$

In other words, every number $\geq$ the true area is the limit of the areas of approximating inscribed polyhedra!

Once this example was discovered, Lebesgue proposed to define the area to be the limit inferior of the areas of approximating inscribed polyhedra. This definition turned out to be satisfactory—but so difficult to work with that it took nearly 50 years to prove that it was satisfactory. As a matter of fact, any definition of area is hard to work with unless the function φ is pretty nice.

3   THE JACOBIAN

DEFINITION 3.1   *If $\varphi : \mathbf{R}^m \to \mathbf{R}^n$, $m \leq n$, then the Jacobian of φ at the point a is the number*

$$J_\varphi(a) = \lim_{r \to 0} \frac{|\varphi(B(a; r))|_m}{|B(a; r)|_m}$$

at any point where the limit exists.

It is not required that φ be defined everywhere on the ball $B(a;r)$. If φ is defined on a set E, and A is any set, then $\varphi(A)$ means simply $\varphi(A \cap E)$. For instance, if φ is the identity function on an interval E in $\mathbf{R}^1$, then $J_\varphi(a)$ is 1 if a is in the interior of E, 0 if a is outside the closure, and $\frac{1}{2}$ if a is one of the end points. As a matter of fact, however, we are interested mainly in the case where a is an interior point of E, in which case φ is defined everywhere on $B(a;r)$ if r is small enough. The purpose of the section is to prove the following basic result—which has already been established when $m = n$ in Theorems 4.2 and 4.3 of Chapter 14.

THEOREM 3.2

If $\varphi : \mathbf{R}^m \to \mathbf{R}^n$ is of class C^1 at a, then

$$J_\varphi(a) = J_{d\varphi(a)} = \sqrt{\det d\varphi(a)^* \, d\varphi(a)}. \tag{1}$$

The proof is rather long. We shall take up first the case where $T = d\varphi(a)$ is nonsingular (i.e., is one to one). This is really the hard case, but it will seem easier because we have some heavy artillery to bring to bear.

Proof When $T = d\varphi(a)$ Is Nonsingular

In this case the range of T has dimension m, and we can choose the coordinates in $\mathbf{R}^n$ so that it is spanned by the first m basis vectors. If $y \in \mathbf{R}^n$, we write $y = (y', y'')$, where y' is the first m coordinates and y'' is the last $n - m$, and also $\varphi(x) = \big(\varphi'(x), \varphi''(x)\big)$, and so on. It is easy to check that with this choice of coordinates we have

$$(T')^* T' = T^* T \qquad \text{and} \qquad T'' = 0. \tag{2}$$

Moreover, we already know the theorem for φ' and T' (Theorems 4.2 and 4.3 of Chapter 14), so we have

$$J_T = J_{T'} = J_{\varphi'}(a). \tag{3}$$

Therefore, it will suffice to prove that

$$\lim_{r \to 0} \frac{\big|\varphi(B(a;r))\big|_m}{\big|\varphi'(B(a;r))\big|_m} = 1, \tag{4}$$

for this will give that $J_\varphi(a) = J_{\varphi'}(a)$.

If P is the projection of $\mathbf{R}^n$ on $\mathbf{R}^m$ defined by $Py = y'$, then by definition we have $\varphi' = P \circ \varphi$. Since P clearly satisfies $|Px - Py| \le |x - y|$, Theorem 1.3 gives

$$|\varphi'(E)|_m \le |\varphi(E)|_m \qquad \text{for any set } E,$$

and this proves half of (4)—that the limit inferior is ≥ 1.

To prove the other half we shall use the same idea—that is, we shall show that if r is small enough, then the function f that takes $\varphi'(x)$ to $\varphi(x)$

satisfies $|f(z) - f(w)| \leq (1 + \epsilon)|z - w|$ on $\varphi'(B(a; r))$. Once this is done it will follow again from Theorem 1.3 that

$$|\varphi(B(a; r))|_m \leq (1 + \epsilon)^m |\varphi'(B(a; r))|_m,$$

which will prove the other half of (4). The basis for this is Theorem 5.3 of Chapter 10. If $\epsilon > 0$ is given, then we can find $\delta > 0$ such that

$$|\varphi(x) - \varphi(y) - T(x - y)| \leq \epsilon|x - y| \qquad \text{if } x, y \in B(a; \delta). \qquad (5)$$

By the choice of the coordinates, T' is nonsingular, so there is a number $m > 0$ such that $|T'x| \geq m|x|$; hence, [by (5)],

$$|\varphi'(x) - \varphi'(y)| \geq (m - \epsilon)|x - y|, \qquad (6)$$

and, of course, we take the precaution to pick $\epsilon < m$. This shows that φ' is one to one on $B(a; \delta)$ and, therefore, that f is well defined. Again by the choice of the coordinates, $T'' = 0$; so, [by (5)],

$$|\varphi''(x) - \varphi''(y)| \leq \epsilon|x - y| \qquad \text{if } x, y \in B(a; \delta). \qquad (7)$$

Combining (6) and (7) we get

$$\begin{aligned}
|\varphi(x) - \varphi(y)|^2 &\leq |\varphi'(x) - \varphi'(y)|^2 + |\varphi''(x) - \varphi''(y)|^2 \\
&\leq |\varphi'(x) - \varphi'(y)|^2 + \epsilon^2|x - y|^2 \\
&\leq |\varphi'(x) - \varphi'(y)|^2 \left(1 + \frac{\epsilon^2}{(m - \epsilon)^2}\right),
\end{aligned}$$

which is equivalent to the required inequality for f. When $m = 2, n = 3$ and $a = \varphi(a) = 0$, the picture looks as shown in Figure 2.

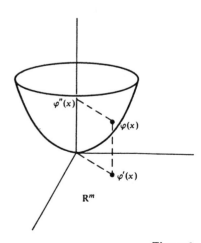

Figure 2

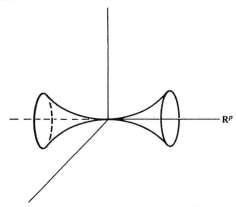

Figure 3

Proof When $T = d\varphi(a)$
Is Singular

In this case the range of T has dimension $p < m$, and the picture looks perhaps as shown in Figure 3. This time we choose the coordinates so that the range of T is spanned by the first p basis vectors, and use primes to denote the first p coordinates and double primes to denote the last $n - p$. If $\epsilon > 0$ is given, we choose $\delta > 0$ so that (5) holds. If $M = \|T\| + 1$, this gives in particular that

$$|\varphi(x) - \varphi(y)| \le M|x - y| \qquad \text{if } x, y \in B = B(a; \delta); \tag{8}$$

and it follows from Theorem 1.3 that

$$\text{if } N \subset B \text{ and } |N|_m = 0, \text{ then } |\varphi(N)|_m = 0. \tag{9}$$

The next step is to show that if $B(b; r) \subset B$, then

$$\varphi(B(b; r)) - \varphi(b) \subset B'(0; Mr) \times B''(0; \epsilon r), \tag{10}$$

where the primes indicate the balls in the respective spaces. Formula (10) means that if $|x - b| < r$, then

$$|\varphi'(x) - \varphi'(b)| < Mr \qquad \text{and} \qquad |\varphi''(x) - \varphi''(b)| < \epsilon r.$$

The first of these follows from (8) and the second from (7). Formula (10) will let us estimate $|\varphi(B(b; r))|_m$—but note that Fubini's theorem does not help when $m < n$.

It is a little easier to deal with cubes than balls, so we replace the ball $B'(0; Mr)$ by the (larger) cube $Q'(0; 2Mr)$ with center 0 and side length $2Mr$, and the ball $B''(0; \epsilon r)$ by the cube $Q''(0; 2\epsilon r)$. What we need to estimate, then, is $|Q|_m^\eta = \mu_m^\eta(Q)/\gamma_m$ for

$$Q = Q'(0; s) \times Q''(0; t) \qquad \text{with } s = 2Mr \text{ and } t = 2\epsilon r.$$

Note that $t \leq s/2$ since $M \geq 1$ and $\epsilon \leq \frac{1}{2}$.

Let l be the integer such that $(l - 1)t < s \leq lt$, and divide Q' in l^p cubes Q_k' of side length s/l. If $Q_k = Q_k' \times Q''$, then each Q_k is contained in a cube of side length t, so $\delta(Q_k) \leq \sqrt{n}\, t$; therefore, if $\eta > \sqrt{n}\, t$, then

$$\mu_m^\eta(Q) \leq l^p n^{m/2} t^m.$$

Using the fact that $l - 1 \leq s/t = M/\epsilon$ (hence $l \leq (M + \epsilon)/\epsilon$), and the fact that $t = 2\epsilon r$, we get

$$\mu_m^\eta(Q) \leq n^{m/2} 2^m (M + \epsilon)^p \epsilon^{m-p} r^m \qquad \text{for } \eta > 2\sqrt{n}\, \epsilon r.$$

Combining this with formula (10) we get the following lemma.

<table>
<tr><td>LEMMA
3.3</td><td>If $B(b; r) \subset B = B(a; \delta)$ and $\eta > 2\sqrt{n}\,\epsilon r$, then

$$\left|\varphi(B(b; r))\right|_m^\eta \leq c\epsilon^{m-p} |B(b; r)|_m,$$

where the constant c depends only on n, m, p, and M.</td></tr>
</table>

The crucial point here is that the constant c does not depend on ϵ, r, η, and so on, but only on the quantities n, m, p, and $M = \|T\| + 1$ that are fixed from the very beginning.

Now take any fixed $\rho < \delta$ and any fixed η. Consider the family $\mathcal{F}$ of all balls $B(b; r)$ such that $2\sqrt{n}\,\epsilon r < \eta$ and $B(b; r) \subset B(a; \rho)$. This family covers $B(a; \rho)$ in the sense of Vitali, so there is a disjoint sequence $\{B_j\}$ that covers $B(a; \rho)$ except for a set N of measure 0. Taking account of the lemma and formula (9), we have

$$\left|\varphi(B(a; \rho))\right|_m^\eta \leq \Sigma |\varphi(B_j)|_m^\eta \leq c\epsilon^{m-p} \Sigma |B_j|_m = c\epsilon^{m-p} |B(a; \rho)|_m.$$

Since this holds for each η, we get

$$\left|\varphi(B(a; \rho))\right|_m \leq c\epsilon^{m-p} |B(a; \rho)|_m.$$

As this holds for every ρ, it follows that the limit superior in Definition 3.1 is $\leq c\epsilon^{m-p}$, and hence that the limit superior is $0 = J_\varphi(a)$.

4   ABSOLUTE CONTINUITY

The notion of absolute continuity that was defined initially for functions from $\mathbf{R}^n$ to $\mathbf{R}^n$ in Section 4 of the last chapter must be extended to the present setting.

**DEFINITION
4.1**

A function $\varphi:\mathbf{R}^m \to \mathbf{R}^n$ is absolutely continuous on a set A if it is defined and continuous on A and if $|\varphi(N)|_m = 0$ whenever $N \subset A$ and $|N|_m = 0$.

Exercise 1 If φ is continuous on $A = \bigcup_{k=1}^{\infty} A_k$ and absolutely continuous on each A_k, then φ is absolutely continuous on A.

Exercise 2 If φ is locally absolutely continuous on A (i.e., if each point of A has a neighborhood G such that φ is absolutely continuous on $A \cap G$), then φ is absolutely continuous on A. (*Hint:* The neighborhood G can be taken to be a ball with rational center and radius.)

**DEFINITION
4.2**

A function $\varphi:\mathbf{R}^m \to \mathbf{R}^n$ is Lipschitzian on a set A if there is a constant M such that $|\varphi(x) - \varphi(y)| \leq M|x - y|$ for any two points x and y of A. It is locally Lipschitzian if each point has a neighborhood G such that it is Lipschitzian on $A \cap G$.

Exercise 3 Consider $\varphi:X \to Y$, where X and Y are metric spaces, and show that if φ is locally Lipschitzian, then it is Lipschitzian on any compact subset of X.

**THEOREM
4.3**

If $\varphi:\mathbf{R}^m \to \mathbf{R}^n$ is C^1 on A, then it is locally Lipschitzian on A. If it is locally Lipschitzian on A, then it is absolutely continuous on A.

Proof Theorem 5.3 of Chapter 10 shows that if φ is C^1 on A, then it is locally Lipschitzian on A. Exercise 1 and Theorem 1.3 show that if φ is locally Lipschitzian, then it is absolutely continuous on A.

The usual way to show that a continuous function φ is absolutely continuous on a set A is to show that φ is C^1 on $A - N$, while $|\varphi(N)|_m = 0$.

Exercise 4 Consider the function $\varphi(t) = \big(t, t \sin (1/t)\big)$, which is the natural parametric representation of the graph of $x \sin 1/x$. Show that φ is absolutely continuous on $\mathbf{R}^1$ but not locally Lipschitzian at 0. Show that the length of φ is infinite on any interval containing 0. (Use the old definition of length.)

It is tempting to think that if φ is absolutely continuous, then $|\varphi(K)|_m < \infty$ for any compact set K. The example in the exercise shows that this is not correct, for in this case the basic theorem of Section 6 will show that $|\varphi(I)|_1$ is the length of φ on I.

**THEOREM
4.4**

Let $\varphi:\mathbf{R}^m \to \mathbf{R}^n$ be absolutely continuous on a set E_0. If $E \subset E_0$ is Lebesgue measurable, then $\varphi(E)$ is α_m measurable. If $A \subset E_0$ is arbitrary, there is a G_δ set $E \supset A$ such that $|\varphi(E \cap E_0)|_m = |\varphi(A)|_m$.

Proof   If E is Lebesgue measurable, then E is the union of a sequence of compact sets and a set of measure 0. Then $\varphi(E)$ is also the union of a sequence of compact sets and a set of measure 0. Indeed, φ takes compacts into compacts because it is continuous, and sets of measure 0 into sets of measure 0 by the definition of absolute continuity.

To prove the other half we shall need the following result on the Hausdorff measures.

Exercise 5   For any $B \subset \mathbf{R}^n$ there is a G_δ set $H \supset B$ with $|H|_m = |B|_m$. [*Hint:* The coverings that are used to define $\mu_m^\epsilon(B)$ can be required to be open. This can be seen as follows: Let $\epsilon > 0$ and $\rho > 1$ be given, and let A be any set with $\delta(A) < \epsilon$. Fix a positive δ and let $G = \{x : d(x, A) < \delta\}$. Then G is open and $\delta(G) < \delta(A) + 2\delta$. Consequently, if δ is small enough so that

$$\delta(A) + 2\delta < \epsilon \qquad \text{and} \qquad \delta(A) + 2\delta < \rho\delta(A),$$

then $\delta(G) < \epsilon$ and $\delta(G) < \rho\delta(A)$. If we replace each A_k in the covering by a G_k of this kind, then at worst we multiply the sum by ρ.]

To prove the other half of the theorem choose a G_δ set $H \supset \varphi(A)$ with $|H|_m = |\varphi(A)|_m$. Let $H = \cap H_k$, where H_k is open. Then $\varphi^{-1}(H_k)$ is open in E_0 because φ is continuous, so $\varphi^{-1}(H_k) = G_k \cap E_0$, where G_k is open in $\mathbf{R}^m$. The set $E = \cap G_k$ does the job, since $\varphi(E \cap E_0) \subset H$.

5   VARIATION

The area of a parametric surface φ can be expressed in terms of the variation of the function φ. There are several ways to define the variation. The one we shall use was discovered by the Polish mathematician Stefan Banach.

DEFINITION 5.1   *The (Banach) variation of a function $\varphi : \mathbf{R}^m \to \mathbf{R}^n$ on a set $E \subset \mathbf{R}^m$ is the number*

$$V(\varphi; E) = \sup \Sigma |\varphi(E_j)|_m, \tag{1}$$

the upper bound being taken over all disjoint sequences $\{E_j\}$ of measurable subsets of E.

THEOREM 5.2   *If $\varphi : \mathbf{R}^m \to \mathbf{R}^n$ is absolutely continuous on the measurable set E, then*

$$V(\varphi; E) = \text{area } \varphi = \int N(E; y) \, d\alpha_m.$$

Proof   Recall that $N(E; y)$ is the number of points $x \in E$ with $\varphi(x) = y$ and that the second equality is just the definition of area that was given in Section 2.

It is obvious that if $\{E_j\}$ is any disjoint sequence of measurable subsets of E, then

$$N(E; y) \geq \Sigma N(E_j; y) \qquad \text{and} \qquad \int N(E_j; y)\, d\alpha_m \geq |\varphi(E_j)|_m.$$

Since any nonnegative series can be integrated term by term, this gives

$$\int N(E; y)\, d\alpha_m \geq \Sigma |\varphi(E_j)|_m;$$

hence

$$\text{area } \varphi \geq V(\varphi; E).$$

[There is one minor point left to be shown—that the multiplicity function $N(E; y)$ is α_m measurable.   This will be taken care of in the second half of the proof.]

In the second half of the proof we shall make use of a construction that will be needed again in Section 6.   For each positive integer k, let $\mathfrak{F}_k$ be a disjoint sequence of measurable subsets of E of diameter $<1/k$ that covers almost all of E.   (Here $\mathfrak{F}_k$ can be any such sequence, but in Section 6 it will be a special one.)   With the family $\mathfrak{F}_k$ we can form another multiplicity function $N_k(E; y)$ that approximates the initial one by letting $N_k(E; y)$ be the number of sets $F \in \mathfrak{F}_k$ such that $y \in \varphi(F)$.   It is easy to see that

$$N_k(E; y) \leq N(E; y) \qquad \text{and} \qquad N_k(E; y) \to N(E; y) \text{ a.e. } \alpha_m. \qquad (2)$$

The first part is self-evident.   As for the second part, if $\mathfrak{F}_k$ covers all of E except for a set M_k of measure 0 and $M = \cup M_k$, then the convergence takes place except on the set $\varphi(M)$ which has α_m measure 0 by the absolute continuity.

There is another way to look at $N_k(E; y)$.   If c_F denotes the characteristic function of the set $\varphi(F)$, then

$$N_k(E; y) = \sum_{F \in \mathfrak{F}_k} c_F(y), \qquad (3)$$

for there is a 1 in the sum on the right each time the point y belongs to the set $\varphi(F)$.   This formula shows that the $N_k(E; y)$ are α_m measurable, and formula (2) shows that $N(E; y)$ is α_m measurable.   Indeed, F is Lebesgue measurable; so by Theorem 4.4, $\varphi(F)$ is α_m measurable.

The conditions in formula (2) guarantee that we can integrate term by term.   Indeed, if $N(E; y)$ is integrable, then the dominated convergence theorem does the job.   If it is not, then Fatou's lemma gives

$$\infty = \int N(E; y)\, d\alpha_m \leq \liminf \int N_k(E; y)\, d\alpha_m.$$

Integrating term by term in (2) and then using (3) we get

$$\int N(E;y)\,d\alpha_m = \lim_{k\to\infty}\int N_k(E;y)\,d\alpha_m = \lim_{k\to\infty}\sum_{F\in\mathcal{F}_k}|\varphi(F|_m)|.$$

Each sum on the right is at most $V(\varphi;E)$, so we have

$$\int N(E;y)\,d\alpha_m \le V(\varphi;E),$$

and the theorem is proved.

Exercise 1 A function $\varphi:\mathbf{R}^1\to\mathbf{R}^1$ on a closed interval $I=[a,b]$ has finite variation on I if and only if it is the difference of two increasing functions. [*Hint:* For one of the two take the function $v(x)=V(\varphi;[a,x])$.]

Exercise 2 Let $\varphi:\mathbf{R}^1\to\mathbf{R}^1$ be absolutely continuous and have finite variation on an interval $I=[a,b]$. Show that for each $\epsilon>0$ there is a $\delta>0$ such that if $E\subset I$ and $|E|<\delta$, then $V(\varphi;E)<\epsilon$. (*Hint:* Use Exercise 1 and then Exercise 6, Section 3 of Chapter 14. The property given here is the classical property of absolute continuity.)

6   THE JACOBIAN FORMULA FOR SURFACE AREA

Now we are ready for the main theorem.

THEOREM 6.1 *If $\varphi:\mathbf{R}^m\to\mathbf{R}^n$, $m\le n$, is absolutely continuous on the measurable set E and of class C^1 at almost every point, then*

$$area\ \varphi = \int N(E;y)\,d\alpha_m = \int_E J_\varphi(x)\,dx. \tag{1}$$

Proof The first equality is the definition of the area, so it is the second one that has to be proved. Note first that if E is replaced by any subset D such that $|E-D|_m=0$, then both sides of (1) remain unchanged. It is obvious that the right side remains unchanged. As for the left side, we have $N(D;y)=N(E;y)$ at each point y not in the set $\varphi(E-D)$, which has α_m measure 0 by the absolute continuity. Thus, $N(D;y)=N(E;y)$ a.e. α_m, so the left side remains unchanged too. Now we shall replace E by the subset on which φ is of class C^1 and suppose from now on that φ is of class C^1 at each point of E. In this case Theorem 5.3 of Chapter 10 shows that each point of E is the center of a ball on which φ is differentiable and Lipschitzian. If Ω is the union of these balls, then Ω is an open set containing E, and φ is differentiable at each point of Ω and is locally Lipschitzian on Ω.

Now we shall prove the theorem when φ is one to one on Ω. (In a moment we shall apply this result to various subsets of Ω on which φ is actually one to one.) If φ is one to one on Ω, then the measure

$$\nu(E) = |\varphi(E)|_m$$

is an absolutely continuous regular Borel measure on Ω. The fact that φ is locally Lipschitzian (hence Lipschitzian on compact sets) gives that compact sets have finite measure and that ν is absolutely continuous. Theorem 4.4 shows that each set is contained in a G_δ of the same measure. Theorem 3.4 of the last chapter gives the formula

$$|\varphi(E)|_m = \int_E D\nu \, dx,$$

and Theorem 3.2 shows that $D\nu(x) = J_\varphi(x)$ at each point $x \in E$. This proves the theorem when φ is one to one on Ω.

Next we shall show that if $M = \{x \in E : J_\varphi(x) = 0\}$, then $|\varphi(M)|_m = 0$, which will prove the theorem for the set M, for it will show that $N(M; y) = 0$ a.e. α_m. In doing this we can assume that M is bounded, for if the assertion is true for each bounded part of M, then it is true for M itself. Choose an open G with $M \subset G \subset \Omega$ and $|G|_m < \infty$, and let $\epsilon > 0$ be given. Let $\mathfrak{F}$ be the family of balls that are contained in G and satisfy $|\varphi(B)|_m < \epsilon|B|_m$. This family covers M in the sense of Vitali; so there is a disjoint sequence $\{B_j\}$ in $\mathfrak{F}$ that covers almost all of M, and we have

$$|\varphi(M)|_m \leq \Sigma|\varphi(B_j)|_m \leq \epsilon\Sigma|B_j|_m \leq \epsilon|G|_m.$$

Since $\epsilon > 0$ is arbitrary, it follows that $|\varphi(M)|_m = 0$.

What we have just done shows that $N(M; y) = 0$ a.e. α_m. Therefore, neither side of formula (1) changes if we replace the set E by the set $E - M$, and we can assume from now on that $J_\varphi(x) \neq 0$ for each point $x \in E$.

Fix a positive integer k and consider the family of balls B such that $B \subset \Omega$, $\delta(B) < 1/k$, and φ is one to one on B. Since $J_\varphi(x) \neq 0$ for each $x \in E$, this family covers E in the sense of Vitali, and we can choose a disjoint countable subfamily $\mathfrak{F}'_k$ that covers almost all of E. We shall apply the results of the last section to the family $\mathfrak{F}_k$ of sets of the form $B \cap E$, where $B \in \mathfrak{F}'_k$. What has been done in the one-to-one case (applied to $\Omega = B$) shows that for each $F \in \mathfrak{F}_k$ we have

$$\int c_F \, d\alpha_m = |\varphi(F)|_m = \int_F J_\varphi(x) \, dx.$$

Summing on the $F \in \mathfrak{F}_k$ and using formula (3) of the last section, we get

$$\int N_k(E; y) \, d\alpha_m = \int_E J_\varphi(x) \, dx.$$

Letting $k \to \infty$ and using formula (2) of the last section, we get the required formula (1). [It is not really necessary to let $k \to \infty$ here. The fact that φ is one to one on each $F \in \mathfrak{F}_k$ implies that $N(E; y) = N_k(E; y)$ a.e. α_m.]

The same argument leads to a Jacobian formula for surface integrals. Note that when $m = n$ and φ is one to one it is just the old formula for changing variables.

THEOREM 6.2

If $\varphi : \mathbf{R}^m \to \mathbf{R}^n$, $m \leq n$, is absolutely continuous on the measurable set E and of class C^1 at almost every point, then for every nonnegative α_m measurable function f,

$$\int f(y) N(E; y) \, d\alpha_m = \int_E f \circ \varphi J_\varphi \, dx.$$

Proof

In the same way as before we discard the points where φ is not C^1 and then find an open set $\Omega \supset E$ on which φ is locally Lipschitzian. Also in the same way as before we discard the points where $J_\varphi = 0$ and then assume that $J_\varphi \neq 0$ everywhere on E. Again this leads us to treat first the case where φ is one to one on Ω, and it is only in this part that there is a slight addition to the previous argument.

We have seen that the measure $\nu(E) = |\varphi(E)|_m$ is an absolutely continuous regular Borel measure on Ω, so by Theorem 3.4 of Chapter 14 we have

$$\int g \, d\nu = \int g \, D\nu \, dx$$

for every nonnegative ν measurable function g. What we have to establish is that if f is nonnegative and α_m measurable, then $g = f \circ \varphi$ is ν measurable and

$$\int f \, d\alpha_m = \int f \circ \varphi \, d\nu. \tag{2}$$

When we apply this formula and the last one to the function $f(y) N(E; y)$ we get exactly the theorem, for $N(E; y)$ is the characteristic function of the set $\varphi(E)$ and $D\nu = J_\varphi$ on E.

It is evident that if the set $\varphi(E)$ is α_m measurable, then the set E is ν measurable. This shows that if f is α_m measurable, then $f \circ \varphi$ is ν measurable. To get formula (2) take first the case where f is the characteristic function of an α_m measurable set F. Then $f \circ \varphi$ is the characteristic function of the set $E = \varphi^{-1}(F)$, and formula (2) is just the definition of ν. The general case results at once by taking first linear combinations of characteristic functions to get all simple functions, and then increasing limits of simple functions.

Now the proof is finished as before. What we have just proved shows that for each set F in the family $\mathcal{F}_k$ we have

$$\int f c_F \, d\alpha_m = \int_F f \circ \varphi J_\varphi \, dx,$$

and summation over the $F \in \mathcal{F}_k$ gives

$$\int f(y) N_k(E; y) \, d\alpha_m = \int_E f \circ \varphi J_\varphi \, dx.$$

Let us review Sard's theorem briefly now that we have plenty of information about surface area. Let $\mathbf{N}$ and $\mathbf{M}$ be smooth manifolds of dimensions n and m, and let $f : \mathbf{N} \to \mathbf{M}$ be of class C^1 on $\mathbf{N}$. The differential $df(a)$ at a point $a \in \mathbf{N}$ is a linear transformation from the tangent space $T_a(\mathbf{N})$ into the tangent space $T_b(\mathbf{M})$, $b = f(a)$. The point a is a regular point if $df(a)$ has the maximum possible rank, which means that $df(a)$ is one to one if $n \leq m$ and that $df(a)$ is onto if $n \geq m$. It is a critical point otherwise. The point $y \in f(\mathbf{N})$ is a regular value of f if every point of the set

$$\mathbf{N}_y = \{x \in \mathbf{N} : f(x) = y\}$$

is a regular point.

THEOREM 6.3

(*Sard's Theorem*) If $f : \mathbf{N} \to \mathbf{M}$ is of sufficiently high class C^r, then almost every value of f is a regular value, where almost every refers of course to the area measure α_m on $\mathbf{M}$.

Exercise 1 Prove the theorem. (*Hint:* This general case reduces immediately to the original one by taking local parametric representations on $\mathbf{N}$ and $\mathbf{M}$. All you have to know is the following easy exercise.)

Exercise 2 Let E be a subset of the smooth m-dimensional manifold $\mathbf{M}$. The following are equivalent:

(a) $|E|_m = 0$.
(b) For every local parametric representation φ, $|\varphi^{-1}(E)|_m = 0$.
(c) For each point $a \in \mathbf{M}$ there is some local parametric representation φ at a such that $|\varphi^{-1}(E)|_m = 0$.

Exercise 3 We have mentioned that Sard's theorem is interesting only when $n \geq m$. If $n < m$, then Theorem 6.1 gives a much better result.

7 EXAMPLES

Consider the length of a path $\varphi : \mathbf{R}^1 \to \mathbf{R}^n$ on an interval $[a, b]$. The Jacobi matrix has the coordinates of φ' arranged in a column. Its adjoint has the coordinates of φ' arranged in a row.   The product has just the single entry $|\varphi'|^2$. Hence $J_\varphi = |\varphi'|$ and the result of Theorem 6.1 is that

$$\text{arc length} = \int_a^b |\varphi'(t)|\, dt, \tag{1}$$

provided, of course, that φ is absolutely continuous and of class C^1 a.e. on $[a, b]$.

Remark   If φ has finite variation on $[a, b]$, then so does each coordinate function.   Hence each coordinate function is the difference of two increasing functions.   Therefore, each coordinate function, and, consequently, φ itself, is differentiable a.e. We have seen (Cantor function) that the arc length is not always given by formula (1).   It is given by formula (1) if φ is absolutely continuous as well as of finite variation, i.e., if φ is absolutely continuous in the classical sense.   In the case of paths the hypothesis that φ is of class C^1 a.e. can be dropped.

As a second example, consider a surface given by an equation

$$y_n = F(y'), \qquad y' \in E \subset \mathbf{R}^{n-1}. \tag{2}$$

The parametric surface is the function $\varphi : \mathbf{R}^{n-1} \to \mathbf{R}^n$ given by

$$\varphi(x) = (x, F(x)), \qquad x \in E \subset \mathbf{R}^{n-1}. \tag{3}$$

At a point x where F is of class C^1 the Jacobi matrix is

$$\frac{\partial \varphi}{\partial x} = \begin{pmatrix} 1 & 0 & \cdots & 0 \\ 0 & 1 & \cdots & 0 \\ \cdots & \cdots & \cdots & \cdots \\ 0 & 0 & \cdots & 1 \\ \dfrac{\partial F}{\partial x_1} & \dfrac{\partial F}{\partial x_2} & & \dfrac{\partial F}{\partial x_{n-1}} \end{pmatrix}$$

The product $(\partial\varphi/\partial x)^*(\partial\varphi/\partial x)$ is the matrix $\{a_{ij}\}$ with

$$a_{ii} = 1 + \left(\frac{\partial F}{\partial x_i}\right)^2, \quad a_{ij} = \left(\frac{\partial F}{\partial x_i}\right)\left(\frac{\partial F}{\partial x_j}\right) \qquad \text{if } i \neq j.$$

This turns out to be a case where we can calculate the Jacobian very nicely by finding some eigenvalues.   To shorten the notation let $a_i = \partial F/\partial x_i$, and let $a = (a_1, \ldots, a_{n-1})$.   Let $T : \mathbf{R}^{n-1} \to \mathbf{R}^{n-1}$ be the linear transformation defined by

$$Tx = x + (x, a)a.$$

We have $\langle Te_i, e_j \rangle = \langle e_i, e_j \rangle + a_i a_j$, so the matrix of T is exactly the one above, that is, $T = d\varphi^* d\varphi$.

The eigenvalues of T are almost in plain sight. If $x \perp a$, then $Tx = x$, so x is an eigenvector with eigenvalue 1, while $Ta = (1 + |a|^2)a$; so a itself is an eigenvector with eigenvalue $1 + |a|^2$. Thus, the product of the eigenvalues is $1 + |a|^2$, and we have

$$J_\varphi = \sqrt{1 + |a|^2} = \sqrt{1 + |\nabla F|^2}. \tag{4}$$

Exercise 1 Is everything all right if $\nabla F = 0$? Note that in this case a is not an eigenvector.

Theorem 6.1 gives

$$\text{area } \varphi = \int_E \sqrt{1 + |\nabla F|^2} \, dx \tag{5}$$

provided, of course, that φ is absolutely continuous and of class C^1 at almost every point.

Consider the unit sphere in $\mathbf{R}^n$, the top half of which has the equation

$$y_n = \sqrt{1 - |y'|^2}, \qquad |y'| \leq 1.$$

In this case $F(x) = \sqrt{1 - |x|^2}$ on $|x| \leq 1$,

$$\frac{\partial F}{\partial x_i} = -x_i(1 - |x|^2)^{-1/2}$$

and

$$|\nabla F|^2 = \frac{|x|^2}{1 - |x|^2};$$

so

$$J_\varphi = \frac{1}{\sqrt{1 - |x|^2}}.$$

Consequently,

$$|S|_{n-1} = 2 \int_{B^{n-1}} (1 - |x|^2)^{-1/2} \, dx, \tag{6}$$

where S is the unit sphere in $\mathbf{R}^n$, and B^{n-1} is the unit ball in $\mathbf{R}^{n-1}$. (Usually the unit sphere in $\mathbf{R}^n$ is written S^{n-1} to indicate that it is a sphere and that its dimension as a surface is $n - 1$.)

Exercise 2 Calculate the integral in (6) by using the formulas in Section 9 of Chapter 13 and the formulas for the gamma function.

Exercise 3 Strictly speaking, the integral in (6) gives the area of the sphere with the section through $x_n = 0$ removed. Show that the area of this section is 0.

Exercise 4 Show that formula (4) holds by calculating determinants instead of eigenvalues.

8   POLAR COORDINATES

Each point $x \neq 0$ in $\mathbf{R}^n$ can be expressed uniquely as $x = r\theta$, where r is a positive real number and θ is a point on the unit sphere S^{n-1}. The pair $(r, \theta) \in \mathbf{R}^1_+ \times S^{n-1}$ is called the polar coordinates of x. ($\mathbf{R}^1_+$ is the set of positive real numbers.) This representation suggests the possibility of expressing an integral over $\mathbf{R}^n$ as an integral with respect to the product measure $dr\, d\alpha_{n-1}$. Such an expression is particularly convenient when the integrand has some spherical symmetry.

THEOREM 8.1   *If $f: \mathbf{R}^n \to \mathbf{R}^1$ is nonnegative and measurable, then*

$$\int f(y)\, dy = \int_{S^{n-1}} \int_0^\infty f(r\theta) r^{n-1}\, dr\, d\alpha_{n-1}. \tag{1}$$

In other words, y should be replaced by $r\theta$ and dy by $r^{n-1}\, dr\, d\alpha_{n-1}$. Note that while α_{n-1} is not σ-finite on $\mathbf{R}^n$, its restriction to S^{n-1} is finite; so Fubini's theorem is applicable here and shows that the integral on the right can be written in either order.

Exercise 1   Use the results of the last section to show that if $E \subset S^{n-1}$, then $\alpha_{n-1}(rE) = r^{n-1}\alpha_{n-1}(E)$, and then provide a heuristic argument for the fact that dy should be replaced by $r^{n-1}\, dr\, d\alpha_{n-1}$.

Proof   In order to get a real proof (as opposed to the heuristic one of the exercise), we shall first use Theorem 6.2 to transfer the surface integral on the right to an integral on $\mathbf{R}^{n-1}$. Then the whole integral on the right will become an integral on $\mathbf{R}^n$ in which we can change variables by Theorem 4.5 of Chapter 14. In order to transfer the surface integral to $\mathbf{R}^{n-1}$, we shall want to use just the top half of the sphere, $S^{n-1}_+ = \{\theta : \theta \in S^{n-1}, \theta_n > 0\}$, rather than the whole thing. The reason is that S^{n-1}_+ is the surface $\varphi : \mathbf{R}^{n-1} \to \mathbf{R}^n$ given by

$$\varphi(x) = (x, \sqrt{1 - |x|^2}), \qquad x \in B^{n-1}$$

which we have studied already in Section 7. Then the formula we shall prove is that

$$\int_{\mathbf{R}^n_+} f(y)\, dy = \int_{S^{n-1}_+} \int_0^\infty f(r\theta) r^{n-1}\, dr\, d\alpha_{n-1}. \tag{2}$$

The same formula holds for the bottom half, and the two together give (1). The function φ is clearly one to one and of class C^1 everywhere on

B^{n-1}, so Theorem 6.2 gives

$$\int_{S_+^{n-1}} \int_0^\infty f(r\theta) r^{n-1} \, dr \, d\alpha_{n-1}$$
$$= \int_{B^{n-1}} \int_0^\infty f(rx, r\sqrt{1-|x|^2}) \, r^{n-1} J_\varphi \, dr \, dx. \quad (3)$$

From the last section we have

$$J_\varphi = \frac{1}{\sqrt{1-|x|^2}}. \quad (4)$$

Now, in the integral $\int_{\mathbf{R}_+^n} f(y) \, dy$, make the change of variable $y = \psi(x, r)$, with

$$\psi(x, r) = (rx, r\sqrt{1-|x|^2}), \qquad x \in B^{n-1}, \quad 0 < r < \infty.$$

According to Theorem 4.5 of Chapter 14, we have

$$\int_{\mathbf{R}_+^n} f(y) \, dy = \int_{B^{n-1}} \int_0^\infty f(rx, r\sqrt{1-|x|^2}) \, J_\psi(x, r) \, dr \, dx \quad (5)$$

Therefore, what remains is to show that

$$J_\psi(x, r) = \frac{r^{n-1}}{\sqrt{1-|x|^2}}. \quad (6)$$

Let M be the Jacobi matrix, and to shorten the notation let $D = \sqrt{1-|x|^2}$. Then

$$M = \begin{pmatrix} r & 0 & \cdots & 0 & x_1 \\ 0 & r & \cdots & 0 & x_2 \\ \cdots\cdots\cdots\cdots\cdots\cdots\cdots \\ 0 & 0 & \cdots & r & x_{n-1} \\ -\dfrac{rx_1}{D} & -\dfrac{rx_2}{D} & & -\dfrac{rx_{n-1}}{D} & D \end{pmatrix}.$$

If you know something about determinants, you can establish formula (6) as follows: Multiply the first row by x_1/D and add to the last row. Then multiply the second row by x_2/D and add to the last row, and so on. The result is a matrix M' with the same determinant, and M' looks like this:

$$M' = \begin{pmatrix} r & 0 & 0 & x_1 \\ 0 & r & 0 & x_2 \\ 0 & 0 & r & x_3 \\ 0 & 0 & 0 & \dfrac{1}{D} \end{pmatrix},$$

for which the determinant is clearly r^{n-1}/D, as required.

However, the Jacobian can be obtained by the same trick as in Section 7 and without any recourse to determinants, for $M^*M = \{a_{ij}\}$, where

$$a_{ij} = r^2\delta_{ij} + \frac{r^2 x_i x_j}{D^2} \qquad \text{if } i < n \text{ and } j < n,$$

$$a_{in} = a_{ni} = 0 \qquad \text{if } i < n,$$

$$a_{nn} = 1,$$

$$\delta_{ij} = \begin{cases} 1 & \text{if } i = j, \\ 0 & \text{if } i \neq j. \end{cases}$$

If we set $b = (x_1/D, \ldots, x_{n-1}/D, 0)$, and define $Ty = y + (y, b)b$, and S by $Se_i = r^2 e_i$ if $i < n$ and $Se_n = e_n$, then it is immediately checked that M^*M is the matrix of ST. Now, $J_S = r^{2(n-1)}$, since S has $n - 1$ eigenvectors with eigenvalue r^2 and one with eigenvalue 1; and

$$J_T = 1 + |b|^2 = 1 + \frac{|X|^2}{D^2} = \frac{1}{D^2}$$

since T (as in Section 7) has one eigenvector, which is b, with eigenvalue $1 + |b|^2$ and $n - 1$ with eigenvalue 1. Thus, we get

$$J_\psi^2 = J_{M \cdot M} = J_S J_T = \left(\frac{r^{n-1}}{D}\right)^2.$$

This establishes formula (6) and, therefore, Theorem 8.1—except for one point that we have let slip by, and shall continue to let slip by unless you want to try the following exercise:

Exercise 2   The function f on $\mathbf{R}^n$ is measurable if and only if the function $g(r, \theta) = f(r\theta)$ is measurable on $\mathbf{R}_+^1 \times S^{n-1}$ with respect to the measure $dr \, d\alpha_{n-1}$.

Ordinarily, we shall write $d\theta$ for the restriction of α_{n-1} to S^{n-1}. In this case the formula for integration in polar coordinates becomes

$$\int_{\mathbf{R}^n} f(y) \, dy = \int_{S^{n-1}} \int_0^\infty f(r\theta) r^{n-1} \, dr \, d\theta. \tag{7}$$

In the case $n = 2$ there is some ambiguity in the usual notations, which may be perplexing at first, but should not cause real confusion. As a curve (i.e., one-dimensional surface), S^1 is the function $\varphi : \mathbf{R}^1 \to \mathbf{R}^2$ given by

$$\varphi(t) = (\cos t, \sin t), \qquad 0 \leq t < 2\pi.$$

As we have seen in Section 7, $J_\varphi = |\varphi'| = 1$. Therefore, when $\theta \in S^1$ and $t \in [0, 2\pi)$ are related by $\theta = \varphi(t)$, we have $d\theta = dt$. Often θ and t are simply identified, and the polar coordinates of a point in the plane are considered interchangeably as a pair $(r, \theta) \in \mathbf{R}_+^1 \times S^1$, or as a pair (r, θ) $(r, \theta) \in \mathbf{R}_+^1 \times$

$[0, 2\pi)$.   Since $d\theta = dt$, this confusion does not cause difficulty.   The polar coordinates described in Section 9 of Chapter 13 are the pair $(r, \theta) \in \mathbf{R}_+^1 \times [0, 2\pi)$.

Exercise 3   If the volume of the unit ball in $\mathbf{R}^n$ is calculated by Theorem 8.1, the result is immediately seen to be

$$|B^n|_n = \frac{|S^{n-1}|_{n-1}}{n}. \tag{8}$$

Along with formula (14), Section 9 of Chapter 13, this gives

$$|S^{n-1}|_{n-1} = \frac{2\pi^{n/2}}{\Gamma(n/2)}. \tag{9}$$

16 | The Brouwer Degree

1 INTRODUCTION

The degree of a continuous function $f: S^{n-1} \to S^{n-1}$ is an integer that measures the number of times the parametric surface f "wraps around" the unit sphere S^{n-1}. The way to define this integer and the fact that it reflects fundamental properties of the function f itself were discovered about 1900 by the Dutch mathematician L. E. J. Brouwer.

The first idea for getting the number of times that f wraps around S^{n-1} would be simply to pick a point y and count the number of points x with $f(x) = y$. This would give just the old multiplicity function $N(S^{n-1}; y)$ of Chapter 15. Consider the function $f: S^1 \to S^1$ representing the path that starts at the top of the circle and goes counterclockwise to the bottom, then returns

L. E. J Brouwer

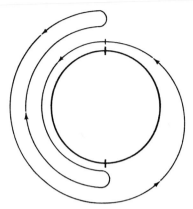

Figure 1

clockwise to the top, and finally goes counterclockwise all the way around (Figure 1). If y is on the right half of the circle, then $N(S^1; y)$ is 1; if y is on the left half, then $N(S^1; y)$ is 3, while at the top and bottom it is 2. On the other hand, it is clear that the path wraps around once. This suggests that we cannot simply count the number of points x with $f(x) = y$, but rather that a given point x should be counted $+1$ times if we are going in a counterclockwise direction at that point, and -1 times if we are going in a clockwise direction. If the counting is done this way and the result is called $\deg(f; y)$, then we have $\deg(f; y) = 1$ at every point y of the circle, except the two points at the top and bottom where we do not know quite what to think.

Exercise 1 Draw several more paths $f: S^1 \to S^1$ and notice that the function $\deg(f; y)$ takes the same value at all points y except a few where you do not know what to think. This constant value is the number $\deg f$ that we are looking for.

Exercise 2 Let $f: S^1 \to S^1$ be of class C^1. Try to find an analytic way to express the fact that at a point $x \in S^1$ the motion is counterclockwise.

In Section 2 we shall carry out the details of this construction for functions $f: S^{n-1} \to S^{n-1}$ that are highly differentiable. In the following section we shall extend the results to continuous functions by approximating them with C^∞ functions. In the case of an arbitrary continuous function there is no way to do a direct counting procedure. It may well be, for example, that for each point y there are infinitely many points x with $f(x) = y$.

2 THE DEGREE FOR C^∞ FUNCTIONS

This section contains the construction of the degree for functions $f: S^{n-1} \to S^{n-1}$ of class C^∞. [The construction is the same for functions of class C^r with $r > n(n+1)/2$, but we can obtain the results for continuous functions by using C^∞ approximations, so there is no point in worrying about class C^r.]

The number $\deg(f; y)$ is defined as follows: First set

$$\tilde{f}(x) = |x| f\left(\frac{x}{|x|}\right) \qquad \text{for } x \neq 0; \tag{1}$$

then define

$$\deg(f; y) = \begin{cases} \displaystyle\sum_{\tilde{f}(x) = y} \text{sign det } d\tilde{f}(x) & \text{if } y \text{ is a regular value of } f \\ 0 & \text{otherwise.} \end{cases} \tag{2}$$

The number sign det $d\tilde{f}(x)$ is (by definition) 1 if det $d\tilde{f}(x) > 0$ and -1 if det $d\tilde{f}(x) < 0$. This is the expression that tells whether to count a given point x plus 1 times or minus 1 times.

Exercise 1   Show that sign det $d\tilde{f}(x)$ answers the question raised in Exercise 2 of the last section.

Note that if y is a regular value of f, then the set $\{x : f(x) = y\}$ is a compact smooth manifold of dimension 0, hence a finite set of points; so the sum in (2) is just a finite sum of plus and minus 1's.

Exercise 2   Show that $\tilde{f}$ is of class C^∞ on $\mathbf{R}^n - \{0\}$ and that for each $y \in S^{n-1}$ the sets $\{x; \tilde{f}(x) = y\}$ and $\{x : f(x) = y\}$ are the same.

Exercise 3   Show that a point $y \in S^{n-1}$ is a regular value of $\tilde{f}$ if and only if it is a regular value of f.

Exercise 4   If K is the set of critical points of f and $L = f(K)$ is the set of critical values, then L is a compact set of α_{n-1} measure 0, and $\deg(f; y)$ is locally constant on $S^{n-1} - L$; that is, each point of $S^{n-1} - L$ has a neighborhood G in $\mathbf{R}^n$ that does not meet L and such that $\deg(f; y)$ is constant on $G \cap S^{n-1}$.

The main fact to be proved is that $\deg(f; y)$ is constant on $S^{n-1} - L$, not just locally constant. Since L is small (has α_{n-1} measure 0), what pops into mind is a connectedness argument—but $S^{n-1} - L$ need not be connected.

Exercise 5   Give an example.

The proof of the fact that $\deg(f; y)$ is constant on $S^{n-1} - L$ is based on the following theorem, which also contains other important information.

THEOREM 2.1

Let $F: S^1 \times S^{n-1} \to S^{n-1}$ be of class C^∞, and let

$$f(x) = F(t_0, x) \qquad and \qquad g(x) = F(t_1, x).$$

If $y \in S^{n-1}$ is not a critical value of either f or g, then $\deg(f; y) = \deg(g; y)$.

[Think of the theorem as saying that if f can be "deformed" into g—the functions $f_t(x) = F(t, x)$ effecting the deformation—then $\deg(f; y) = \deg(g; y)$.]

Proof

In the first place it can be assumed that y is not a critical value of any of the three f, g, or F. Indeed, by Exercise 4 there is a neighborhood G of y in $\mathbf{R}^n$ containing no critical value of f and no critical value of g and such that $\deg(f; z)$ and $\deg(g; z)$ are both constant on $G \cap S^{n-1}$. The set L of critical values of F has α_{n-1} measure 0 by Sard's theorem, so in particular it cannot fill up the whole set $G \cap S^{n-1}$. If we choose a point $z \in G - L$, and if we know the theorem for such z, then we have $\deg(f; y) = \deg(f; z) = \deg(g; z) = \deg(g; y)$. Henceforth, we assume that y is not a critical value of f, g, or F. Furthermore, we assume that y is actually a value of F; for if it is not, then it is not a value of f, nor of g, in which case the definition gives $\deg(f; y) = 0 = \deg(g; y)$. Thus, y is a regular value of F.

Since y is a regular value of F, the set

$$\mathbf{M}_y = \{(t, x) : F(t, x) = y\}$$

is a compact smooth manifold of dimension 1. It is plain that

$$\deg(f; y) = \sum_{(t_0, x) \in \mathbf{M}_y} \text{sign det } \frac{\partial \tilde{F}}{\partial x},$$

$$\deg(g; y) = \sum_{(t_1, x) \in \mathbf{M}_y} \text{sign det } \frac{\partial \tilde{F}}{\partial x},$$

where

$$\tilde{F}(t, x) = |x| \, F\left(t, \frac{x}{|x|}\right), \qquad x \neq 0.$$

What we shall show is that if $\mathbf{M}$ is any component of $\mathbf{M}_y$, then

$$\sum_{(t_0, x) \in \mathbf{M}} \text{sign det } \frac{\partial \tilde{F}}{\partial x} = \sum_{(t_1, x) \in \mathbf{M}} \text{sign det } \frac{\partial \tilde{F}}{\partial x} \tag{3}$$

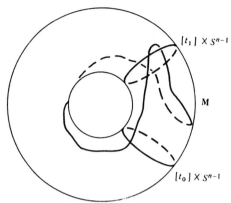

$[t_1] \times S^{n-1}$

M

$[t_0] \times S^{n-1}$

Figure 2

This will prove the theorem because the compact smooth manifold $\mathbf{M}_y$ is the disjoint union of its components and the components are finite in number.

The advantage in dealing with the component $\mathbf{M}$ is that it is a compact connected smooth one-dimensional manifold, so it has a parametric representation by arc length. That is, there is a C^∞ function $\varphi : \mathbf{R}^1 \to \mathbf{M}$ such that $\varphi(\mathbf{R}^1) = \mathbf{M}$, $\varphi(s + \alpha) = \varphi(s)$ for all s, $|\varphi'| = 1$, and φ is one to one on the interval $[0, \alpha)$. (Theorem 3.14 of Chapter 11.) In order to make use of φ we shall have to find its connection with the numbers sign det $\partial \tilde{F} / \partial x$. When $n = 2$, the picture looks as shown in Figure 2.

In doing the calculations we shall be just a little bit illogical and shall identify the circle S^1 with the interval $[0, 2\pi]$ so that we can speak of $\partial \tilde{F} / \partial t$, of the idea that $t_1 > t_0$, and so on. To be perfectly correct about this we should put $t(r) = (\cos r, \sin r)$ and replace $\tilde{F}(t, x)$ by $\tilde{F}(t(r), x)$, but this complicated notation introduces more confusion than it eliminates. If S^1 is identified with $[0, 2\pi]$, then automatically $S^1 \times S^{n-1}$ is identified with a subset of $\mathbf{R}^{n+1}$. We shall use the subscript 0 to indicate the first coordinate. With this notation the connection between φ and sign det $\partial \tilde{F} / \partial x$ is that

$$\text{sign } \varphi_0'(s) = \text{sign det } \frac{\partial \tilde{F}}{\partial x}, \qquad (4)$$

the right side being calculated at the point $\varphi(s)$.

To prove formula (4) we calculate the tangent vector to $\mathbf{M}$ at $\varphi(s)$ in two different ways. On the one hand, the tangent vector is just $\varphi'(s)$. On the other, since $\mathbf{M}$ is defined by the equation $\tilde{F}(t, x) = y$, the normal is spanned by the vectors $\nabla \tilde{F}_j$, $j = 1, \ldots, n$. What we need is a vector orthogonal to each of these. Note that the Jacobi matrix of $\tilde{F}$ is the $n \times n + 1$ matrix

$$\begin{pmatrix} \nabla \tilde{F}_1 \\ \nabla \tilde{F}_2 \\ \cdot \\ \cdot \\ \cdot \\ \nabla \tilde{F}_n \end{pmatrix}.$$

Let A_j be the square matrix obtained by skipping the jth column (starting with $j = 0$), and let $a_j = (-1)^j \det A_j$. At each point of $\mathbf{M}$ the vector a is $\neq 0$ because each point of $\mathbf{M}$ is a regular point of $\tilde{F}$. Furthermore, a is orthogonal to each $\nabla \tilde{F}_j$, as can be seen by putting $\nabla \tilde{F}_j$ at the bottom of the Jacobi matrix and calculating the resulting $n + 1 \times n + 1$ determinant by Cramer's rule. The determinant is 0 because two rows of the matrix are the same. It is also the inner product of a with $\nabla \tilde{F}_j$ by Cramer's rule. Since the tangent space to $\mathbf{M}$ is one dimensional, we conclude that

$$a(\varphi(s)) = \rho(s)\varphi'(s), \qquad \rho(s) \neq 0. \tag{5}$$

Thus, ρ is either always positive or always negative, and, replacing $\varphi(s)$ by $\varphi(-s)$ if necessary, we can suppose that ρ is always positive. In this case sign $a_0(\varphi(s)) = \text{sign } \varphi_0'(s)$, while obviously $a_0 = \det \partial \tilde{F}/\partial x$. This proves formula (4), and now we will use formula (4) to prove the theorem.

If the sections $\{t_0\} \times S^{n-1}$ and $\{t_1\} \times S^{n-1}$ are removed from $S^1 \times S^{n-1}$, then what is left consists of two connected components. With the identification of S^1 with $[0, 2\pi]$ and $t_0 < t_1$, they are

$$\begin{aligned} C &= \{(t, x) : t_0 < t < t_1\} \\ D &= \{(t, x) : t < t_0 \text{ or } t > t_1\}. \end{aligned}$$

D is connected because we are really on the circle S^1 rather than the interval $[0, 2\pi]$. The points s with $\varphi_0(s) = t_0$ are the ones that contribute to $\deg(f; y)$, and the points s with $\varphi_0(s) = t_1$ are the ones that contribute to $\deg(g; y)$. At each such point we have $\varphi_0'(s) \neq 0$ because of formula (5) and the regularity of f and g. For the same reason there are only a finite number of these points in the interval $[0, \alpha)$. Consider two adjacent ones, s_0 and s_1 with $s_0 < s_1$. Since $\varphi((s_0, s_1))$ is connected, it must be contained in either C or D. For definiteness let us suppose that it is contained in C and that $\varphi_0(s_0) = t_0$. For $s > s_0$ we have $\varphi(s) \in C$; hence $\varphi_0(s) > t_0$; therefore, $\varphi_0'(s_0) > 0$. [Remember that $\varphi_0'(s_0) \neq 0$!] Now there are two cases to consider.

Case a. Suppose that $\varphi_0(s_1) = t_0$. In this case $\varphi_0'(s_1) < 0$; for if $s < s_1$, then $\varphi(s) \in C$ so that $\varphi_0(s) > t_0$. Thus, the contribution of the two terms

$$\text{sign } \varphi_0'(s_0) + \text{sign } \varphi_0'(s_1)$$

to $\deg(f; y)$ is 0.

Case b. Suppose that $\varphi_0(s_1) = t_1$. In this case $\varphi_0'(s_1) > 0$; for if $s < s_1$, then $\varphi(s) \in C$ so that $\varphi_0(s) < t_1$. Consequently, the two points s_0 and s_1 together contribute $+1$ to $\deg(f; y)$ and $+1$ to $\deg(g; y)$.

Now the theorem is proved, for it has been shown that any pair of adjacent points s_0 and s_1 with $\varphi_0(s) = t_0$ or t_1 contribute the same thing to both $\deg(f; y)$ and $\deg(g; y)$.

Exercise 6   To pair up the points in this way it is necessary to know that the number of points $s \in [0, \alpha)$, with $\varphi_0(s) = t_0$ or t_1, is even. Why is this true?

Exercise 7   In what situations would the contribution be -1 to both degrees?

THEOREM 2.2   *If neither y nor z is a critical value of f, then $\deg(f; y) = \deg(f; z)$.*

Proof   Let U_t be an orthogonal transformation on $\mathbf{R}^n$ that effects a rotation through an angle t in the plane spanned by y and z and leaves the orthogonal complement of this plane alone, and set $F(t, x) = U_t f(x)$.

Exercise 8   Write a formula for U_t to show that F is of class C^∞.

Choose t_1 so that $U_{t_1} z = y$, and set $g(x) = F(t_1, x)$. Apply Theorem 2.1 to g and to $f = U_0 f$. It is immediately checked that $\{x : f(x) = z\} = \{x : g(x) = y\}$ and that $\deg(f; z) = \deg(g; y)$. Theorem 2.1 gives $\deg(g; y) = \deg(f; y)$.

DEFINITION 2.3   *If $f : S^{n-1} \to S^{n-1}$ is of class C^∞, then $\deg f$ is the common value of $\deg(f; y)$ on the set of points y that are not critical values of f.*

THEOREM 2.4   *If $S^{n-1} \xrightarrow{f} S^{n-1} \xrightarrow{g} S^{n-1}$, then $\deg(g \circ f) = \deg g \deg f$.*

Exercise 9   Prove the theorem.

This construction of Brouwer's degree comes from an elegant little book of John Milnor called *Topology from the Differential Point of View.*

John Milnor

3   THE DEGREE FOR CONTINUOUS FUNCTIONS

The degree of a continuous function $f: S^{n-1} \to S^{n-1}$ can be obtained by approximation in the norm

$$\|u\| = \sup_{x \in S^{n-1}} |u(x)|. \tag{1}$$

The basis for this is the following lemma.

LEMMA 3.1   *If $f, g: S^{n-1} \to S^{n-1}$ are of class C^∞ and $\|f - g\| < 2$, then $\deg f = \deg g$.*

Proof   If $t = (t_1, t_2)$ is a point of S^1, set

$$G(t, x) = t_1^2 f(x) + t_2^2 g(x);$$

then set $F(t, x) = G(t, x)/|G(t, x)|$.   Note that $G(t, x)$ is never 0.   If it were, then we would have $t_1^2 f(x) = -t_2^2 g(x)$, and taking absolute values we would get $t_1^2 = t_2^2$; hence $f(x) = -g(x)$, which is impossible if $\|f - g\| < 2$.   Since G is of class C^∞ and never 0, it follows that F is of class C^∞ and we can apply Theorem 2.1.

DEFINITION 3.2   *If $f: S^{n-1} \to S^{n-1}$ is continuous, then $\deg f = \deg f_0$, where $f_0: S^{n-1} \to S^{n-1}$ is any C^∞ function with $\|f - f_0\| < 1$.*

If f_0 and g_0 are any two C^∞ functions satisfying $\|f - f_0\| < 1$ and $\|f - g_0\| < 1$, then $\|f_0 - g_0\| < 2$; the lemma shows that $\deg f_0 = \deg g_0$, so the definition does not depend on the particular C^∞ function f_0.   In order to show that there exist C^∞ functions satisfying $\|f - f_0\| < 1$, we use the Stone–Weierstrass approximation theorem as follows: Given $\epsilon > 0$ we use Stone–Weierstrass to find a C^∞ function $g_0 : S^{n-1} \to \mathbf{R}^n$ such that $\|f - g_0\| < \epsilon$, and we set $f_0(x) = g_0(x)/|g_0(x)|$. Since $|f(x)| = 1$, we have

$$\big|1 - |g_0(x)|\big| \leq \|f - g_0\| < \epsilon.$$

If $\epsilon < 1$, it follows that $g_0(x) \neq 0$ (so the definition of f_0 makes sense), and also that

$$f_0(x) - g_0(x) = \left(\frac{1}{|g_0(x)|} - 1\right) g_0(x),$$

which gives

$$|f_0(x) - g_0(x)| = \big|1 - |g_0(x)|\big|.$$

Hence, $\|f_0 - g_0\| < \epsilon$, and finally

$$\|f - f_0\| < 2\epsilon. \tag{2}$$

This shows that a continuous function from S^{n-1} to S^{n-1} can be approximated arbitrarily well by a C^∞ function from S^{n-1} to S^{n-1} in the norm (1).

THEOREM 3.3

If $f, g : S^{n-1} \to S^{n-1}$ are continuous and $\|f - g\| < 2$, then $\deg f = \deg g$.

Proof

Take $\epsilon < 2 - \|f - g\|$ and use (2) to find f_0 and g_0 of class C^∞ so that $\|f - f_0\| < \epsilon/2$ and $\|g - g_0\| < \epsilon/2$. Then $\|f_0 - g_0\| < 2$, and Lemma 3.1 gives

$$\deg f = \deg f_0 = \deg g_0 = \deg g.$$

DEFINITION 3.4

Let $f, g : S^{n-1} \to S^{n-1}$ be continuous. We write $f \sim g$ and say that f is homotopic to g if there is a continuous function $F : [t_0, t_1] \times S^{n-1} \to S^{n-1}$ with

$$f(x) = F(t_0, x) \quad and \quad g(x) = F(t_1, x).$$

(This is just our old notion of a deformation, but we no longer have to stay within the framework of C^∞ functions and manifolds.)

THEOREM 3.5

Let $f, g : S^{n-1} \to S^{n-1}$ be continuous. If $f \sim g$ then $\deg f = \deg g$.

Proof

Set $f_t(x) = F(t, x)$. Since $[t_0, t_1] \times S^{n-1}$ is compact, F is uniformly continuous. From this it follows that there is a $\delta > 0$ such that if $|t - s| < \delta$, then $\|f_t - f_s\| < 2$. This fact and Theorem 3.3 give that the function $d(t) = \deg f_t$ is continuous on the interval $[t_0, t_1]$. Since it takes only integer values, it must be constant.

THEOREM 3.6

If $S^{n-1} \xrightarrow{f} S^{n-1} \xrightarrow{g} S^{n-1}$, then $\deg(g \circ f) = \deg g \deg f$.

Proof

Choose f_0 and g_0 of class C^∞ and approximating f and g. Then we have

$$|g(f(x)) - g_0(f_0(x))| \le |g(f(x)) - g_0(f(x))| + |g_0(f(x)) - g_0(f_0(x))|.$$

Now, g_0 is of class C^∞, so it is Lipschitzian; if we write M for the Lipschitz constant, then we have

$$\|g \circ f - g_0 \circ f_0\| \le \|g - g_0\| + M\|f - f_0\|.$$

First we choose g_0 so that $\|g - g_0\| < 1$. This fixes M, and we choose f_0 so that $M\|f - f_0\| < 1$. The result is that $\|g \circ f - g_0 \circ f_0\| < 2$, so by Theorems 3.3 and 2.4 we have

$$\deg(g \circ f) = \deg(g_0 \circ f_0) = \deg g_0 \deg f_0 = \deg g \deg f.$$

Theorems 3.3, 3.5, and 3.6 allow the computation of the degree in many important cases.

Exercise 1   If $f:S^{n-1} \to S^{n-1}$ does not map S^{n-1} onto S^{n-1}, then $\deg f = 0$. [*Hint:* It is plain from the initial definition that if g is constant, then $\deg g = 0$. If a is a point outside the range of f, b the diametrically opposite point, and g the constant $g(x) = b$, then $\|f - g\| < 2$.]

Exercise 2   If $f:S^{n-1} \to S^{n-1}$ is both one to one and onto, then $\deg f = \pm 1$. (*Hint:* In this case there exists g with $g \circ f = I$, and it is plain from the initial definition that $\deg I = 1$.)

Exercise 3   If $f:S^{n-1} \to S^{n-1}$ is of class C^1, then

$$\deg f = \frac{1}{|S^{n-1}|_{n-1}} \int_{S^{n-1}} \det d\vec{f}(x) \, d\alpha_{n-1}.$$

[*Hint:* Do first the case where f is of class C^∞. In this case $\deg (f; y)$ is similar to $N(S^{n-1}; y)$, except that each point x is counted with either a $+1$ or a -1. This fact and the Jacobian formula for surface area suggest the formula

$$\int_{S^{n-1}} \deg(f; y) \, d\alpha_{n-1} = \int_{S^{n-1}} \det d\vec{f}(x) \, d\alpha_{n-1},$$

which is just what is to be proved because $\deg(f; y)$ is constant almost everywhere on S^{n-1}. Now approximate a C^1 function by C^∞ functions. (This is not a completely trivial exercise. You will have to use the ideas of Chapter 15 and notice that the Jacobian J_φ is essentially $|\det d\vec{f}|$.)]

4   SOME APPLICATIONS OF THE DEGREE

In most applications we are not presented directly with a function from S^{n-1} to S^{n-1}. One common situation is to have a function from S^{n-1} to $\mathbf{R}^n - \{0\}$. For such functions the degree is defined as follows:

DEFINITION 4.1   *If $f: S^{n-1} \to \mathbf{R}^n - \{0\}$, then* $\deg f = \deg \tilde{f}$, *where*

$$\tilde{f}(x) = \frac{f(x)}{|f(x)|}.$$

Exercise 1   If $S^{n-1} \to S^{n-1} \overset{g}{\to} \mathbf{R}^n - \{0\}$, then $\deg (g \circ f) = \deg g \deg f$.

DEFINITION 4.2   *If $f, g: S^{n-1} \to \mathbf{R}^n - \{0\}$, then we write $f \sim g$ in $\mathbf{R}^n - \{0\}$ and say that f is homotopic to g in $\mathbf{R}^n - \{0\}$, if there is a continuous function $F: [t_0, t_1] \times S^{n-1} \to \mathbf{R}^n - \{0\}$ such that*

$$f(x) = F(t_0, x) \qquad and \qquad g(x) = F(t_1, x).$$

Exercise 2   If $f \sim g$ in $\mathbf{R}^n - \{0\}$, then $\deg f = \deg g$.

Now let us calculate the degree of the restriction of a linear transformation T to S^{n-1}. T must be nonsingular, of course, for otherwise there is a point $x \in S^{n-1}$ with $Tx = 0$, and T is not a function from S^{n-1} to $\mathbf{R}^n - \{0\}$.

THEOREM 4.3   *If T is a nonsingular linear transformation from $\mathbf{R}^n$ to $\mathbf{R}^n$, then*

$$\deg T\Big|_{S^{n-1}} = \operatorname{sign} \det T.$$

Proof   Suppose first that $T = U$ is orthogonal. In this case U maps S^{n-1} into S^{n-1}, and we can go back to the initial definition of the degree in Section 2. It is plain that $\tilde{U}x = Ux$ and that for a given point y there is only one point x with $Ux = y$, so the initial definition gives the required formula directly.

Suppose next that $T = H$ is strictly positive definite. In this case we have in $\mathbf{R}^n - \{0\}$ the homotopy

$$H_t x = (1 - t)Hx + tx, \qquad 0 \leq t \leq 1,$$

which gives $H|_{S^{n-1}} \sim I|_{S^{n-1}}$ in $\mathbf{R}^n - \{0\}$; hence $\deg H|_{S^{n-1}} = 1$. On the other hand, $\operatorname{sign} \det H = 1$ because the determinant is the product of the eigenvalues, which are all positive.

In the general case we have $T = HU$ and can apply Exercise 1 and the formula det $T = $ det H det U.

THEOREM
4.4

Let $f: S^{n-1} \to S^{n-1}$ be continuous. If n is odd, then there is at least one point $x \in S^{n-1}$ with $f(x) = \pm x$.

Proof

If $f(x) \neq x$ for all $x \in S^{n-1}$, then $\|f + I\| < 2$, and Theorem 3.3 gives deg $f = $ deg $-I$, which is -1 by Theorem 4.3. If $f(x) \neq -x$ for all $x \in S^{n-1}$, then $\|f - I\| < 2$, so deg $f = $ deg $I = 1$. Not both are possible.

Let $\mathbf{M}$ be a smooth manifold of dimension m in $\mathbf{R}^n$. A tangent vector field to $\mathbf{M}$ is a continuous function $v: \mathbf{M} \to \mathbf{R}^n$ such that for each point $x \in \mathbf{M}$, $v(x)$ lies in the tangent space to $\mathbf{M}$ at x. Theorem 4.4 has the following immediate corollary.

COROLLARY
4.5

There is no nonvanishing tangent vector field to S^{n-1} if n is odd.

Exercise 3

Prove the corollary and show that there are nonvanishing tangent vector fields to S^{n-1} if n is even.

Remark 1

A very interesting and difficult question is whether it is possible to choose at each point $x \in S^{n-1}$ vectors $e_1(x), \ldots, e_{n-1}(x)$ that form a basis of the tangent space at x, and to do this so that the functions $e_j(x)$ are continuous. Corollary 4.5 shows that it certainly is not possible (even to choose one single nonzero tangent vector) unless n is even. Quite recently J. F. Adams proved the very difficult theorem that it is possible only if $n = 2, 4,$ or 8.

THEOREM
4.6

Let $F: \bar{B}(0; 1) \to \mathbf{R}^n$ be continuous, let f be its restriction to S^{n-1}, and let y be a point of $\mathbf{R}^n$. If deg $(f - y) \neq 0$, then there is a point $x \in B(0; 1)$ with $F(x) = y$.

Proof

If F does not take the value y, then we have in $\mathbf{R}^n - \{0\}$ the homotopy

$$h(t, x) = F(tx) - y, \qquad 0 \le t \le 1,$$

from the constant $F(0) - y$ to the function $f - y$; so the degree of $f - y$ is equal to the degree of the constant, which is 0.

Exercise 4

Given $f: S^{n-1} \to \mathbf{R}^n - \{0\}$, you can find $\delta > 0$ such that if $\|f - g\| < \delta$, then $f \sim g$ in $\mathbf{R}^n - \{0\}$; hence deg $f = $ deg g.

Exercise 5

Let $f: S^{n-1} \to \mathbf{R}^n$. The function $d(y) = $ deg $(f - y)$ is constant on each connected component of $\mathbf{R}^n - f(S^{n-1})$. [*Hint:* The previous exercise shows that $d(y)$ is continuous.]

The use of the degree provides a very simple proof of the inverse-function theorem and at the same time an improvement.

THEOREM 4.7

Let $f: \mathbf{R}^n \to \mathbf{R}^n$ be continuous on a neighborhood of a and differentiable at a with $df(a)$ nonsingular. Then f maps each neighborhood of a on a neighborhood of $f(a)$.

Proof

It simplifies the notation and is clearly no restriction to assume that $a = f(a) = 0$. Let $T = df(0)$, and for each $r > 0$ set

$$f_r(x) = f(rx) \quad \text{and} \quad g_r(x) = T(rx) = rTx, \qquad x \in S^{n-1}.$$

The idea is to show that if r is small, then $f_r \sim g_r$ in $\mathbf{R}^n - \{0\}$.

To see this choose $m > 0$ so that

$$|Tx| > m|x|,$$

then ϵ with $0 < \epsilon < m$, and finally r_0 so that

$$|f(x) - Tx| < \epsilon|x| \qquad \text{if } |x| < r_0.$$

Now consider the homotopy between $f_r - y$ and $g_r - y$ given by

$$\begin{aligned}
F(t, x) &= t(f_r(x) - y) + (1 - t)(g_r(x) - y) \\
&= t(f_r(x) - g_r(x)) + g_r(x) - y, \qquad 0 \le t \le 1.
\end{aligned}$$

If $r < r_0$ and $|y| < (m - \epsilon)r$, then $F(t, x) \ne 0$, for we have

$$\left| t(f_r(x) - g_r(x)) \right| < \epsilon r \qquad \text{and} \qquad |g_r(x)| \ge mr.$$

Thus, if $r < r_0$ and $|y| < (m - \epsilon)r$, then $(f_r - y) \sim (g_r - y)$ in $\mathbf{R}^n - \{0\}$; consequently, $\deg(f_r - y) = \deg(g_r - y)$.

Next we show that $\deg(g_r - y) \ne 0$. Since $|g_r(x)| \ge mr$, the ball $B(0; mr)$ is contained in $\mathbf{R}^n - g_r(S^{n-1})$; so by Exercise 5 we have $\deg(g_r - y) = \deg g_r$. And $\deg g_r \ne 0$ by Exercise 1 and Theorem 4.3.

Now the proof is finished, for Theorem 4.6 shows that

$$f(B(0; r)) \supset B(0; (m - \epsilon)r) \qquad \text{if } r < r_0. \tag{1}$$

Remark 2

Note that formula (1) [in the case where $df(0) = I$ and hence $m = 1$] is just what was needed in proving the formulas for changing variables in multiple integrals. We could now go back to these formulas and replace the assumption that f is of class C^1 almost everywhere by the assumption that f is just differentiable almost everywhere. (But we still need to assume that f is absolutely continuous, of course.) One case of particular interest is when f is Lipschitzian. It is plain that a Lipschitz function is absolutely continuous, and the important theorem of H. Rademacher says that a Lipschitz function is differentiable almost everywhere. This is done in Section 5.

Exercise 6 Theorem 4.7 gives half of the inverse-function theorem, but it does not assert that f is one to one on sufficiently small neighborhoods of a. What about this assertion?

A retraction of a metric space X on a subset A is a continuous function $R:X \to A$ such that $R(a) = a$ for each $a \in A$.

THEOREM 4.8

There is no retraction of the ball $\bar{B}(a; r)$ on the sphere $S(a; r)$.

Proof

It is enough to deal with the unit ball and sphere. To get a contradiction suppose that there is a retraction R of $\bar{B}(0; 1)$ on $S(0; 1)$. The restriction of R to $S(0; 1)$ is the identity and its degree is 1. Therefore, by Theorem 4.6 there must be a point $x \in B(0; 1)$ with $R(x) = 0$, which is impossible since all values of R lie on $S(0; 1)$.

THEOREM 4.9

*(**Brouwer Fixed-Point Theorem**)   If $F:\bar{B}(0; 1) \to \bar{B}(0; 1)$ is continuous, then there is a point $x \in \bar{B}(0; 1)$ with $F(x) = x$.*

Proof

We shall suppose that there is no fixed point and get a contradiction by producing a retraction of the ball on the sphere. Geometrically, $R(x)$ is the point where the half-line from $F(x)$ through x meets the sphere S^{n-1}. Analytically, it is defined by

$$R(x) = F(x) + t(x - F(x)), \tag{2}$$

where t is the positive number such that

$$\left|F(x) + t(x - F(x))\right| = 1. \tag{3}$$

Exercise 7 Find the positive solution of (3), put it back into (2), and deduce that R really is continuous. [*Hint:* Square both sides of (3) and write things out with inner products. You will get a quadratic equation for t.]

THEOREM 4.10

*(**Fundamental Theorem of Algebra**)   Every nonconstant complex polynomial has a complex zero.*

Proof

If $p(z) = \sum_{k=0}^{m} a_k z^k$, $a_m \neq 0$, we can divide through by a_m and consider

$$P(z) = z_m + \sum_{k=0}^{m-1} b_k z^k$$

instead. The job is to show that $P(z) = 0$ for some z. Let

$$f_r(x) = P(rx) \quad \text{and} \quad g_r(x) = (rx)^m, \qquad x \in S^1,$$

and consider the homotopy

$$F(t, x) = tf_r(x) + (1 - t)g_r(x) = t(f_r(x) - g_r(x)) + g_r(x), \quad 0 \le t \le 1.$$

The problem is to show that if r is large enough then F does not vanish. If $F(t, x) = 0$, then $f_r(x) - g_r(x) = -g_r(x)/t$; therefore, since $t \le 1$,

$$|f_r(x) - g_r(x)| \ge |g_r(x)|. \tag{4}$$

On the one hand, $|g_r(x)| = r^m$. On the other, if M is the largest of the $|b_k|$ and $r \ge 1$, then $|f_r(x) - g_r(x)| \le Mmr^{m-1}$; so if $r \ge 1$ and $r > Mm$, then (4) is obviously impossible.

What we have is that if $r \ge 1$ and $r > Mm$, then $f_r \sim g_r$ in $\mathbf{R}^2 - \{0\}$; so deg f_r = deg g_r = deg g_1, the last equality coming from Exercise 1.

Exercise 8   Go back to the initial definition of the degree and show that deg $g_1 = m$, where $g_1(x) = x^m$.

Now what we have is that if $r \ge 1$ and $r > Mm$, then deg $f_r = m \ne 0$, so Theorem 4.6 shows that there is a point z with $|z| < r$ and $P(z) = 0$.

In many of the theorems of this section it is not essential to have a true ball or sphere but suffices to have a set that is equivalent to a ball or sphere in the following sense:

DEFINITION 4.11   *The metric space X is homeomorphic to the metric space Y if there is a one to one function φ from X onto Y such that φ and φ^{-1} are both continuous. Such a φ is called a homeomorphism from X onto Y.*

Exercise 9   Any compact connected smooth one-dimensional manifold is homeomorphic to S^1.

Exercise 10   Let $\|x\| = \max|x_j|$, and let

$$\varphi(x) = \frac{\|x\|}{|x|} x, \qquad \varphi(0) = 0.$$

Show that φ is a homeomorphism of the unit cube $\bar{Q}(0; 1)$ on the unit ball $\bar{B}(0; 1)$ and of the surface of Q on S^{n-1}.

Exercise 11   Prove the Brouwer fixed-point theorem for any metric space X that is homeomorphic to the ball $\bar{B}(0; 1)$.

Exercise 12 Prove Theorem 4.4 for the surface of the cube Q. (Note that this theorem would not make sense for an arbitrary metric space homeomorphic to S^{n-1}.)

Exercise 13 Let $\mathbf{M}$ be a smooth manifold of dimensions $n-1$ and suppose that there is a homeomorphism φ of $\mathbf{M}$ on S^{n-1} that is regular at each point. Show that $\mathbf{M}$ has no nonvanishing tangent vector field if n is odd and that it does have one if n is even.

Exercise 14 Use Exercise 13 to show that there is no regular homeomorphism of the torus on the sphere S^2. (As a matter of fact, there is no homeomorphism at all. Unlike the one-dimensional case, there are infinitely many essentially different compact connected smooth two-dimensional manifolds.)

Exercise 15 Let X and Y be metric spaces that are homeomorphic to S^{n-1}, and let $\varphi:X\to S^{n-1}$ and $\psi:Y\to S^{n-1}$ be homeomorphisms. If f is a continuous function from X to Y, set $f_{\varphi\psi}=\psi\circ f\circ\varphi^{-1}$. The natural way to try to define the degree of f is to define it to be the degree of $f_{\varphi\psi}$. To what extent does this depend on φ and ψ?

5   CHANGE OF VARIABLE REVISITED

The Brouwer degree, via Theorem 4.7, allows the elimination of the C^1 hypothesis in the formula for changing variables in multiple integrals.

THEOREM 5.1 *If $\varphi:\mathbf{R}^n\to\mathbf{R}^n$ is absolutely continuous and differentiable a.e. on the open set Ω, and $\det d\varphi$ is locally integrable, then*

$$\int N(E,y)f(y)\,dy = \int_E f\circ\varphi(x)|\det d\varphi(x)|\,dx \qquad (1)$$

for every nonnegative measurable function f defined a.e. on Ω and every measurable set $E\subset\Omega$.

In the proof it can be assumed that E is bounded and that $\bar{E}\subset\Omega$. Moreover, if (1) holds for a sequence of disjoint sets E, then clearly it holds for the union. This will make it possible to discard various sets along the way. The main step in the proof is to show that (1) holds when $f=1$, i.e., to show that

$$\int N(E,y)\,dy = \int_E|\det d\varphi(x)|\,dx. \qquad (2)$$

The general statement follows directly, and by the usual argument, from the case where f is the characteristic function of a measurable set Y. In this case, $f\circ\varphi$ is the characteristic function of $\varphi^{-1}(Y)$ and $N(E,y)f(y)=N(E',y)$, where $E'=E\cap\varphi^{-1}(Y)$, so (1) reduces to (2) with E replaced by E'. There is a rub, however, that makes necessary a little maneuvering: $\varphi^{-1}(Y)$ is not necessarily measurable. More generally, $f\circ\varphi$ is not necessarily measurable, though, as follows from the proof, $f\circ\varphi|\det d\varphi|$ is measurable.

LEMMA 5.2

If $d\varphi$ is singular a.e. on the set A, then $|\varphi(A)| = 0$.

Proof

Because of the absolute continuity of φ it can be assumed that $d\varphi$ is singular everywhere on A, and it can be assumed that A is bounded (but it is not assumed that A is measurable). Let $G \supset A$ be open with $|G| \leq |A| + \epsilon$. In the half of the proof of Theorem 4.3, Chapter 14, dealing with $d\varphi(a)$ singular, only the differentiability at a is used. Consequently, that theorem shows that each point of A is the center of a ball B with arbitrarily small radius such that $|\varphi(B)| \leq \epsilon|B|$. The family of such balls, which are also contained in G, covers A in the sense of Vitali. By Vitali's theorem there is a disjoint sequence B_k so that

$$A \subset \bigcup_{k=1}^{\infty} B_k \cup N, \qquad |N| = 0.$$

Therefore,

$$|\varphi(A)| \leq \epsilon \sum_{k=1}^{\infty} |B_k| + |\varphi(N)| \leq \epsilon|G| \leq \epsilon(|A| + \epsilon),$$

by the absolute continuity of φ. Since ϵ is arbitrary, the lemma follows.

Proof of (2).

We begin by throwing out some subsets of E. By Theorem 5.2, Chapter 15, we have

$$\text{If } |\varphi(E)| = 0, \quad \text{then } \int N(E, y)\, dy = 0. \tag{3}$$

Therefore, (2) holds whenever E has measure 0. We throw out the set where φ is not differentiable and the set where $|\det d\varphi|$ is not the derivative of its integral. By Lemma 5.2 and statement (3), (2) holds on the set where $d\varphi$ is singular, and we throw out that set also. Now we will prove (2) under the additional assumptions that φ is differentiable everywhere on E, $d\varphi$ is nonsingular everywhere on E, and $|\det d\varphi|$ is the derivative of its integral everywhere on E. E of course is bounded and measurable with $\bar{E} \subset \Omega$.

Let $\epsilon > 0$ be given. Use the local integrability of $|d\varphi|$ to find an open $G \supset E$ with $|G| \leq |E| + \epsilon$ and

$$\int_G |\det d\varphi(x)|\, dx - \int_E |\det d\varphi(x)|\, dx \leq \epsilon. \tag{4}$$

Let $a \in E$, and let $\psi = d\varphi(a)^{-1}\circ\varphi$, so that $d\psi(a) = I$. If r is sufficiently small, then by Theorem 4.7 and the definition of the differential,

$$B(b, (1 - \epsilon)r) \subset \psi(B(a, r)) \subset B(b, (1 + \epsilon)r), \qquad b = \psi(a).$$

Therefore,

$$(1 - \epsilon)^n \leq \frac{|\psi(B(a, r))|}{|B(a, r)|} \leq (1 + \epsilon)^n,$$

and since $|\varphi(A)| = |\det d\varphi(a)| \, |\psi(A)|$,

$$|\det d\varphi(a)|(1 - \epsilon)^n|B(a, r)| \leq |\varphi(B(a, r))|$$
$$\leq |\det d\varphi(a)|(1 + \epsilon)^n|B(a, r)|. \qquad (5)$$

Since $|\det d\varphi(a)|$ is the derivative of the integral of $|\det d\varphi(x)|$, it is also true that for small r

$$\left| \, |\det d\varphi(a)| \, |B(a, r)| - \int_{B(a, r)} |\det d\varphi(x)| \, dx \, \right| \leq \epsilon|B(a, r)|.$$

Combining the last two inequalities we find that

$$|\varphi(B(a, r))| \qquad \text{lies between}$$

$$(1 \pm \epsilon)^n \int_{B(a, r)} |\det d\varphi(x)| \, dx \pm \epsilon|B(a, r)| \qquad (6)$$

for all sufficiently small r.

This shows that for any positive integer k, the balls of diameter less than $1/k$ with center in E, contained in G, and satisfying (6), cover E in the sense of Vitali. For each k, let $\mathscr{F}_k$ be a disjoint sequence of such balls covering almost all of E. According to formula (3), Section 5, Chapter 15, we have

$$\int N_k(E, y) \, dy = \sum_{B \in \mathscr{F}_k} |\varphi(B)|.$$

According to formula (2) of that same section, the integral of $N(E, y)$ lies between

$$(1 - \epsilon)^n \int_E |\det d\varphi(x)| \, dx - \epsilon|G|$$

and

$$(1 + \epsilon)^n \int_G |\det d\varphi(x)| \, dx + \epsilon|G|,$$

which proves formula (2), because of the conditions on G.

LEMMA 5.3   *If $|\varphi(A)| = 0$, then $N(A, y) = 0$ a.e. and $d\varphi$ is singular a.e. on A.*

Proof   If A is measurable, the first conclusion follows from formula (3), and the second from the first and formula (2). If A is not measurable, let B' be a G_δ of measure 0 containing $\varphi(A)$, and let $A' = \varphi^{-1}(B')$. Then A' is measurable and $\varphi(A') = B'$, so the conclusions hold for A', and therefore for the smaller set A.

End of Proof of the
Theorem

As stated at the beginning, it is sufficient to prove formula (2) with E replaced by $E' = E \cap \varphi^{-1}(Y)$, where Y is measurable. If A is the set where either φ is not differentiable or $d\varphi$ is singular, then

$$N(E', y) = N(E' \cap A, y) + N(E' - A, y). \tag{7}$$

By the absolute continuity of φ and Lemma 5.2 $|\varphi(A)| = 0$, and then by Lemma 5.3 $N(E' \cap A, y) = 0$ a.e.

Now, $E' - A$ is measurable. Let Y' be a G_δ containing Y with $|Y' - Y| = 0$. If

$$X' = \varphi^{-1}(Y') \qquad \text{and} \qquad A' = E \cap \varphi^{-1}(Y' - Y) - A$$

then $E' - A = (E \cap X') - A - A'$. Now $\varphi(A') \subset Y' - Y$, so $|\varphi(A')| = 0$. By Lemma 5.3, $d\varphi$ is singular a.e. on A', but A' is contained in the set where $d\varphi$ is nonsingular, so $|A'| = 0$; and obviously $(E \cap X') - A$ is measurable.

Since $E' - A$ is measurable, formula (2) holds with E replaced by $E' - A$, and we have just seen that neither side changes when the latter is replaced by E'.

THEOREM 5.4

(*H. Rademacher*) *If* $\varphi: \mathbf{R}^n \to \mathbf{R}^n$ *is locally Lipschitzian on the open set* Ω, *then*

$$\int N(E, y)f(y) \, dy = \int_E f \circ \varphi(x) |\det d\varphi(x)| \, dx$$

for every nonnegative measurable function f defined a.e. on Ω *and every measurable set* $E \subset \Omega$.

This is immediate from Rademacher's differentiability theorem, as the absolute continuity of φ and local integrability of det $d\varphi$ are evident.

Remark

When φ is one to one the proof of Theorem 5.1 is much simpler—almost identical to that of the earlier theorem on change of variable, Theorem 4.5, Chapter 14. In this case, $\nu(E) = |\varphi(E)|$ is an absolutely continuous regular Borel measure, so all that is needed is the fact that $D\nu = |\det d\varphi|$ a.e. which can be obtained from Lemma 5.2 and formula (5). The local integrability of det $d\varphi$ comes automatically from the local integrability of $D\nu$. In consequence, it also comes automatically when φ is locally one to one. ($\varphi(x) = x \sin (1/x)$ is an example where det $d\varphi$ is not locally integrable.)

One reason for not assuming that φ is one to one appears in the following results on Lipschitz changes of variable. In his famous articles on the extension of differentiable functions, H. Whitney gave in passing a simple formula for extending Lipschitz functions.

Exercise 1   Let f be a real-valued Lipschitz function defined on a subset A of a metric space X and with Lipschitz constant M. Then

$$F(x) = \sup\{f(a) - Md(x, a): a \in A\}$$

is a Lipschitz extension of f to X with the same Lipschitz constant. Consequently, a Lipschitz function from any subset of $\mathbf{R}^n$ to $\mathbf{R}^m$ has a Lipschitz extension to $\mathbf{R}^n$.

THEOREM 5.4   *If $\varphi: \mathbf{R}^n \to \mathbf{R}^n$ is Lipschitz on a measurable set E, then for any nonnegative measurable f defined a.e. on E*

$$\int N(E, y) f(y) \, dy = \int_E f \circ \varphi(x) |\det d\varphi(x)| \, dx,$$

where $d\varphi$ is the differential of any Lipschitz extension of φ.

Remark   Even if φ is one to one on E, there may be no one-to-one Lipschitz extension, so this theorem requires Theorem 5.1 without a one-to-one assumption. On the other hand, the theorem shows that $\det d\varphi$ is independent a.e. of the extension of φ.

Exercise 2   Discuss whether or not the last statement is surprising.

17 Extension of Differentiable Functions

1   INTRODUCTION

Extension problems take the following general form. Let X and Y be metric spaces and let $F(X, Y)$ be a class of functions from X to Y. Let A be a subset of X, and let $F(A, Y)$ be a class of functions from A to Y. The extension problem is to determine whether each function in $F(A, Y)$ has an extension in $F(X, Y)$.

The first useful extension theorem was probably the classical theorem of Tietze.

THEOREM 1.1   *Let A be a closed subset of the metric space X and let Y be an interval of real numbers. Every continuous function from A to Y has a continuous extension from X to Y.*

Exercise 1   If Y is the closed interval $0 \leq y \leq b$, one extension is the sum of the infinite series in the last part of the proof of Stone–Weierstrass.

Exercise 2   If Y is the half open interval $0 \leq y < b$, let F be the extension given by Exercise 1, let B be the set where $F = b$, and let e be continuous with values 1 on A, 0 on B, and between 0 and 1 elsewhere. Then eF is the required extension.

Exercise 3   If Y is the open interval $-b < y < b$, the positive and negative parts of the function can be extended separately by Exercise 2.

The theorem results from the fact that every interval is homeomorphic to one of the above.

In one of his pioneering papers on extension, H. Whitney gave in passing a simple formula for extending real-valued Lipschitz functions.

THEOREM 1.2
Let f be a real-valued Lipschitz function with Lipschitz constant M. Then the function

$$F(x) = \sup\{f(a) - Md(x, a) : a \in A\}$$

is an extension of f with the same Lipschitz constant. If f has values in $[a, b]$, then $G(x) = \min(\max(F, a), b)$ is an extension with the same Lipschitz constant and with values in $[a, b]$.

Exercise 4 Show that Whitney's formula works.

Throughout the chapter, as in the above examples, the functions will be real-valued, but there are important extension theorems for functions with values in other spaces Y. Here are a couple of them. The first, a result of M. Kirszbraun, is given without proof. The second is a special case of a classical result in dimension theory.

THEOREM 1.3
Every Lipschitz function from a subset of $\mathbf{R}^m$ to $\mathbf{R}^n$ has a Lipschitz extension with the same Lipschitz constant.

The Euclidean spaces are quite special in this respect. Usually, Lipschitz functions do not extend with the same constant. (Whitney's formula gives a Lipschitz extension, but not one with the same constant unless $n = 1$.)

THEOREM 1.4
If $m \leq n$, then every continuous function from a closed subset of the sphere S^m to the sphere S^n has a continuous extension with values in S^n. If $m > n$, this is not so.

Proof
If f is defined on A, first use Tietze to extend each coordinate function in order to obtain an extension F_1 with values in $\mathbf{R}^{n+1}$. If F_1 does not take the value 0, then $F = F_1/|F_1|$ is the required extension. If F_1 does take the value 0, fix ϵ (how small to be determined presently), and use Stone–Weierstrass to find F_2 of class C^1 with $\|F_2 - F_1\| < \epsilon$, the norm being the maximum. By Corollary 2.9 of Chapter 13 the range of F_2 has measure 0, so there exist points y not in the range satisfying $|y| < \epsilon$. Choose such a y and set $F_3(x) = F_2(x) - y$, and $F_4 = F_3/|F_3|$. It is easy to see that for any $\theta \in S^n$ and any $z \neq 0$, $|\theta - z/|z|\,| < 2|\theta - z|$, so that $\|F_4 - f\|_A < 4\epsilon$. Thus, if $g = F_4 - f$ on the set A, then the maximum of $|g|$ on A is 4ϵ, and we can use Tietze again to get an extension G with $\|G\| \leq 4\epsilon\sqrt{n+1}$. Now $F_5 = F_4 - G$ is an extension of f, and it cannot take the value 0 if $4\epsilon\sqrt{n+1} < 1$, since F_4 has absolute value 1 everywhere. Thus $F = F_5/|F_5|$ is the required extension.

If $m = n + 1$, take $A = S^n$ to be the equator in S^{n+1}, and f to be the identity function. An extension of f would give a retraction of the upper hemisphere in S^{n+1}, which is the same as B^{n+1}, onto S^n, which is impossible by Theorem 4.8 of Chapter 16.

Exercise 5   The sphere S^m in Theorem 1.4 can be replaced by $\mathbf{R}^m$. (Note that although $\mathbf{R}^m$ is homeomorphic to a subset of S^m, the image of a closed unbounded set cannot be closed, so this result is not a completely trivial consequence of Theorem 1.4.)

There are many different classes of differentiable functions that play important roles in the various branches of analysis. The most immediate of these are the classes C^m themselves. These have been defined only for functions on open sets, while for the purposes of extension it is necessary to expand the definition. There will be a basic open set Ω, and A will be the union of Ω and a part of the boundary.

DEFINITION   *Suppose that $A \subset \overline{A^0}$. Then $C^m(A)$ consists of the functions in $C^m(A^0)$ with*
1.5   *the following property: each point of A has a neighborhood G such that all derivatives of order $\leq m$ are uniformly continuous on $G \cap A^0$.*

Exercise 6   If $u \in C^m(A)$, then each derivative of order $\leq m$ has a unique continuous extension to A.

The symbol $D_i u$ is used for the extension to A as well as for the original derivative.

The extension problem for the classes C^m was solved in a very satisfying way in the work of Whitney mentioned above. Whitney gave a general definition of "class C^m on an arbitrary closed set in $\mathbf{R}^n$", showed that every function of class C^m on a closed set has an extension of class C^m on $\mathbf{R}^n$, and showed that if Ω is open and has the Property W below, then $C^m(\overline{\Omega})$ in his sense is the same as $C^m(\overline{\Omega})$ in the sense of Definition 1.5.

PROPERTY W   *For each point a of $\overline{\Omega}$ there are positive numbers r and L such that any two points x and y of $\Omega \cap B(a, r)$ can be joined by an arc in Ω of length $\leq L|x - y|$.*

Remark   These beautiful results of Whitney were largely ignored until Whitney himself showed how they could be used to answer other difficult questions of interest, at which point they became appreciated. Proofs and applications of Whitney's theorems are now readily accessible, and will not be given here.

Some other classes of differentiable functions that play important roles in analysis are as follows.

The space $BC^m(A)$ consists of the functions $u \in C^m(A)$ such that

$$\|u\|_m = \sup\{|D_i u(x)| : x \in A, |i| \leq m\} < \infty.$$

For $0 < s \leq 1$, set

$$\|u\|_s = \sup \frac{|u(x) - u(y)|}{|x - y|}, \qquad x, y \in A, \qquad x \neq y.$$

The space $BC^{m,s}(A)$ consists of the functions in $BC^m(A)$ such that

$$\|u\|_{m,s} = \max\{\|u\|_m, |D_i u|_s : |i| = m\} < \infty.$$

These are called Hölder or Lipschitz spaces.

The Sobolev space $L_m^p(A)$ is the completion of the functions in $C^m(A)$ such that

$$\|u\|_{L_m^p} = \left\{ \sum_{|i| \leq m} \int_A |D_i u|^p \, dx \right\}^{1/p} < \infty, \qquad 1 \leq p < \infty.$$

When the role of the set A must be stressed it is included as a subscript, e.g., $\|u\|_{m,A}$.

Exercise 7 $BC^m(A)$ and $BC^{m,s}(A)$ are complete.

All of the above have useful variants, but for present purposes these will suffice. The objective of the chapter is to construct an extension operator E, depending only on A, which extends locally integrable functions on A to locally integrable functions on $\mathbf{R}^n$ in such a way that if u belongs to some space of the above type, then Eu belongs to the corresponding space on $\mathbf{R}^n$. This kind of result (even for the Sobolev spaces alone) requires a stronger condition on A than the Whitney property, but a condition that is still relatively mild. The basic condition is that the boundary is locally the graph of a Lipschitz function in some set of coordinates.

DEFINITION 1.6 *An open set Ω is a Lipschitz graph domain if for each point $a \in \partial\Omega$ there exist a neighborhood G of a, a choice of the coordinates, and a Lipschitz function f on $\mathbf{R}^{n-1}$ such that*

$$\Omega \cap G = \{x : x_n < f(x')\} \cap G.$$

THEOREM 1.7 *The open set Ω is a Lipschitz graph domain if and only if for each point $a \in \partial\Omega$ there exist $r > 0$ and an open convex cone Γ such that if $\Gamma_r = \Gamma \cap B(0, r)$, then*

$$b + \Gamma_r \subset \Omega \quad \text{and} \quad b - \Gamma_r \subset R^n - \overline{\Omega} \quad \text{for } b \in \partial\Omega, |a - b| < r.$$

Exercise 8 Prove the theorem.

The differentiable extensions will be accomplished by first localizing and then performing a generalized reflection across the graph of a Lipschitz function, and it is the second step that will involve difficulty. The set-up for this second step is as follows. A' is an open set in $\mathbf{R}^{n-1}$, f is a Lipschitz function, which by Whitney's formula is defined throughout $\mathbf{R}^{n-1}$,

$$A_+ = \{x : x' \in A', \, x_n \geq f(x')\}, \quad A_- = \{x : x' \in A', \, x_n \leq f(x')\}$$

and $A = A_+ \cup A_-$. The function u is defined on A_-, a generalized reflection Ru is defined on A_+, and the extension Eu is defined essentially to be u on A_- and Ru on A_+. This requires a differentiable matching of u and Ru. The following lemma shows that differentiable matching is enough for differentiability.

LEMMA 1.8   *Let f be Lipschitz and let u and Ru be of class C^m on A_- and A_+. If $D_i Ru = D_i u$ on $A_- \cap A_+$ for $|i| \leq m$, then Eu is of class C^m on A.*

Exercise 9   Prove the lemma. Hint: It is enough to treat the case $m = 1$. If θ is a direction such that lines with direction θ meet the graph in only one point, then $D_\theta Eu$ exists on the graph and has the common value of $D_\theta u$ and $D_\theta Ru$. Therefore $D_\theta Eu$ is continuous. Since this is true for all θ in a certain open cone (note Theorem 1.7), u is C^1.

The next lemma has a certain technical use.

LEMMA 1.9   *Let $u \in C^m(A_-)$ vanish for $x_n \leq f(x') - d$, and let $B' \subset A'$ be compact. There is a sequence $u_k \in C^\infty(\mathbf{R}^n)$ such that $\|u - u_k\|_{m, B-} \to 0$.*

Proof   For $h > 0$, let $\tau_h u(x) = u(x', \, x_n - h)$. Since the derivatives of u of order $\leq m$ are uniformly continuous on B_-, it follows that

$$\|u - \tau_h u\|_{m, B-} \to 0 \text{ as } h \to 0.$$

For each fixed h, the regularization procedure of Theorems 11.5 and 12.1 of Chapter 13 provides $u_h \in C^\infty(\mathbf{R}^n)$ with $\|\tau_h u - u_h\|_{m, B-} < h$.

Exercise 10   Lemma 1.9 only requires continuity of f.

Remark   There are two standard meanings for the symbol $D_i u$. In this book $i = (i_1, \ldots, i_m)$ with $1 \leq i_j \leq n$, $|i| = m$, and

$$D_i u = D_{i_1} \ldots D_{i_m} u.$$

2   REFLECTION ACROSS HYPERPLANES

The first differentiable reflection across hyperplanes was given by L. Lichtenstein for functions of class C^1. Suppose u is of class C^1 on $A_- = \{x : x' \in A', \, x_n \leq 0\}$, let b_0 and b_1 be distinct positive numbers, and set

$$Ru(x) = a_0 u(x', -b_0 x_n) + a_1 u(x', -b_1 x_n) \quad \text{on } A_+.$$

Then Ru is of class C^1 on A_+, and Ru matches u, along with the first derivatives, if and only if a_0 and a_1 satisfy the equations

$$a_0 + a_1 = 1, \qquad -a_0 b_0 - a_1 b_1 = 1.$$

If a_0 and a_1 do satisfy these equations, then by Lemma 1.8 the function Eu which is u on A_- and Ru on A_+ is of class C^1 on A.

The Lichtenstein reflection for class C^1 was extended to C^m, $m < \infty$, by M. Hestenes. If u is of class C^m on A_- and $b_0, \ldots, b_m$ are distinct positive numbers, set

$$Ru(x) = \sum_{\mu=0}^{m} a_\mu u(x', -b_\mu x_n) \qquad \text{on } A_+.$$

Then Ru is of class C^m on A_+, and

$$D_i Ru(x) = \sum_{\mu=0}^{m} a_\mu D_i u(x', -b_\mu x_n)(-b_\mu)^\nu,$$

where ν is the order of differentiation in x_n. Thus Ru matches u, along with all derivatives of order $\leq m$, if and only if $a_0, \ldots, a_m$ satisfy the equations

$$\sum_{\mu=0}^{m} a_\mu(-b_\mu)^\nu = 1, \qquad \text{for } \nu = 0, \ldots, m. \tag{1}$$

The matrix for this system of equations is the matrix $V(-b)$, where $V(z) = V(z_0, \ldots, z_n)$ has z_μ^ν in row ν, column μ, starting with row 0 column 0. Such a matrix is called a Vandermonde matrix.

Exercise 1   If there is a linear relation between the rows of $V(z)$ with coefficients $c_0, \ldots, c_m$, then the polynomial

$$p(x) = \sum_{\nu=0}^{m} c_\nu x^\nu$$

vanishes at the points $z_0, \ldots, z_m$, so all c_ν are 0 if these points are distinct.

By Exercise 1 the equations (1) have a unique solution. If this solution is $a_0, \ldots, a_m$, then by Lemma 1.8 the function Eu which is u on A_- and Ru on A_+ is of class C^m on A.

R. Seeley had the very nice idea of choosing the numbers $b_0, \ldots, b_m$ in a special way so that, as m goes to ∞, the solution to the system (1) converges to a solution to the infinite system

$$\sum_{\mu=0}^{\infty} a_\mu(-b_\mu)^\nu = 1, \qquad \nu = 0, 1, \ldots . \tag{2}$$

To carry out Seeley's idea, an explicit formula for the solution to (1) is needed, and this can be obtained through Cramer's Rule (Chapter 9, Section 11, Exercise 9).

The determinant of the Vandermonde matrix $V(z)$ is calculated by successively multiplying row $\nu - 1$ by z_0 and subtracting from row ν, starting from the bottom. In column μ this produces the entries $z_\mu^{\nu-1}(z_\mu - z_0)$. Thus column 0 has a nonzero entry only in row 0, and in the minor of that entry column μ has the factor $z_\mu - z_0$. The removal of these factors leaves a Vandermonde matrix $V(z_1, \ldots, z_m)$. Therefore

$$\det V(z_0, \ldots, z_n) = \prod_{\mu > 0} (z_\mu - z_0) \det V(z_1, \ldots, z_n).$$

It follows by induction that

$$\det V(z) = \prod_{\nu > \mu} z_\nu - z_\mu. \tag{3}$$

When one of the columns in the Vandermonde matrix is replaced by a column of 1's, the matrix remains Vandermonde. From equation (3) and Cramer's rule it follows that the solution to the system (1) is

$$a_\mu = a_m(\mu) = \prod_{\nu \neq \mu} \frac{b_\nu + 1}{b_\nu - b_\mu}.$$

Seeley took

$$b_\mu = b^{\mu+1} - 1 \qquad \text{with } b \geq 2.$$

In this case the solution to the system (1) becomes

$$a_\mu = a_m(\mu) = (-1)^\mu \prod_{\nu=1}^{\mu} \frac{1}{b^\nu - 1} \prod_{\nu=1}^{m-\mu} \frac{1}{1 - b^{-\nu}}.$$

Since $b \geq 2$ it follows that $b^\nu - 1 > b^{\nu-1}$, and therefore

$$\prod_{\nu=1}^{\mu} \frac{1}{b^\nu - 1} \leq b^{-\mu(\mu-1)/2}.$$

Since $-\log(1 - x) \leq x$ for $0 \leq x < 1$ it follows that

$$\prod_{\nu=1}^{m-\mu} \frac{1}{1 - b^{-\nu}} \leq e^{1/(b-1)} \leq e.$$

It is plain that for fixed μ the sequence $(-1)^\mu a_m(\mu)$ is increasing, so we have

LEMMA 2.1   *If $a_m(\mu)$ is the unique solution to the system (1) then*

$$0 \leq (-1)^\mu a_m(\mu) \nearrow (-1)^\mu a_\mu \leq e b^{-\mu(\mu-1)/2}$$

and the a_μ satisfy the equations (2).

Proof   It remains to be proved that the $a_\mu = a(\mu)$ satisfy the equations (2). Think of a and the a_m as functions on the integers and the sums as integrals with respect to counting measure. The estimate in the Lemma allows use of the dominated convergence theorem.

DEFINITION 2.2

If u is a function on A_-, then

$$Ru(x) = \sum_{\mu=0}^{\infty} a_\mu u(x', -b_\mu x_n) \qquad \text{on } A_+$$

wherever the series converges.

LEMMA 2.3

If u is continuous on A_- and for each compact $K' \subset A'$ there are constants C, d, and ν such that

$$|u(x)| \leq C|x_n|^\nu \qquad \text{for } x' \in K' \quad \text{and} \quad x_n < -d,$$

then the series for Ru converges uniformly on compact sunsets of A_+, and Ru is continuous on A_+ and matches u to give a continuous function on A.

Exercise 2   Prove the lemma.

In order to obtain a single extension operator that can be applied to all functions, we first cut off the function below $x_n = -2$. (If the operator is to be applied only to functions that satisfy the condition in the lemma, along with all relevant derivatives, then the cut off is not needed.) For the cut off we fix once and for all a function $\psi \in C^\infty(\mathbf{R}^n)$ which is 1 for $x_n \geq -1$ and is 0 for $x_n \leq -2$.

DEFINITION 2.4

If u is a function on A_-, then Eu is the function which is u on A_- and $R(\psi u)$ on A_+.

THEOREM 2.5

If u is of class C^m on A_-, then the extension Eu is of class C^m on A. For each of the common norms in Section 1 there is a constant C such that $\|Eu\| \leq C\|u\|$.

Proof

On any compact subset of A_+^0 the series for $R(\psi u)$ is finite, so $R(\psi u)$ is of class C^m on A_+^0 and the derivatives are given by

$$D_i R(\psi u) = \sum_{\mu=0}^{\infty} a_\mu D_i \psi u(x', -b_\mu x_n)(-b_\mu)^\nu \qquad \text{for } |i| \leq m,$$

where ν is the order of differentiation in x_n. Since the derivatives of ψu are bounded on compact subsets of A_-, the estimate in Lemma 2.1 gives uniform convergence of the above series on compact subsets of A_+, and this proves that $R(\psi u)$ is of class C^m on A_+. The required matching then follows from the equations (2), and Lemma 1.8 shows that Eu is of class C^m on A.

As far as the norms go, the estimate in Lemma 2.1 gives, for example,

$$\|Eu\|_m \leq C\|\psi u\|_m$$

with

$$C = e \sum_{\mu=0}^{\infty} b^{-\mu(\mu-1)/2}(b^{\mu+1} - 1)^m.$$

Exercise 3   Show that the above gives the required norm estimate for the norm on BC^m.

Exercise 4   Use the above to obtain the norm estimates for the Hölder and Sobolev cases. (These estimates are proved for general Lipschitz graph reflections in Sections 5 and 6, but they are very simple here.)

3   REGULARIZED DISTANCE

For many years I had the good luck to work closely with N. Aronszajn, a wonderful mathematician who left a permanent mark in many areas of modern analysis. The results in the rest of the chapter were joint work with Aronszajn and R. Adams, but they are due mainly to Aronszajn.

At issue is to reflect, not across a hyperplane, but across the graph of a function $x_n = f(x')$ defined on $\mathbf{R}^{n-1}$. It is reasonable to expect that $x_n - f(x')$, the vertical distance to the graph, should take over the previous role of x_n, the vertical distance to the hyperplane. In this case, the analog of Seeley's reflection formula is

$$Ru(x) \ = \ \sum_{\mu=0}^{\infty} a_\mu u(x', f(x') - b_\mu(x_n - f(x'))).$$

It can be shown that this works when f has enough differentiability. However, that differentiability also allows transformation of the graph to a hyperplane, so nothing new is obtained. Aronszajn's idea was to rewrite the formula so that f appears only via the vertical distance:

$$Ru(x) \ = \ \sum_{\mu=0}^{\infty} a_\mu u(x', x_n - b^{\mu+1}(x_n - f(x'))),$$

to replace the vertical distance by a "regularized vertical distance" ρ, and to use the reflection formula

$$Ru(x) \ = \ \sum_{\mu=0}^{\infty} a_\mu u(x', x_n - b^{\mu+1}\rho(x)).$$

This idea works whenever f is a Lipschitz function. The theorem on regularized distance is as follows.

THEOREM 3.1

Let f be Lipschitz on $\mathbf{R}^{n-1}$ with Lipschitz constant M, and let $0 < \epsilon < 1$. There are constants C_m and a function ρ of class C^∞ on $x_n > f(x')$ such that

(a) $(1 - \epsilon)(x_n - f(x')) \leq \rho(x) \leq x_n - f(x')$,
(b) $|D_i\rho(x)| \leq C_m(x_n - f(x'))^{1-m}$ *for* $|i| = m$,
(c) $D_n\rho(x) \geq 1 - \epsilon$.

Remark   Properties (*a*) and (*b*) are sufficient for extension in the classes C^m, BC^m, etc. but (*c*) is needed for integral estimates such as those in the Sobolev norms.

Proof   Fix a function e_1 of class C^∞ on $(0, \infty)$ which is decreasing, constant on a neighborhood of 0, 0 for $t \geq 1$, and satisfying

$$\int_0^\infty e_1(t) t^{n-2}\, dt \; = \; \frac{1}{|S^{n-2}|}. \tag{1}$$

Then $e(x') = e_1(|x'|)$ is of class C^∞ on $\mathbf{R}^{n-1}$, is a decreasing function of $|x'|$, vanishes for $|x'| \geq 1$, and has integral 1. It will be shown that numbers c and k can be chosen so that

$$\rho(x) \; = \; k \int (x_n - f(y'))^{2-n} e\!\left(\frac{M(x' - y')}{c(x_n - f(y'))} \right) dy' \tag{2}$$

has the required properties. For the moment c and k remain undetermined, but subject to $c < 1$.

Set $r = x_n - f(x')$ and $r_1 = x_n - f(y')$. In the range where e is not 0, $M|x' - y'| \leq cr_1$ so that

$$r \leq r_1 + |f(y') - f(x')| \leq r_1 + M|x' - y'| \leq r_1 + cr_1.$$

Hence $r \leq (1 + c)r_1$, and similarly $r \geq (1 - c)r_1$. Therefore

$$0 < \frac{r}{1 + c} \leq r_1 \leq \frac{r}{1 - c}. \tag{3}$$

In particular, r_1 is not 0 on the range of integration, so formula (2) makes sense. If we set

$$z' \; = \; \frac{M(x' - y')}{c(x_n - f(y'))} \; = \; \frac{M(x' - y')}{cr_1}, \tag{4}$$

formula (2) becomes

$$\rho(x) \; = \; k \int r_1^{2-n} e(z')\, dy'. \tag{5}$$

Continuation of the proof requires formulas for the derivatives of z' and the Jacobian of the transformation (4) (as a change of variable from y' to z').

Exercise 1   The derivatives of z' are given by

$$\frac{\partial z_j}{\partial x_k} \; = \; \delta_{jk} \frac{M}{cr_1} - \delta_{nk} \frac{z_j}{r_1}, \tag{6}$$

$$\frac{\partial z_j}{\partial y_k} \; = \; \frac{M}{cr_1}\!\left(-\delta_{jk} + \frac{c}{M} z_j \frac{\partial f}{\partial y_k} \right). \tag{7}$$

Exercise 2   If A is the $n \times n$ matrix $\{-\delta_{jk} + a_j b_k\}$, then $\det A = (-1)^n(1 - \langle a, b \rangle)$. Consequently, the Jacobian of the transformation (4) is given by

$$\mathbf{J} = (M/cr_1)^{n-1}|1 - (c/M)\langle z', \nabla f \rangle| = (M/cr_1)^{n-1} \mathbf{J}_1. \tag{8}$$

Hint: A is the matrix of the linear transformation $\mathbf{T}x = -x + \langle x, a \rangle b$. Use a new orthonormal basis in which $e_1 = a/|a|$.

Proof of (a)   The transformation (4) is Lipschitz with the Jacobian given by (8). Since $|\nabla f| \leq M$, and since $|z'| \leq 1$ on the range of integration in (5), it follows that

$$1 - c \leq J_1 \leq 1 + c. \tag{9}$$

The Rademacher Theorem 5.4 of Chapter 16 justifies using (4) as a change of variable in (5). The result is

$$\rho(x) = k\left(\frac{c}{M}\right)^{n-1} \int \frac{r_1 e(z')}{J_1} \, dz'. \tag{10}$$

Using the right half of (3) and the left half of (9) we get

$$\rho(x) \leq k(c/M)^{n-1}(1 - c)^{-2} r(x).$$

Consequently, the right half of (a) holds if

$$k = (M/c)^{n-1}(1 - c)^2. \tag{11}$$

The left half of (a) will follow from (c).

Proof of (b)   For $|i| = m$, $D_i \rho$ is a sum of terms of the form

$$I_m = k \int r_1^{2-n-m} (M/c)^{|p|-|q|} D_p e(z') z'^q \, dy', \qquad |q| \leq |p| \leq m.$$

To verify this, assume it for a given value of m and apply an additional derivative in accordance with (6). The result is a sum of terms of the same form with m increased by 1. Making the change of variable (4) and inserting the value of k from (11), we get

$$I_m = (1 - c)^2 \int r_1^{1-m} (M/c)^{|p|-|q|} D_p e(z') \frac{z'^q}{J_1} \, dz'.$$

Using the left half of (3) and the left half of (9), we get

$$|I_m| \leq |B^{n-1}|(1 - c)(1 + c)^{m-1}(M/c)^{|p|-|q|}\|e\|_m r(x)^{1-m}. \tag{12}$$

Proof of (c)   A simple calculation gives

$$D_n \rho(x) = -k \int r_1^{1-n}(|z'|e_1'(|z'|) + (n - 2)e_1(|z'|)) \, dy'.$$

The change of variable (4) gives

$$D_n \rho(x) = -(1 - c)^2 \int \frac{|z'|e_1'(|z'|) + (n - 2)e_1(|z'|)}{J_1} \, dz'.$$

Since $e_1 \geq 0$, and $e_1' \leq 0$, it follows from (9) that

$$D_n \rho(x) \geq -\frac{(1 - c)^2}{1 + c} \int |z'|e_1'(|z'|) \, dz'$$

$$-(1 - c)(n - 2) \int e_1(|z'|) \, dz'.$$

According to (1) and integration by parts, the first integral is $1 - n$, and the second one is 1. Inserting these values we get

$$D_n\rho(x) \geq (1 - c(2n - 3))(1 - c)/(1 + c) \geq 1 - (2n - 1)c.$$

Therefore, (c) holds, hence also the left half of (a), if $c = \epsilon/(2n - 1)$.

DEFINITION 3.2

Fix once and for all a function $e_1 \in C^\infty$ on $(0, \infty)$ which is decreasing, constant on a neighborhood of 0, 0 for $t \geq 1$, and satisfying

$$\int_0^\infty e_1(t)t^{n-2}\, dt = 1/|S^{n-2}|,$$

and set $e(x') = e_1(|x'|)$. For $0 < \epsilon < 1$, let $c = \epsilon/(2n - 1)$. If f is Lipschitz with Lipschitz constant $M \neq 0$, the regularized distance to the graph of f is the function

$$\rho(x) = (M/c)^{n-1}(1 - c)^2 \int (x_n - f(y'))^{2-n} e\left(\frac{M(x' - y')}{c(x_n - f(y'))}\right) dy'.$$

Exercise 3 If $f'(x') = f(x'/M)$, then f' has Lipschitz constant 1. If ρ' is the corresponding regularized distance, then $\rho(x) = \rho'(Mx', x_n)$.

The final form of Theorem 3.1 is as follows.

THEOREM 3.3

There are constants C_m depending only on m and on the dimension n such that if f is Lipschitz with Lipschitz constant $M \neq 0$, then the regularized distance to the graph of f is of class C^∞ on $x_n > f(x')$ and has the following properties:

(a) $(1 - \epsilon)(x_n - f(x')) \leq \rho(x) \leq x_n - f(x')$.

(b) $|D_i\rho(x)| \leq C_m M^{m'}\epsilon^{-m}(x_n - f(x'))^{1-m}$
where $m = |i|$ and m' is the order in x'.

(c) $D_n\rho(x) \geq 1 - \epsilon$.

Proof All that needs discussion is the new version of (b) in which the form of the constants is made explicit. For $M = 1$, this is clear from (12), and then for any $M \neq 0$ it follows from Exercise 3.

Before going to the next section on reflections we record a couple of lemmas for future use, but first, for given points x and y above the graph of f we construct a path joining them with length comparable to $|x - y|$, and no closer to the graph than the points themselves. For any direction θ' in $\mathbf{R}^{n-1}$, set

$$x(s) = (x' + s\theta', x_n + sM). \tag{13}$$

It is easily checked that

$$x_n(s) - f(x'(s)) \geq x_n - f(x') \qquad \text{for } s \geq 0. \tag{14}$$

If $y' \neq x'$, let $\theta' = (y' - x')/|y' - x'|$, let $s_0 = |y' - x'|$, and let $z = x(s_0)$, so that

$$z = (y', x_n + M|x' - y'|).$$

If $y' = x'$, set $z = x$. According to (14), the path consisting of the segments $[x, z]$ and $[z, y]$ lies above the graph, and it is clear that

$$|x - z| \leq \sqrt{M^2 + 1}\,|y' - x'|, \tag{15}$$

and that

$$|z - y| \leq |y_n - x_n - M|y' - x'|\,|. \tag{16}$$

Consequently,

$$|x - z| + |z - y| < 3\sqrt{M^2 + 1}\,|x - y|. \tag{17}$$

LEMMA 3.4   *The regularized distance $\rho = \rho_\epsilon$ is Lipschitz, satisfying*

$$|\rho_\epsilon(y) - \rho_\epsilon(x)| \leq C\epsilon^{-1}M_0^2|x - y|, \qquad M_0 = \max(M, 1),$$

where C depends only on n.

Proof   By Theorem 3.3(b), any directional derivative of ρ_ϵ is bounded by $C_1 M_0 \epsilon^{-1}$, so, by (17), the integral over the path above is bounded as in the lemma.

LEMMA 3.5   *If $\varphi(x) = D_{q_1}\rho_\epsilon \ldots D_{q_r}\rho_\epsilon$, $q = \sum|q_j|$, then*

$$|\varphi(x)| \leq CM_0^q \epsilon^{-q}(x_n - f(x'))^{r-q}. \tag{18}$$

If $x_n - f(x') \leq y_n - f(y')$, then

$$|\varphi(y) - \varphi(x)| \leq CM_0^{q+2}\epsilon^{-q-1}(x_n - f(x'))^{r-q-1}|y - x|. \tag{19}$$

C depends only on q, r, and n.

Proof   Formula (18) follows directly from Theorem 3.3(b), which gives also that any directional derivative of φ is bounded by the same expression with q replaced by $q + 1$. Formula (19) then follows from integration of the directional derivative over the two segments described above and the inequalities (14) and (17).

4   REFLECTION ACROSS LIPSCHITZ GRAPHS

Throughout the section f is a Lipschitz function with Lipschitz constant M, $M_0 = \max(M, 1)$, and ρ is a regularized distance to the graph with some fixed $b \geq 2$ and $\epsilon \leq \frac{1}{4}$. The symbol C_m is used to designate a constant depending only on m, the dimension n, and quantities that are assumed to be fixed once and for all such as b, ϵ, the function e_1 used in constructing ρ, etc. A' is an open set in $\mathbf{R}^{n-1}$,

$$A_+ = \{x : x' \in A', \ x_n \ge f(x')\}, \qquad A_- = \{x : x' \in A', \ x_n \le f(x')\}$$

and $A = A_+ \cup A_-$.

DEFINITION 4.1

If u is a function on A_-, then

$$Ru(x) = \sum_{\mu=0}^{\infty} a_\mu u(x', x_n - c_\mu \rho(x)) \qquad \text{on } A_+, \qquad c_\mu = b^{\mu+1}, \tag{1}$$

wherever the series converges.

Exercise 1 If $x \in A_+$, then $(x', \ x_n - c_\mu \rho(x)) \in A_-$ and its vertical distance to the graph lies between $b^\mu(x_n - f(x'))$ and $(b^{\mu+1} - 1)(x_n - f(x'))$.

LEMMA 4.2

If u is continuous on A_- and for each compact $\mathrm{K}' \subset A'$ there are constants C, d, and ν such that

$$|u(x)| \le C|x_n|^\nu \qquad \text{for } x' \in \mathrm{K}' \quad \text{and} \quad x_n \le -d,$$

then the series for Ru converges uniformly on compact subsets of A_+, and Ru is continuous on A_+ and matches u to give a continuous function on A.

Exercise 2 Prove the lemma.

In order to obtain a single extension operator that can be applied to all functions, we cut off below $f(x') - 2$. Fix once and for all a function $\psi \in C^\infty(\mathbf{R}^n)$ which is 1 for $x_n \ge f(x') - 1$ and 0 for $x_n \le f(x') - 2$ and is bounded along with all derivatives.

Exercise 3 Produce such a ψ. Hint: As in Lemma 12.2 of Chapter 13, regularize the characteristic function of the set where $x_n \ge f(x') - \frac{3}{2}$. Note that $\|\psi\|_m$ depends only on the regularizing function, which can be fixed ahead of time and once and for all.

Exercise 4 There are constants C_m such that

$$\|\psi u\|_m \le C_m \|u\|_m.$$

DEFINITION 4.3

If u is a function on A_-, then Eu is the function which is u on A_- and $R(\psi u)$ on A_+.

THEOREM 4.4

If $u \in C^m(A_-)$, then $Eu \in C^m(A)$. There are constants C_m such that if $u \in BC^m(A_-)$, then $Eu \in BC^m(A)$ and

$$\|Eu\|_m \le C_m M_0^m \|u\|_m.$$

By Exercise 4 the function ψ can be dropped, and it can be assumed that u vanishes for $x_n \leq f(x') - 2$. In this case, by Exercise 1, the series for Ru is finite on any compact subset of A_+^0. Therefore, $u \in C^m(A_+^0)$ and the main problem is to show C^m on A_+ itself. In the present situation the derivatives are less simple, and some preparatory lemmas are needed. For purposes of induction we define

$$R_\nu u(x) = \sum_{\mu=0}^{\infty} a_\mu u(x', x_n - c_\mu \rho(x))(-c_\mu)^\nu \qquad \text{on } A_+ \tag{2}$$

wherever the series converges.

It is immediately verified that

$$\frac{\partial R_\nu u}{\partial x_j} = R_\nu \frac{\partial u}{\partial x_j} + \frac{\partial \rho}{\partial x_j} R_{\nu+1} \frac{\partial u}{\partial x_n}. \tag{3}$$

In the sequel we will encounter integers ν, r, and q, and systems of indices i, p, and q_j. They will always satisfy the relations

$$q = \sum |q_j|, \qquad |p| + q = |i| + r, \tag{4}$$

and

$$|q_j| \geq 1, \qquad 1 \leq \nu \leq |i|, \qquad 1 \leq r \leq q < |i|. \tag{5}$$

LEMMA 4.5 $D_i R_\lambda u = R_\lambda D_i u$ *plus terms of the form*

$$D_{q_1}\rho \ldots D_{q_r}\rho R_{\lambda+\nu} D_p u.$$

Proof   The lemma is obvious for $|i| = 0$, so we assume it is true for $|i| = m$ and apply an additional derivative in accordance with formula (3). For example, when we differentiate one of the factors $D_{q_j}\rho$, r and $|p|$ remain the same, while $|i|$ and q both jump 1. When we differentiate $R_{\lambda+\nu} D_p u$, there are two terms. In one, $|i|$ and $|p|$ both jump 1, while the other indices remain the same. In the other, r, $|i|$, $|p|$, and q all jump 1. Consequently the relations (4) and (5) are preserved.

LEMMA 4.6 $D_i R u = R D_i u$ *plus terms of the form*

$$D_{q_1}\rho \ldots D_{q_r}\rho R_\nu D_p u. \tag{6}$$

This is just the previous lemma with $\lambda = 0$.

Exercise 5   Since $c_\mu = b^{\mu+1} = b_\mu + 1$, the equations (2) in Section 2 become

$$\sum_{\mu=0}^{\infty} a_\mu = 1, \qquad \sum_{\mu=0}^{\infty} a_\mu c_\mu^\nu = 0 \qquad \text{for } \nu \geq 1.$$

Consequently,

$$\sum_{\mu=0}^{\infty} (c_\mu - c_0)^k c_\mu^\nu = 0 \qquad \text{for } k \geq 0, \quad \nu \geq 1. \tag{7}$$

LEMMA 4.7

For each $v \geq 1$ and $k \geq 0$ there is a constant C such that

$$|R_v u(x)| \leq C\|u\|_k \rho(x)^k. \tag{8}$$

Proof

By Taylor's formula, centered at $x_n - c_0\rho(x)$,

$$u(x', x_n - c_\mu\rho(x)) = \sum_{j=0}^{k-1} D_n^j u(x', x_n - c_0\rho(x))(c_0 - c_\mu)^j \rho(x)^j/j!$$
$$+ r_k(x, \mu)(c_0 - c_\mu)^k \rho(x)^k$$

with

$$|r_k(x, \mu)| \leq \|u\|_k/k!.$$

When any individual term from the polynomial part is inserted in the series for $R_v u$, the sum is 0 because of the equations (7). When the remainder is inserted, the result is the inequality (8) with

$$C \leq \frac{1}{k!} \sum_{\mu=0}^{\infty} |a_\mu| b^{(v+k)(\mu+1)}.$$

This lemma provides an estimate for the unwanted terms in Lemma 4.6.

LEMMA 4.8

There are constants C_m such that if $1 \leq v \leq |i| \leq m$ and $|i| \leq k \leq m + 1$, then

$$|D_{q_1}\rho(x) \ldots D_{q_r}\rho(x) R_v D_p u(x)| \leq C_m M_0^m \|u\|_k \rho(x)^{k-|i|}. \tag{9}$$

Proof

According to Lemma 3.5,

$$|D_{q_1}\rho(x) \ldots D_{q_r}\rho(x)| \leq CM_0^m \rho(x)^{r-q}.$$

According to Lemma 4.7,

$$|R_v D_p u(x)| \leq C\|D_p u\|_{k-|p|}\rho(x)^{k-|p|}.$$

Putting the two together and using the relation (4) between the indices we get (9).

LEMMA 4.9

There are constants C_m such that if $u \in BC^m(A_-)$, then

$$\|Ru\|_m \leq C_m M_0^m \|u\|_m,$$

hence

$$\|Eu\|_m \leq C_m M_0^m \|u\|_m,$$

provided $Eu \in C^m(A)$.

Proof

Consider the decomposition in Lemma 4.6. First,

$$\|RD_i u\|_0 \leq \|D_i u\|_0 \sum_{\mu=0}^{\infty} |a_\mu|.$$

Then the remainder terms are covered by Lemma 4.8 with $k = |i|$.

To complete the proof of Theorem 4.4 it remains to be shown that if $u \in C^m(A_-)$, then $Eu \in C^m(A)$. This is done in two steps.

<table>
<tr><td>

LEMMA
4.10

</td><td>

If $u \in C^{m+1}(A_-)$, then $Eu \in C^m(A)$.

</td></tr>
<tr><td>

Proof

</td><td>

By Lemma 4.2, $RD_i u$ matches $D_i u$ to give a continuous function on A, so what has to be shown is that the remainder terms in Lemma 4.6 do not spoil the matching. If $\|u\|_{m+1}$ is finite, this is shown by Lemma 4.8 with $k = |i| + 1$. If B' is any open subset of A' whose closure is compact and contained in A', then on B' this norm is finite and we conclude that Eu is of class C^m on B. Since this is true of every B', Eu is of class C^m on A.

</td></tr>
<tr><td>

End of Proof of
Theorem 4.4

</td><td>

Let $u \in C^m(A_-)$, and let B' be open in $\mathbf{R}^{n-1}$ with $B' \subset A'$ and B' compact. By Lemma 1.9 there is a sequence $u_k \in C^\infty(\mathbf{R}^n)$ with $\|u - u_k\|_{m,B_-} \to 0$. By Lemmas 4.9 and 4.10 the sequence Eu_k is Cauchy in $BC^m(B)$, so by Exercise 7 of Section 1 Eu_k has a limit v in $BC^m(B)$, and it is plain that $Eu = v$. Therefore $Eu \in C^m(B)$. Since this is true of every B', $Eu \in C^m(A)$, and the proof is finished.

</td></tr>
</table>

5   REFLECTION OF HÖLDER FUNCTIONS

As defined in Section 1,

$$|u|_s = \sup\{|u(x) - u(y)|/|x - y|^s, \quad x \neq y\}, \qquad 0 < s \le 1,$$

and the Hölder space $BC^{m,s}(A)$ consists of the functions in $BC^m(A)$ such that

$$\|u\|_{m,s} = \max \{\|u\|_m, |D_i u|_s : |i| = m\} < \infty.$$

The main theorem is as follows.

<table>
<tr><td>

THEOREM
5.1

</td><td>

There are constants C_m such that if $u \in BC^{m,s}(A_-)$, then $Eu \in BC^{m,s}(A)$, and

$$\|Eu\|_{m,s} \le C_m M_0^{m+2} \|u\|_{m,s}.$$

</td></tr>
<tr><td>

Exercise 1

</td><td>

There are constants C_m such that

$$\|\psi u\|_{m,s} \le C_m \|u\|_{m,s}.$$

This exercise makes it possible to drop ψ in the definition of Eu, and to assume instead that u vanishes for $x_n \le f(x') - 2$, an assumption that is made from now on. The proof begins with a couple of lemmas.

</td></tr>
<tr><td>

LEMMA
5.2

</td><td>

There is a constant C, depending only on v and n, such that

$$|R_v u(x)|_s \le C M_0^{2s} |u|_s.$$

</td></tr>
<tr><td>

Exercise 2

</td><td>

Use Lemma 3.4 to prove this lemma.

</td></tr>
</table>

LEMMA 5.3

For each $\nu \geq 1$ and $k \geq 0$, there is a constant C such that

$$|R_\nu u(x)| \leq C\|u\|_{k,s}\rho(x)^{k+s}.$$

Proof

This is the replacement for Lemma 4.7, and it is proved in just about the same way. In Taylor's formula we write the remainder after the term of degree $k - 1$, then add and subtract the term of degree k to get

$$u(x', x_n - c_\mu\rho) = \sum_{j=0}^{k} D_n^j u(x', x_n - c_0\rho)(c_0 - c_\mu)^j\rho^j/j!$$
$$+ r_k(x, \mu)(c_0 - c_\mu)^k\rho^k,$$

where

$$r_k(x, \mu) = (D_n^k u(x', t_\mu) - D_n^k u(x', x_n - c_0\rho))/k!$$

with t_μ between $x_n - c_0\rho$ and $x_n - c_\mu\rho$. Therefore,

$$|r_k(x, \mu)| \leq \|u\|_{k,s}(c_\mu - c_0)^s\rho(x)^s.$$

As in the proof of Lemma 4.7, the polynomial terms sum to 0 when put in the series for R_ν, so we have Lemma 5.3 with

$$C \leq \sum_{\mu=0}^{\infty} |a_\mu| b^{(k+\nu+s)(\mu+1)}.$$

Proof of Theorem 5.1

From Theorem 4.4 we know already that $Eu \in BC^m(A)$ and have an evaluation of $\|u\|_m$, so all that is needed is an inequality of the form

$$|D_i Ru|_s \leq C_m M_0^{m+2}\|u\|_{m,s}, \qquad |i| = m. \tag{1}$$

For the term $RD_i u$ in the decomposition of Lemma 4.6 this inequality is given by Lemma 5.2. Therefore, what is needed is an inequality of the form

$$|\varphi(y)R_\nu D_p u(y) - \varphi(x)R_\nu D_p u(x)| \leq C_m M_0^{m+2}\|u\|_{m,s}|x - y|^s$$

with

$$\varphi(x) = D_{q_1}\rho \ldots D_{q_r}\rho. \tag{2}$$

The indices satisfy the relations (4) and (5) preceding Lemma 4.5, with $|i| = m$. We assume the notation is chosen so that

$$\rho(x) \leq \rho(y), \tag{3}$$

and distinguish two cases.

Case 1

$\rho(x) \leq |x - y|$. According to Lemma 5.3 with $k = m - |p|$ we have

$$|R_\nu D_p u(x)| \leq C\|u\|_{m,s}\rho(x)^{m-|p|+s}.$$

According to Lemma 3.5 we have

$$|\varphi(x)| \leq C M_0^q \rho(x)^{r-q}.$$

Combining the two and using the relations between the indices we get

$$|\varphi(x)R_\nu D_p u(x)| \leq C M_0^q \|u\|_{m,s}\rho(x)^s. \tag{4}$$

Similarly,

$$|\varphi(y)R_\nu D_p u(y)| \le CM_0^q \|u\|_{m,s}\rho(y)^s. \qquad (5)$$

Now, $\rho(x) \le |x-y|$, and by Lemma 3.4 $\rho(y) \le (1+CM_0^2)|x-y|$, so the inequality (2) is proved.

Case 2 $\rho(x) \ge |x-y|$. Suppose first that $|p| = m$. In this case, by virtue of the relations between the indices, $q = r$, so, by Lemma 3.5, $|\varphi|$ is bounded by CM_0^m. We write the left side of (2) in the form

$$\varphi(y)(R_\nu D_p u(y) - R_\nu D_p u(x)) + (\varphi(y) - \varphi(x))R_\nu D_p u(x).$$

By Lemma 5.2 the first term is bounded in the way required by (2). By Lemmas 3.5, 5.3 (with $k = 0$), the second term is bounded by

$$CM_0^{m+2}\|u\|_{m,s}\rho(x)^{s-1}|x-y|.$$

Since $\rho(x) \ge |x-y|$ and $s \le 1$, this gives (2).

Now let $|p| < m$. In this case we produce the left side of (2) by integrating the directional derivative of $\varphi R_\nu D_p u$ along the path described just before Lemma 3.4, a path of length $\le CM_0|x-y|$. By Lemma 4.5 any first derivative of $\varphi R_\nu D_p$ is a sum of terms of the form

$$D_{q_1}\rho \ldots D_{q_{r'}}\rho R_\nu D_p u$$

with indices satisfying $|p'| + q' = m + 1 + r'$. By Theorem 3.3 and Lemma 5.3 any such derivative is therefore bounded by

$$C_m M_0^{m+1}\|u\|_{m,s}\rho(x)^{s-1}$$

when account is taken of the relation between the indices. (Recall that the vertical distance from the graph to any point on the path is at least the vertical distance to x.) These bounds for the derivatives and the path length yield (2).

6   REFLECTION OF SOBOLEV FUNCTIONS

THEOREM 6.1   *There are constants C_m such that if $u \in L_m^s(A_-^0)$ and $u \in C^m(A_-)$, then $Eu \in L_m^s(A)$ and*

$$\|Eu\|_{L_m^s} \le C_m M_0^m \|u\|_{L_m^s}. \qquad (1)$$

Remark   The theorem is true without the restriction $u \in C^m(A_-)$. With an elementary background on Sobolev spaces it can be removed immediately. The problem is that the functions in $L_m^s(A)$ are not defined everywhere, but are defined more precisely than almost everywhere. In particular, if $u \in L_m^s(A_-^0)$, then by the current definition Eu is not defined anywhere on the graph of f, while, for $s \ge \frac{1}{2}$, the Sobolev functions cannot be undefined on that large a set. This particular problem is solved by extending slightly the definition of E.

DEFINITION 6.2 *Let E_0 be the current extension operator. For any locally integrable function u on A_-*

$$Eu(x) = \lim_{r \to 0} \frac{1}{|B(x, r)|} \int_{B(x, r)} E_0 u(y) \, dy \qquad (2)$$

at any point x where the right side exists.

If u is locally integrable, then $E_0 u$ is locally integrable, and $Eu = E_0 u$ almost everywhere, including every point of continuity. Consequently, the extended definition is consistent with the old one. This removes the problem with the definition, but a few elementary facts are needed to use the new definition. With these facts the restriction $u \in C^m(A_-)$ disappears immediately.

Since, however, we are assuming $u \in C^m(A_-)$, we know already that $Eu \in C^m(A)$, so all that needs proof is the inequality. As usual, we begin with an exercise to get rid of ψ, and thenceforth assume that $u = 0$ for $x_n \leq f(x') - 2$.

Exercise 1 There are constants C_m such that

$$\|\psi u\|_{L^s_m} \leq C_m \|u\|_{L^s_m}.$$

As before, the proof depends on the decomposition in Lemma 4.6, where the first term causes no difficulty, and the others require evaluation. As usual, the evaluation is based on Taylor's formula and on the fact that the polynomial terms disappear when inserted in the series for R_ν because of the relations (7) just below Lemma 4.6. Therefore, the needed estimate is one for the remainder. Since it is an integral estimate, we use the integral form of the remainder.

$$g(t) = \sum_{j=0}^{k-1} \frac{1}{j!} g^j(a)(t - a)^j + r_k(t)(t - a)^k,$$

$$r_k(t) = \frac{1}{(k-1)!} \int_0^1 g^k((1 - \tau)a + \tau t)(1 - \tau)^{k-1} \, d\tau. \qquad (3)$$

Exercise 2 Use the integral form of the remainder given in Exercise 5, Section 2, Chapter 5 to establish this one.

Because of the form of (3), for a given function v we will have to consider

$$T_{\mu, \tau} v(x) = v(x', x_n - ((1 - \tau)c_0 + \tau c_\mu)\rho(x)), \qquad (4)$$

and will need the following estimate for the L^s norm.

$$\|T_{\mu, \tau} v\|_{L^s} \leq 2^{1/s} \|v\|_{L^s}. \qquad (5)$$

To see this, in the integral defining the norm make the change of variable

$$y' = x', \qquad y_n = x_n - ((1 - \tau)c_0 + \tau c_\mu)\rho(x). \qquad (6)$$

The Jacobian is given by

$$J = (\tau c_\mu + (1 - \tau)c_0)D_n\rho(x) - 1 \geq \tfrac{1}{2}, \tag{7}$$

the inequality coming from

$$c_\mu \geq c_0 \geq 2 \quad \text{and} \quad D_n\rho(x) \geq 1 - \epsilon \geq \tfrac{3}{4}.$$

The more elementary Theorem 4.5 of Chapter 14 is sufficient for the change of variable here, since the transformation (6) is C^∞.

Proof of the Theorem

For the term RD_iu in Lemma 4.6 the inequality (5) gives

$$\|RD_iu\|_{L^s} \leq \sum_{\mu=0}^{\infty} |a_\mu| \, \|T_{\mu,1}D_iu\|_{L^s} \leq 2 \sum_{\mu=0}^{\infty} |a_\mu| \, \|D_iu\|_{L^s}.$$

For the others, expansion by Taylor's formula in the form (3) gives

$$D_pu(x', x_n - c_\mu\rho(x)) =$$
$$\sum_{j=0}^{k=1} \frac{1}{j!} D_n^j D_pu(x', x_n - c_0\rho(x))(c_0 - c_\mu)^j\rho(x)^j \tag{8}$$
$$+ r_k(x, \mu)(c_0 - c_\mu)^k\rho(x)^k$$

with

$$r_k(x, \mu) = (1/(k-1)!)\int_0^1 T_{\mu,\tau}D^kD_pu(1 - \tau)^{k-1} \, d\tau.$$

Hölder's inequality gives

$$|r_k(x, \mu)|^s \leq \int_0^1 |T_{\mu,\tau}D_n^kD_pu|^s \, d\tau.$$

For $k = |i| - |p|$, integration with respect to x and the inequality (5) give

$$\|r_k(x, \mu)\|_{L^s} < 2\|u\|_{L^s_m}. \tag{9}$$

When put in the series for R_ν, the polynomial terms sum to 0, so we have

$$R_\nu D_pu(x) = \rho(x)^k \sum_{\mu=0}^{\infty} a_\mu(c_0 - c_\mu)^k(-c_\mu)^\nu r_k(x, \mu).$$

Because of the standard relation between the indices and the fact that $k = |i| - |p|$, the power of ρ that dominates the derivatives of ρ cancels with ρ^k, and we have

$$|D_{q_1}\rho \ldots D_{q_r}\rho R_\nu D_pu(x)| \leq C_m M_0^m \sum_{\mu=0}^{\infty} |a_\mu|(c_\mu - c_0)^k c_\mu^\nu \, |r_k(x, \mu)|.$$

Thus,

$$\|D_{q_1}\rho \ldots D_{q_r}\rho R_\nu D_pu\|_{L^s} \leq C_m M_0^m\|u\|_{L^s_m} \sum_{\mu=0}^{\infty} |a_\mu|(c_\mu - c_0)^m c_\mu^m,$$

which proves the theorem.

7 EXTENSION FROM LIPSCHITZ GRAPH DOMAINS

Although uniform extension theorems can be proved in a more general setting, the Lipschitz graph domains and the uniform Lipschitz graph domains defined below provide enough generality for most purposes, and they avoid serious difficulties in the proofs. Some remarks on the more general setting are given at the end.

**DEFINITION
7.1**

An open set Ω is a Lipschitz graph domain if for each point $a \in \partial\Omega$ there exist a neighborhood G of a, a choice of the coordinates, and a Lipschitz function f on $\mathbf{R}^{n-1}$ such that

$$\Omega \cap G = \{x : x_n < f(x')\} \cap G.$$

The problem of defining an operator E which extends differentiable functions on Ω to differentiable functions on $\mathbf{R}^n$ is localized and brought back to a reflection across a Lipschitz graph by the use of a partition of unity, which we proceed to describe.

**LEMMA
7.2**

Let X be a metric space. If the compact set K is covered by the open sets $G_1, \ldots, G_m$, then there are open sets $\Omega_1, \ldots, \Omega_m$ covering K and satisfying $\bar{\Omega}_\nu \subset G_\nu$.

Proof

Let $d(x) = \max\{d(x, \mathbf{R}^n - G_\nu)\}$. Since $d(x) > 0$ on K, there is a positive ϵ so that $d(x) \geq \epsilon$ on K. Take

$$\Omega_\nu = \{x : d(x, \mathbf{R}^n - G_\nu) > \tfrac{1}{2}\epsilon\}.$$

Exercise 1   If B is an open ball containing K, the Ω_ν can be taken to lie inside B.

**LEMMA
7.3**

Let O be a family of open sets in $\mathbf{R}^n$. There is a sequence of open sets Ω_ν with the same union, such that the closure of each Ω_ν is contained in some set in O, and such that every infinite subsequence has empty intersection.

Proof

If G is the union, let $\{K_j\}$ be a sequence of compact sets with union G and with $K_j \subset K_{j+1}^\circ$, and set $K_0 = \varnothing$. Let O_j consist of all intersections of sets in O with $K_{j+3}^\circ - K_j$. O_0 covers K_2, so finitely many of its sets, $G_1, \ldots, G_m$, cover K_2. To these we apply Lemma 7.2 to select corresponding Ω's that cover K_2. For $j \geq 1$, O_j covers $K_{j+2} - K_{j+1}^\circ$, and we apply the same process to get finitely many Ω's covering this set, each one contained in a set in O_j. The totality of Ω's has union G, and since sets in O_{j+3} and O_j must be disjoint, any infinite set of Ω's must have empty intersection.

If Ω is a Lipschitz graph domain, then the lemma provides sequences of open sets Ω_ν and G_ν with the following properties.

(i) The Ω_ν cover $\partial\Omega$.

(ii) $\bar{\Omega}_\nu \subset G_\nu$.

(iii) Any infinite set of G_ν's has empty intersection.

(iv) In a suitable coordinate system depending on ν

$$\Omega \cap G_\nu = \{x : x_n < f_\nu(x')\} \cap G_\nu,$$

where f_ν is Lipschitz.

These properties suffice to construct the extension operator E for the classes $C^m(\bar{\Omega})$. The other classes involve global norm evaluations which require some uniformity in the properties.

DEFINITION 7.4   *A Lipschitz graph domain is uniform if the Ω_ν and G_ν can be chosen so that there are positive numbers ϵ, N, and M such that*

(v) *The distance from Ω_ν to $\mathbf{R}^n - G_\nu$ is $\geq \epsilon$.*

(vi) *Any N + 1 G_ν's have empty intersection.*

(vii) *The Lipschitz constants of the f_ν are $\leq M$.*

Equivalently, an open Ω is a uniform Lipschitz graph domain if $\partial\Omega$ can be covered by a sequence of open sets G_ν so that (iv), (vi), and (vii) hold, and in addition each point of $\partial\Omega$ is at a distance $\geq \epsilon$ from the complement of some G_ν.

Exercise 2   Prove the equivalence stated in the definition.

Remark   A classical theorem in dimension theory asserts that the Ω_ν in Lemma 7.3 can be chosen so that any $n + 2$ have empty intersection. However, it does not provide the required additional properties.

Exercise 3   If F_0 and F_1 are disjoint closed sets, then there are open sets $G_0 \supset F_0$ and $G_1 \supset F_1$ with $\bar{G}_0$ and $\bar{G}_1$ disjoint. Hint: $G_0 = \{x : d(x, F_0) < \frac{1}{3}d(x, F_1)\}$.

Exercise 4   If F_0 and F_1 are disjoint closed sets, there is a function $\psi \in C^\infty(\mathbf{R}^n)$ with $0 \leq \psi(x) \leq 1$ for all x and with $\psi = 0$ on a neighborhood of F_0 and $\psi = 1$ on a neighborhood of F_1. Hint: Use the last exercise along with Exercise 7, Section 2, Chapter 12 and Exercise 10, Section 6, Chapter 7.

First we will construct a partition of unity for any Lipschitz graph domain Ω and use it to produce an operator E that extends functions in $C^m(\bar{\Omega})$ to functions in $C^m(\mathbf{R}^n)$. Then we refine the construction in the case of a uniform Lipschitz graph domain so that E extends functions in the other classes too.

Let $\{\Omega_\nu\}$ and $\{G_\nu\}$ have the properties (i)–(iv). Use Exercise 3 to get an open $U_\nu \supset \bar{\Omega}_\nu$ with $\bar{U}_\nu \subset G_\nu$, and let U be the union of the U_ν. Then $\Omega - U$ is closed, so we can find an open $U_0 \supset \Omega - U$ with $\bar{U}_0 \subset \Omega$. For convenience of notation, set $G_0 = \Omega$. Now use Exercise 4 to produce functions of class C^∞, with values between 0 and 1, as follows:

$\psi_\nu = 1$ on U_ν,

    $= 0$ on a neighborhood of $\mathbf{R}^n - G_\nu$,

$\psi = 1$ on a neighborhood of $\bar{\Omega}$,

    $= 0$ on a neighborhood of $\mathbf{R}^n - (\Omega \cup U)$.

At each point of $\Omega \cup U$ at least one of the functions is positive, and each point has a neighborhood on which all but finitely many are 0. Consequently, the functions

$$\varphi_\nu(x) = \psi_\nu(x)\psi(x)\left(\sum_{\mu=0}^{\infty}\psi_\mu(x)^2\right)^{-\frac{1}{2}} \tag{1}$$

have the following properties.

    (I) $\varphi_\nu \in C^\infty(\mathbf{R}^n)$ and $\varphi_\nu = 0$ outside a closed subset of G_ν.

    (II) Each point of $\mathbf{R}^n$ has a neighborhood on which all but finitely many φ_ν are 0.

    (III) $\sum_{\nu=0}^{\infty}\varphi_\nu^2 = 1$ on a neighborhood of $\bar{\Omega}$.

Remark   The functions φ_ν^2 form a partition of unity for Ω.

For $\nu \geq 1$, let $A_\nu = \{x : x_n < f_\nu(x')\}$. If u is a function on Ω let

$u_0(x) = \varphi_0(x)u(x)$ if $x \in \Omega$, 0 otherwise,

$u_\nu(x) = \varphi_\nu(x)u(x)$ if $x \in A_\nu \cap \Omega$, 0 if $x \in A_\nu - \Omega$.

LEMMA 7.5   *If $u \in C^m(\Omega)$, then $u_0 \in C^m(\mathbf{R}^n)$.*

Proof   If $a \in \Omega$, then on a neighborhood of a u_0 coincides with $\varphi_0 u$, while if $a \notin \Omega$, then on a neighborhood of a u_0 is 0.

LEMMA 7.6   *If $u \in C^m(\bar{\Omega})$, then $u_\nu \in C^m(\bar{A}_\nu)$.*

Proof   If $a \in G_\nu$, then u_ν coincides with $\varphi_\nu u$ on the intersection of A_ν with a neighborhood of a, while if $a \notin G_\nu$, then u_ν is 0 on the intersection of A_ν with a neighborhood of a.

DEFINITION 7.7   *Let E_ν be an extension operator for A_ν. If u is a function on $\bar{\Omega}$, then*

$$Eu = \varphi_0 u_0 + \sum_{\nu=1}^{\infty}\varphi_\nu E_\nu u_\nu. \tag{2}$$

THEOREM 7.8   *Let Ω be a Lipschitz graph domain. If $u \in C^m(\bar{\Omega})$, then $Eu \in C^m(\mathbf{R}^n)$, and Eu is an extension of u.*

Proof   By Lemmas 7.5 and 7.6 and the fact that E_ν is an extension operator, each term in the sum in (2) lies in $C^m(\mathbf{R}^n)$. Since each point has a neighborhood on which the sum is finite, it follows that $Eu \in C^m(\mathbf{R}^n)$.

To show that Eu is an extension of u we show that on Ω

$$\varphi_\nu E_\nu u_\nu = \varphi_\nu^2 u$$

and use the relation (III). If $x \notin G_\nu$, then both sides are 0. If $x \in G_\nu$, then $x \in \Omega \cap G_\nu = A_\nu \cap G_\nu$, and

$$E_\nu u_\nu(x) = E_\nu \varphi_\nu(x) u(x) = \varphi_\nu(x) u(x).$$

A corresponding result for the other function classes requires norm evaluations, which in turn require uniform bounds on the derivatives of the φ_ν, the Lipschitz constants of the f_ν, and the number of G_ν with nonempty intersection.

LEMMA 7.9   *There are constants K_m such that if E_0 and E_1 are measurable sets with $d(E_0, E_1) \geq \epsilon$, then there is a function $\psi \in C^\infty(\mathbf{R}^n)$ with $0 \leq \psi(x) \leq 1$, $\psi = 0$ on the set $\{x : d(x, E_0) \leq \epsilon/3\}$, $\psi = 1$ on the set $\{x : d(x, E_1) \leq \epsilon/3\}$, and*

$$\|\psi\|_m \leq K_m \epsilon^{-m}.$$

Exercise 5   Prove the lemma by expanding on the proof of Lemma 12.2, Chapter 13.

LEMMA 7.10   *If Ω is a uniform Lipschitz graph domain, then the φ_ν can be chosen so that*

$$\|\varphi_\nu\|_m \leq K_m,$$

where the K_m are independent of ν.

Proof   Take $U_\nu = \{x : d(x, \bar{\Omega}_\nu) < \epsilon/3\}$. The distances from $\bar{\Omega}_\nu$ to $\mathbf{R}^n - G_\nu$, $\Omega - U$ to $\mathbf{R}^n - \Omega$, and $\bar{\Omega}$ to $\mathbf{R}^n - (\Omega \cup U)$ are at least ϵ, $\epsilon/3$, and $\epsilon/3$, respectively. By Lemma 7.9, ψ_ν and ψ can be chosen with bounds of the kind specified. Let φ be the sum of the squares of the ψ_ν so that $\varphi_\nu = \psi_\nu \psi \varphi^{-1/2}$. Since the sum defining φ has at most N nonzero terms at each point, φ also has bounds of the kind specified. Any derivative of $\varphi^{-1/2}$ of order $\leq m$ is a sum of products of derivatives of φ of orders $\leq m$ and negative powers of φ of degree at most $m + \frac{1}{2}$. The result therefore follows from the fact that $\varphi \geq 1$ wherever ψ is $\neq 0$.

THEOREM 7.11   *Let Ω be a uniform Lipschitz graph domain. If u lies in one of the spaces $BC^m(\bar{\Omega}), BC^{m,s}(\bar{\Omega}),$ or $L_m^p(\Omega)$, then Eu lies in the corresponding space on $\mathbf{R}^n$, and*

$$\|Eu\| \leq C_m \|u\|,$$

where the norm is the corresponding norm and the constants C_m depend only on Ω.

Proof   We have

$$D_i E u = D_i \varphi_0 u_0 + \sum_{\nu=1}^{\infty} D_i \varphi_\nu E_\nu u_\nu. \tag{3}$$

From the fact that the extension operators E_ν have uniform bounds and Lemma 7.10 we get

$$|D_i \varphi_\nu E_\nu u_\nu(x)| \leq C_m K_m \|u\|_m,$$

and a similar inequality for $D_i \varphi_0 u_0(x)$. For any given x there are at most $N + 1$ nonzero terms. Therefore,

$$\|Eu\|_m \leq (N + 1) C_m K_m \|u\|_m.$$

An almost identical argument works for $BC^{m,s}$, but a little more is needed in the Sobolev case.

If $\{a_\nu\}$ is a sequence with at most $N + 1$ nonzero terms, then Hölder's inequality gives

$$\left(\sum_\nu |a_\nu|\right)^p \leq (N + 1)^{p-1} \sum_\nu |a_\nu|^p.$$

Consequently, at each point we have

$$|D_i E u|^p \leq (N + 1)^{p-1}\left(|D_i \varphi_0 u_0|^p + \sum_{\nu=1}^{\infty} |D_i \varphi_\nu E_\nu u_\nu|^p\right).$$

Therefore,

$$\|Eu\|_{L_m^p(\mathbf{R}^n)}^p \leq (N + 1)^{p-1}\left(\|\varphi_0 u_0\|_{L_m^p(\mathbf{R}^n)} + \sum_{\nu=1}^{\infty} \|\varphi_\nu E_\nu u_\nu\|_{L_m^p(\mathbf{R}^n)}^p\right).$$

From the fact that the extension operators E_ν have uniform bounds and Lemma 7.10 we have

$$\|\varphi_\nu E_\nu u_\nu\|_{L_m^p(\mathbf{R}^n)} \leq C_m K_m \|\varphi_\nu u\|_{L_m^p(\Omega)}.$$

Now, $\|\varphi_\nu u\|_{L_m^p(\Omega)}^p$ is a sum of terms of the form

$$\int_\Omega |D_j \varphi_\nu|^p |D_k u|^p \, dx, \qquad |j| + |k| \leq m,$$

so the result follows from the fact that

$$\sum_{\nu=1}^{\infty} |D_j \varphi_\nu(x)|^p \leq N K_m^p.$$

Remark   The extension procedure described here is quite different from Whitney's, both in its nature and in the situations to which it applies. For example, the Whitney extension does not apply to the Sobolev spaces, and in the other cases it provides increasingly complex extension operators as the order of differentiability increases. On the other hand, it does apply to more general domains than the Lipschitz graph domains.

At least for the Sobolev spaces, a uniform extension operator does exist for domains that are more general, though not as general as Whitney's. They are domains with a sort of polyhedral structure in which the cells are uniform Lipschitz graph domains. When there are a finite number of bounded cells, the condition for extension is that the cells meet nontangentially and that the boundary of Ω does not separate Ω locally. This result appeared first in an article of Adams, Aronszajn, and Smith in University of Kansas Technical Reports, New Series, No. 8, 1964; and subsequently in the Annales de L'Institut Fourier, 1967. The Lipschitz graph case was presented (with a similar but not identical proof) by E. M. Stein in lectures at the University of Paris, Orsay, during 1966–1967. Stein's proof is given in his book *Singular Integrals and Differentiability Properties of Functions* (Princeton University Press, 1970). The first theorems on the extension of Sobolev functions across Lipschitz graphs were given by A. P. Calderón. Calderón's method works for an arbitrary but fixed finite order of differentiability and for norms in which singular integral operators are bounded, e.g. L_m^p with $1 < p < \infty$.

Index

Undergraduate Texts in Mathematics

continued from ii

Malitz: Introduction to Mathematical
Logic: Set Theory - Computable
Functions - Model Theory.
1979. xii, 198 pages. 2 illus.

Martin: The Foundations of Geometry
and the Non-Euclidean Plane.
1975. xvi, 509 pages. 263 illus.

Martin: Transformation Geometry: An
Introduction to Symmetry.
1982. xii, 237 pages. 209 illus.

Millman/Parker: Geometry: A Metric
Approach with Models.
1981. viii, 355 pages. 259 illus.

Prenowitz/Jantosciak: Join Geometrics:
A Theory of Convex Set and Linear
Geometry.
1979. xxii, 534 pages. 404 illus.

Priestly: Calculus: An Historical
Approach.
1979, xvii, 448 pages. 335 illus.

Protter/Morrey: A First Course in Real
Analysis.
1977. xii, 507 pages. 135 illus.

Ross: Elementary Analysis: The Theory
of Calculus.
1980. viii, 264 pages. 34 illus.

Sigler: Algebra.
1976. xii, 419 pages. 27 illus.

Simmonds: A Brief on Tensor
Analysis.
1982. xi, 92 pages. 28 illus.

Singer/Thorpe: Lecture Notes on
Elementary Topology and Geometry.
1976. viii, 232 pages. 109 illus.

Smith: Linear Algebra.
1978. vii, 280 pages. 21 illus.

Smith: Primer of Modern Analysis
1983. xiii, 442 pages. 45 illus.

Thorpe: Elementary Topics in Differential
Geometry.
1979. xvii, 253 pages. 126 illus.

Troutman: Variational Calculus
with Elementary Convexity.
1983. xiv, 364 pages. 73 illus.

Whyburn/Duda: Dynamic Topology.
1979. xiv, 338 pages. 20 illus.

Wilson: Much Ado About Calculus:
A Modern Treatment with Applications
Prepared for Use with the Computer.
1979, xvii, 788 pages. 145 illus.